Reproduction in Mammals

REPRODUCTION IN MAMMALS

The Female Perspective

VIRGINIA HAYSSEN ∎ TERI ORR

Johns Hopkins University Press   *Baltimore*

© 2017 Johns Hopkins University Press
All rights reserved. Published 2017
Printed in the United States of America on acid-free paper
9 8 7 6 5 4 3 2 1

Johns Hopkins University Press
2715 North Charles Street
Baltimore, Maryland 21218-4363
www.press.jhu.edu

Library of Congress Cataloging-in-Publication Data

Names: Hayssen, Virginia, author. | Orr, Teri, 1980– author.
Title: Reproduction in mammals : the female perspective / Virginia Hayssen,
    Teri Orr.
Description: Baltimore : Johns Hopkins University Press, [2017] | Includes
    bibliographical references and index.
Identifiers: LCCN 2016049247 | ISBN 9781421423159 (hardcover : alk. paper) |
    ISBN 9781421423166 (electronic) | ISBN 1421423154 (hardcover : alk. paper) |
    ISBN 1421423162 (electronic)
Subjects: LCSH: Mammals—Reproduction.
Classification: LCC QL739.23 .H39 2017 | DDC 599.156—dc23
    LC record available at https://lccn.loc.gov/2016049247

A catalog record for this book is available from the British Library.

*Special discounts are available for bulk purchases of this book. For more information,
please contact Special Sales at 410-516-6936 or specialsales@press.jhu.edu.*

Johns Hopkins University Press uses environmentally friendly book materials,
including recycled text paper that is composed of at least 30 percent post-consumer
waste, whenever possible.

Contents

Preface

Newborn mammals can weigh as little as a dime or as much as a motorcycle. Some of these babies receive milk for only a few days, whereas others nurse for a few years. Humans have one baby at a time after 9 months of pregnancy, but other mammals have up to 20 or more after only 2 to 3 weeks in utero. What causes this incredible diversity? Do bigger moms have bigger babies (does size matter)? Do primates have longer pregnancies than other groups (does genetics matter)? Do aquatic animals have particular patterns (does habitat matter)? Predatory lions have many young at one time, whereas their herbivorous antelope prey usually have only one (does diet matter)?

This book addresses not only the broad diversity of reproduction in mammals but also the ways in which natural selection has influenced that diversity. We focus on the reproductive biology of a female in relation to her environment, her mates, her offspring, and other females. We aim to illustrate this perspective— the female perspective (hers as well as our own)—in a variety of contexts as reproduction is an emergent property of genes, tissues, environment, and evolution.

Reproductive biology covers widely divergent fields. It includes the complex molecular interactions of hormones and their receptors, genomic comparisons with sophisticated statistical analysis, mark and recapture of mammals in the field, assisted reproduction of endangered wildlife in captive settings, analyses of milk yield in dairy cows on different diets, changes in seasonality with climate change, and amenorrhea associated with anorexia. Each area of investigation has its own methodology and jargon. Often students with a focus in one area may be ignorant of findings that are accepted as fact in another area. Sections of a book that cover such wide areas may seem simplified to those specializing in a specific aspect or overly complicated to those new to a given topic. However, we have tried to write for both audiences, in part by defining terms when we use them, but also by including a detailed glossary of specialist terms. Our language may also be more conversational than may be customary for specialists. We have aimed at maintaining a casual narrative summarizing each area of reproductive biology that will engage the reader (no matter their background). In addition, sidebars provide detail on specific areas or highlight specific work.

The scope of each chapter begins with a hyena story. Charismatic and well-studied, but so seemingly different, even alien, from ourselves and other familiar mammals, the spotted hyena (*Crocuta crocuta*) provides a context for the material of each chapter.

Throughout the book, we also highlight the pivotal work of scientists in reproductive biology. We do so in the text but also in special boxes. These boxes provide examples of researchers who have made major contributions to our understanding of reproductive biology. Our choices reflect the perspective of the book, as well as the unfortunate underrepresentation of women of color in the field of reproductive biology. We are optimistic and hopeful that future books will reflect a greater inclusivity in the sciences.

In a narrow sense, our book is written for animal scientists, reproductive physiologists, mammalogists, physiological ecologists, evolutionary biologists, animal behaviorists, and conservation biologists as well as researchers who study reproduction or who need a book that places mammalian reproduction in a cohesive evolutionary and ecological context. However, we have used the text successfully in both upper-level and first-year undergraduate seminars on mammalian reproduction. In addition, the material was successfully incorporated into an interdisciplinary workshop on "Mothers and Others" that included faculty in art, literature, sociology, legal theory, ethnic and women's studies, as well as professionals from nonprofit organizations. Thus, the book is accessible to the specialist as well as to the educated generalist.

Scope and Organization

This book is a review not only of female mammal reproduction but also of the diversity and scope of reproductive strategies in female mammals. This is in contrast to texts that detail physiological mechanisms or use information only from domesticated or laboratory mammals. We build on earlier explorations of the female perspective in biology such as Bettyann Kevles's *Females of the Species* (1986), which examines all female animals. In some respects, our book updates Kevles's work but with a focus only on mammals. At the other extreme are books that focus mostly or solely on humans. Excellent books in this vein include Evelyn Shaw and Joan Darling's *Female Strategies* (1985); an edited volume by Mary Ellen Morbeck et al. (1997), *The Evolving Female: A Life History Perspective*; and three Sarah Blaffer Hrdy books, *The Woman That Never Evolved* (1991), *Mother Nature* (1999), and *Mothers and Others: The Evolutionary Origins of Mutual Understanding* (2009). We also build on a long history of scholarly works on reproductive physiology and ecology, such as Sydney Asdell's *Patterns of Mammalian Reproduction* (1946); Richard Sadleir's *The Ecology of Reproduction in Wild and Domestic Mammals* (1969); Ari van Tienhoven's *Reproductive Physiology of Vertebrates* (1983); the various editions of F.H.A. Marshall and colleagues, *Marshall's Physiology of Reproduc-*

tion (1994); E.S.E. Hafez's *Reproduction in Farm Animals* (2000); and the multiple volumes of C.R. Austin and R.V. Short's *Reproduction in Mammals* (1982).

In this age of web-based searches, one is inundated with almost unlimited facts on reproduction in mammals, but facts without context are surprisingly uninformative. Facts can tell us the length of gestation for a sloth (three-toed sloth, *Bradypus*, 6–7 months; Hayssen 2009, 2010) but not whether that gestation is long or short compared to other sloths (about half the length of the two-toed sloth, *Choloepus*, 11–12 months; Hayssen 2011), nor how differences in gestation length relate to other aspects of the animal's life. For sloths, low metabolic rates (MR) and low body temperatures (T_b) may lead to slow growth rates and result in long gestations; however, MR and T_b are the same for both sloths, leaving the difference in gestation a mystery. This book provides context on multiple levels—as well as the opportunity to be distracted by specific cases, as with the sloths example! The literature on mammalian reproduction contains many stories specific to individual taxa, but those stories are without synthesis or integration. Here we put individual stories from diverse taxa into a single perspective—a female mammal.

The first major section of the book includes three chapters that outline the individual female: her genetics, anatomy, and physiology. In our conversations with undergraduate students and academics outside of biology, these three chapters were the most challenging as they have the most unfamiliar terminology. After exploring the internal workings of individual females, our focus in the second major section moves that female through a reproductive cycle (oogenesis, ovulation, mating, conception, implantation, birth, lactation, weaning) and includes a female's intimate interactions with males and offspring. Our third major section puts a female's reproductive life in context with the rest of the world, the abiotic environment and the biotic environment (both conspecifics and heterospecifics). Two minor sections bookend the three major ones. The first explores the female perspective (chapter 1) and the diversity and reproductive evolution in mammals (chapter 2). The final two chapters of the book focus on human concerns: a penultimate chapter on conservation and a final chapter on women as mammals.

Our aim is to put a female perspective not only on how reproduction fits into the lives of mammals but also on how the biology of mammals affects their reproductive patterns. We start with an exploration of the subtitle of the book, the female perspective.

Acknowledgments

Specific chapters of the book benefited from comments from Sara Cairns (New Hampshire Natural Heritage Bureau, Division of Forests and Lands), Tom Eiting (University of Utah), Brock Fenton (University of Western Ontario), Casey Gilman (University of Massachusetts), Anson Koehler (University of Melbourne), Peter Lurz (Randersacker, Germany), and Amy Skibiel (Auburn University). Angela Baerwald, Jennifer Marshall Graves, Sarah Blaffer Hrdy, and Eileen Lacey graciously reviewed their respective scientist sections. Similarly, two anonymous reviewers provided useful comments.

Invaluable research or other academic assistance was provided by Ida Hay and Linda Kurowski (5-College book depository); Cristina Ryan and Susan Daily (Smith interlibrary loans); and Isabella Fielding, Abigail Michelson, Paula Noonan, and Siobhan Prout (Smith College). Camryn McCarthy and Lanee Jung compiled the subject index with help of students in the fall 2016, first-year seminar "Mammalian Reproduction: A Female Perspective." Vincent J. Burke, Tiffany Gasbarrini, and Meagan Szekely of Johns Hopkins University Press helped shepherd the book through the editorial labyrinth.

Elizabeth Adkins-Regan (Cornell University) provided valuable advice in the early stages of the project. Barbara Blake (University of North Carolina at Greensboro) and Kimberly Hammond (University of California, Riverside) provided welcome support and encouragement. Marlene Zuk (University of Minnesota) provided inspiration through her many published works on the female perspective. Conversations with Patricia Brennan (Mount Holyoke College) were very insightful. M. Denise Dearing (University of Utah) has served as an excellent mentor (TJO) through some of the later stages of this book.

Emily Fusco (University of Massachusetts), Abigail Michelson (Smith College), and Jennifer Wen (University of Massachusetts) created illustrations for the book. Photographs were donated by Angela Baerwald, Jennifer Marshall Graves (Micheline Pelletier, GAMMA), Kay Holekamp, Sarah Blaffer Hrdy (Anula Jayasuriya), Eileen Lacey, Virpi Lummaa, and Marilyn Renfree. Permissions for images were granted by the Burke Museum of Natural History and Culture, John Wiley and Sons, *Journal of Reproduction and Fertility*, McGraw-Hill Education, Mindenstock.com, and the Royal Society, or were from open-access sources.

Funding from the Kahn Institute and the Blakeslee Grant for Genetics Research (both at Smith College) is gratefully acknowledged (VH). During initial stages working on this book, TJO was supported by a National Science Foundation Postdoctoral Research Fellowship in Biology under Grant No. DBI-297 1202871.

Many undergraduate students contributed to our understanding of the female mammal perspective. We wish to acknowledge these valuable contributions.

TJO: Caitlin Sanchez (University of California Riverside), Jen Silva as well as Jennifer Wen (University of Massachusetts).

VH: Mariam Ali, Nicole Bartlett, Sonya Bhatia, Louren Bridges, Allison Corbosiero, Diane Chen, Talya Davis-Johnson, Eve DeRosa, Katheryn Dickhut, Abbey Fleming, Sarah Gaffney, Kimberly Geisler, Shari Jainuddin, Lanee Jung, Arcadia Kratkiewicz, Joy Lapseritis, Lesley Latinville, Josie Little, Junzhou Liu, Camryn McCarthy, Megan McCusker, Karen Messerschmitt, Ana Moreno-Mesa, Katherine Morris, Sophia Ong, Taryn Pestalozzi, Katherine Pielmeier, Catherine Rafferzeder, Samantha Ross, Maggie Sawdy, Sarah Schulten, Sarah Soss, Cathryn Starr, Britni Steingard, Quinn Tompkins, Alicia VandeVusse, Maryalice Walker, and Julia Yun (all Smith College).

TJO would also like to acknowledge the support of her family including Delbert Orr and Leona Thompson (*nee* Elliott), who instilled in me a love of nature and an appreciation that anything a boy can do I could do (possibly better) as I reinforced through joyfully shared rounds of "I can jump a hurdle even in my girdle" (Irving Berlin, 1946; Annie Get Your Gun) sung around the house.

Reproduction in Mammals

CHAPTER ONE

The Female Perspective

[Mammalia] By privileging a uniquely female characteristic . . . , Linnaeus broke with longstanding traditions that saw the male as the measure of all things.

—Schiebinger 1993:393

Perhaps more than any other class of organisms on Earth, female mammals possess extraordinary control over their reproduction. They regulate major aspects of mating and conception as well as offspring survival, growth, and development. They do this using a combination of internal conception, in utero development, and lactation, all of which provide mammalian females unprecedented influence on their reproductive success. Yet, historically, the female perspective has been given short shrift.

We have learned a great deal using the male perspective. For one, thinking about how sperm might interact after copulation led to the idea of post-copulatory sexual selection. By being "apparent," the male ejaculate led scientists to consider post-mating interactions at a biochemical level. Because the male perspective has arguably been the platform from which we have studied mammalian reproduction for some time, one could say that nearly all the information we have gathered thus far has come from this perspective. However, in the field of life history, the focus has been the exact opposite. Here, the value of females as the key aspect of measuring fitness has long been appreciated, and one often encounters reference to females and "granddaughters" (see chapters 3, 5). Alternative perspectives allow the framing of questions in new ways.

> The emphasis of scientific research until about ten years ago was on the male, and the major research thrust pursued his behaviors, treating the female only as an incidental sperm receptacle. (Shaw, Darling 1985:3)

> Before the mid-twentieth century, the female as an active participant in evolution was largely overlooked. Most scientists interested in animal behavior were men who sometimes displayed bias toward, but more often obliviousness to, the often subtle patterns in the lives of female animals. (Kevles 1986:vii)

Female mammals have long been neglected in biomedical research. (Beery, Zucker 2011:565). [Only in 1993, did the National Institutes of Health require that women be included in human clinical trials (Beery, Zucker 2011).]

The title of this book is *Reproduction in Mammals: The Female Perspective*, but what is the female perspective? Perhaps this is best answered with an example. Kathryn Clancy gave a talk at Smith College in 2015 entitled "A Feminist Perspective on Ovarian Follicular Dynamics." This title raised a few eyebrows among the faculty, who were both puzzled and relieved when Clancy explained that a feminist perspective in this context was a methodological one: using an unconventional (i.e., female-focused) lens through which to examine a familiar topic. For Clancy, this meant examining ovarian and uterine function in rural, Polish women compared with that of urban or suburban women of equal age (Clancy et al. 2009). Her focus was on the luteal phase, when the uterine lining thickens under the influence of progesterone, important for implantation. The rural, Polish women had much lower progesterone levels, an endometrium that changed in thickness over the luteal phase, as well as a shorter luteal phase, but these differences did not lower fertility. In fact, lower hormone levels were associated with higher fertility, as 73% of the rural women had children compared with none from the urban sample. These results challenge the medical practice in fertility regimes, which is to administer hormones at higher than physiological levels (Clancy et al. 2009). In this case, the consequence of asking a standard question of a nonstandard sample was to challenge established medical practice.

In this chapter, we explore the female perspective and associated terminology. Our goal is not as lofty as altering clinical procedures. Instead, we review the standard topics of mammalian reproduction but from a nonstandard, female-focused perspective. We also focus on diverse mammals rather than just domesticated or laboratory species. In doing so, we acknowledge that fundamental insights can be gleaned from a diversity of taxa. With these approaches we hope to open new avenues for understanding the most important aspect in the evolution of mammals: their ability to successfully reproduce. In mammals, the success of reproduction is almost entirely the province of females. Conception, early embryology, gestation, and lactation are primarily under the control of females. This is the most important reason for taking the female perspective.

The Female Perspective Explained Further

In this section, we explore what happens when the female perspective is not part of the discussion. Some readers may not be interested in the terminology and issues we discuss here and would rather jump into the information-rich chapters that follow. However, we anticipate that many readers will be curious to explore sex-biased terminology. Such readers may find the remainder of this chapter thought provoking. Using the female perspective presents us with exciting avenues for future research. Thus, unlike other chapters where we focus on the biology at hand, here we explore a few examples, primarily in terminology but also in methodology, phrasing, organization, and emphasis that illustrate how the "female perspective" might differ from the usual (often male-centric) perspective. Such perspectives are so entrenched in the literature that one is nearly forced to use male-centric terms to avoid confusing reviewers and readers. To this end, we are not criticizing anyone for their use, but we hope pointing out such cases will provide food for thought.

Androcentric Terminology

One of the most striking aspects of male-biased terminology is that features of indeterminate sex may be given male names. A few examples follow. The embryonic genital tubercle gives rise to female and male genital structures (chapter 4) but is often referred to as the primordial phallus, although it could equally be called the primordial clitoris. Similarly, prostate glands are present in both sexes but in females are called the female prostate. Meanwhile, the enlarged clitoris of some females, e.g., spotted hyenas, is called a female phallus. The clitoris is described as masculinized, or virilized, rather than enlarged or prominent. As we will see in chapter 3, the process of sex differentiation is much more complex than having a female or male path. Even as female a structure as the vagina has a male bias, as its etymology is from the Latin for "sheath" or "scabbard," which focuses on its interaction with male genitalia rather than other functions. As an aside, the word *virile* is interesting. It characterizes sexual strength and energy, positive traits in females and males, but its synonyms are manly, masculine, or male. What is the equivalent word for female sexual strength and energy?

Common language in the field of reproductive biology may be loaded with value judgments or not consider the female perspective. For example, the term *miscarriage* suggests the mother is at fault for miscarrying the fetus, when, more likely, the fetus itself is defective. For a female, spending additional resources on an offspring that will not survive to produce young of its own would be a costly mistake with potential evolutionary consequences. *Embryo rejection* is a more apt term than miscarriage. Other medical terms are laden with value-based (generally negative) terminology, such as *luteal-phase deficit* rather than *short luteal phase*. Recalling Clancy's study, the Polish women had short luteal phases but were clearly fertile, and thus not deficient. Another example is using *cervical incompetence* or *cervical insufficiency*, rather than *early cervical dilation*. Early cervical dilation can lead to embryo rejection, which again could benefit the mother.

Females may be "mature" at first ovulation or conception or at first birth or mating. However, sexual maturation in females is often considered from the male perspective, such as when females are capable of being fertilized (Boness et al. 2002).

Female sexual behavior, too, is often described from the male perspective and textbook authors may go to awkward lengths to do so. An example is the tripartite classification of female sexual behavior as attractiveness (aka attractivity): "the stimulus value of a female to a male," proceptivity: "the extent to which a female initiates copulation," and receptivity: "the stimulus value of a female for eliciting an intravaginal ejaculation" (Nelson 2011:289). The sex biases of "attractiveness" and "receptivity" are clear and alternatives exist. For instance, from the female perspective, attractivity is solicitation—behaviors and cues used to attract potential mates—whereas receptivity is facilitation—behaviors used by females to achieve conception. Proceptivity too has a strong male bias.

Sex drive, or libido, is commonly assumed to be equivalent to proceptivity, but as Hrdy (2000:80; sidebar on page 5) points out, comparing the "sex drive" of a potentially fertile male with a non-ovulating female, or "[assuming] that the urge to mate derives from the same "motivation" or evolved for the same reason in both sexes" is the biological equivalent of comparing apples to oranges. Both sexes have heightened libido when their hormones dictate. Males are under more consistently high testosterone levels (albeit with some often ignored fluctuation), whereas females have more obvious hormonal peaks and troughs. Comparing the libido of an estrous mare with that of a gelding would be an equivalently mismatched juxtaposition. Phrasing sexual behavior from the male perspective suggests that females are not soliciting mating nor participating actively in the process. As we explore in chapter 6, females are not passive bystanders in mating.

Between mating and conception, several common perspectives focus on the activity of sperm. Readers may be familiar with the idea that sperm "race" up the female reproductive tract to the "passive" ovum, even though, over 70 years ago, Hartman (1957:419) concluded that "it is highly unlikely that sperm motility has the slightest value for ascent through the oviduct." Still, in 2016, Holt and Fazeli (2016:105) needed to remind their readers "that the 'sperm race' is no longer a tenable hypothesis." Though often retold, this idea is a misconception (pardon the pun). The orgasmic and other contractions of the female reproductive tract propel or impede sperm as appropriate. Female secretions nurture suitable sperm, store sperm, and biochemically alter sperm so that conception is possible. In general, once a female obtains sperm, her physiology manages their activity and function (chapter 6). For the most part, sperm are passive recipients.

Females may ovulate based on cues from the environment or endogenous hormone cycles. However, coitus and penile stimulation are also said to induce ovulation. Here penile stimulation does not refer to stimulation of the penis, but stimulation of the vagina or cervix by the penis. Again, what a male does is used to explain what happens in a female. Strangely, stimulation of the penis by the female (e.g., to solicit or induce ejaculation) is not described as important for mating (another passive female process presumably), whereas stimulation of the vagina or cervix is considered critical (and male active). Tangentially, one might

"Mothers and Others"
Sarah Blaffer Hrdy (1946–)

Sarah Blaffer Hrdy is an anthropologist and sociobiologist famous for her studies of motherhood and group dynamics in primates. Her major contributions connect research on nonhuman primates to humans. She is particularly well known for her theories on infanticide and female sexual strategies.

While pursuing her PhD, Hrdy investigated the evolutionary causes of infanticide among the langurs of Mount Abu, India. Her book *The Langurs of Abu: Female and Male Strategies of Reproduction* (1977) was controversial. Rather than treating infanticide as pathological behavior triggered by overcrowding, Hrdy proposed that infanticide was an adaptive strategy beneficial to the males practicing it. When male langurs take over a preexisting group, they may kill unweaned young, thereby causing the females to ovulate sooner because they no longer have suckling young. Infanticide thus increases the reproductive success of males, albeit at the expense of mothers, infants, and the male predecessors.

Hrdy also hypothesized that female primates developed counterstrategies against infanticide, including situation-dependent sexual receptivity and solicitation of multiple male partners. Because males almost never attack offspring they might have sired, a mother may protect offspring by manipulating information available to males about paternity. Far from passive bystanders, mothers are active agents in determining their lifetime reproductive success.

In *Mother Nature: A History of Mothers, Infants and Natural Selection* (1999), Hrdy explored possible maternal instincts. She pointed out that all mammalian females have innate maternal responses, but this does not mean that a mother automatically nurtures every offspring she produces nor that she does so with the same level of commitment. In humans, initial maternal commitment depends on a range of factors, especially perception of social support. Hrdy argued that the evolution of offspring that are as costly and as dependent on parental care for very long periods as are those in the relatively large-brained, bipedal genus *Homo*, required the concomitant evolution of assistance from conspecifics. This idea became known as "the cooperative breeding hypothesis."

In *Mother and Others: The Evolutionary Origins of Mutual Understanding* (2009), Hrdy examined the effects of prolonged dependence on alloparents (group members other than genetic parents) on offspring and parents. Over the course of development, already clever, manipulative little apes with rudimentary theory of mind became conditioned to read the intentions of others, appeal to them, and elicit their solicitude, thus producing novel ape phenotypes. These more "other regarding" ape youngsters were in turn subjected to directional Darwinian selection favoring those better at reading the mental states of others. These would be the youngsters most likely to be cared for and fed and thus survive. The emergence of apes interested in the thoughts and feelings of others and eager to attract and to ingratiate themselves with these others, laid the cognitive and emotional foundations for such distinctively human capacities as language and morality. In this way, she argued, without any foresight on the part of "Mother Nature" (her metaphor for Darwinian natural selection), regarding the eventual benefits of an improved ability to read the intentions of others and work with them toward joint goals, the peculiar mode of child-rearing in the lineage of apes leading to *Homo sapiens* served as "the prequel" to the "main human feature film."

Hrdy's work dismantled traditional views of motherhood and gender roles. Her studies of motherhood in primates illuminated the many things humans have in common with our relatives, as well as the ways in which we evolved.

(Photo courtesy of Sarah Blaffer Hrdy. The photographer is Dr. Anula Jayasuriya, and the child is the photographer's daughter, Shanika.)

argue that ejaculation is under female control as it often requires the stimulation of the female reproductive tract (or surrogate thereof) to occur. Returning to ovulation, one could argue that orgasm, rather than penile stimulation, influences ovulation in facultative ovulators (chapter 6). Of course, orgasm is a controversial subject when applied to non-human mammals, especially females (Fox, Fox 1971).

Often, terminology is defined differently when a trait occurs in a female versus a male. An example is the terminology for mating with more than one member of the opposite sex: *polyandry* and *polygyny*. One paper on marine mammals defines polygyny as "successful competitors mate and fertilize more than one female" (Boness et al. 2002:287). With this phrasing, any female with singletons has no potential to be polyandrous, even if she mates with multiple males. Thus, polygyny and polyandry cannot be consistently applied.

While not necessarily an aspect of male bias, classification schema used for mating systems can also distort the female perspective. Monogamy, polyandry, polygyny, and promiscuity are broad, commonly used categories that identify numbers of copulatory partners. But for natural selection, the important outcome of reproduction is progeny, not mating. The number of matings and the number of different mates are only relevant if these numbers directly correlate with the genetic contribution to offspring. The usual categorization of mating systems does not take that key issue into account. As a result, the number of categories has expanded. For instance, what was once monogamy is now broken into three categories: (1) social monogamy, (2) sexual monogamy, and (3) genetic monogamy. Mating is not key to fitness, but the genetic contribution to a reproductive effort is. Using number of matings to infer genetic relationships is problematic. Mating does not automatically assure paternity. Mating systems, per se, may be important from the male perspective but are less so from the female perspective. Use of molecular markers, instead of observed copulations, could lead to a categorization system based on the genetic contribution to progeny over the lifetime of individuals.

Similarly, sexual selection, while usually put in the context of mates is actually measured by successful conceptions and the offspring that result. That is, males compete to genetically contribute to conception. This is true. However, after mating, a female may continue to exert sexual selection. For instance, her physiology may select which sperm to use for conception or which embryos will implant.

"Conventional sex roles imply caring females and competitive males" (Kokko, Jennions 2008:919). The presumed biological basis is that, because individual ova are much larger than individual sperm, females put more resources into reproduction from the start and, thus, continue to do so (i.e., pre- and post-conception investment is high in females). That conclusion is problematic. It assumes that equal numbers of gametes are contributed to each mating. That may be reasonable for sea urchins with external conception but not for mammals. With internal conception, the number of sperm per mating is much higher than the number of oocytes. In mammals, the sperm count at any given copulation is in the millions or billions of sperm (84,000,000,000 per ejaculate in the case of boars; Estienne et al. 2008). Thus, the actual gametic cost for males to achieve conception is equal to if not much, much higher than for females. A more important argument focuses on the erroneous assumption that optimal decisions should depend on past costs, not future expectations. In this case, the conclusion that anisogamy (unequal gametes) leads to increased care commits the Concorde fallacy (Kokko, Jennions 2008). Good investment decisions look ahead to the probability of future gains or losses, not past history, which is why the US Security and Exchange Commission requires any investment prospectus to state that past performance is not a reliable indicator of future results. Overall, trying to link pre-mating to post-mating investment is fraught with logical difficulties (Kokko, Jennions 2008). Anisogamy may have led to internal conception as a way for females to prevent wasting large gametes rather than a way to increase parental care. As an aside, polar bodies (box 4.1) are technically an example of anisogamy but never mentioned in this context.

Since Aristotle, studies of conception have had a male bias. The fusion of oocyte and spermatocyte is usually termed *fertilization* and has a male active-female passive undertone. Conception is "a gender-neutral unbiased term for the fusion of gametes to produce a full genome. Unlike fertilization, conception implies two interactive partners, egg and sperm, contributing equally to the formation of the zygote" (Chen 2014:9). Dynamic oocyte activity and function during conception was visually evident in 1895, but those processes were not studied until 80 years later (Schatten, Schatten 1983). As a result, use of the term *fertilization* remains a convention in the field of reproductive biology and development. However, in using *fertilization* we unintentionally present the male perspective. The term *conception* refers to the union of male and female gametes (syngamy) and lacks a gender bias. Thus, in the remainder of the book, we use *conception* rather than *fertilization* to try to make a stride in changing this terminology.

Using Males to Understand Females

In an unusual methodological twist, male traits may be employed to measure female behavior. Strangely, the term *promiscuity* is commonly used for females that mate with more than one male but rarely in the reverse. For example, Google Scholar turned up seven papers when searched for titles including the phrase "male promiscuity" and 51 hits for the phrase "female promiscuity." Apparently, in titles, promiscuity is more remarkable when associated with females. Or perhaps the default assumption for females is monogamy and for males promiscuity? Some of these assumptions are discussed in the excellent book *Sexual Selections* (Zuk 2002).

The male perspective may be used in a variety of other cases to explain exclusively female traits. For instance, Wasser and Waterhouse (1983:23) compiled the following male-oriented explanations for reproductive synchrony, continuous receptivity, concealed ovulation, and orgasm in women.

> Polygynous women synchronized their menstrual cycles to avoid inundating males with "contradictory information" from independent cycles (Burley, 1979); estrus disappeared in women to facilitate male-male bonds (Etkin, 1963; Pfeiffer, 1969); loss of estrus evolved among women because it prolonged their period of sexual attractiveness to men, who provided them with meat in exchange for sex (Symons, 1979); concealed ovulation evolved to increase paternal certainty in humans and to force males into pair bonds (Alexander and Noonan, 1979); the female orgasm evolved to make women quiescent following copulations so as to prevent the male's sperm from leaking out of the vagina (Morris, 1967); the female orgasm evolved as a "by-product of selection for male orgasm (Symons, 1979). (Wasser, Waterhouse 1983:23)

Although the explanations for concealed ovulations have not been completely abandoned, they may be irrelevant as current research demonstrates women can and do make subtle but measurable (and observable) changes to their voice, scent, appearance, and behavior around the time of ovulation (Haselton, Gildersleeve 2016). Thus, ovulation may not be concealed at all. We return to humans in chapter 15.

Language

Pregnant pauses and seminal issues—such reproductive metaphors populate our speech to good effect. When non-reproductive terms are used metaphorically, the result is not always as helpful. For instance, in an article on conception, Bedford et al. (2004) refer to "the ripening follicle" as if it is a fruit to be plucked and eaten. Would one say "the ripening sperm" for gametes in the male ejaculate?

Our short compilation of biases is just a sampling of the myriad ways in which the female perspective is not part of the history or language of mammalian reproductive biology. Changing tone and emphasis to incorporate the female perspective might lead to some awkward phrasing. For instance, we are so accustomed to saying "external fertilization" that using "external conception" may seem odd. Similarly, use of "zygote" rather than "fertilized egg" will also take effort. But as our culture is increasingly accustomed to more gender-neutral language in the public arena, perhaps the time has come to also make our science more gender neutral.

With that thought, we leave the details of terminologies and gender perspectives behind and dive into an exploration of the female mammal as she reproduces. We will revisit biases as relevant, but the focus of this book is female reproduction, not only gender bias. Before we start in earnest, we review the evolution and taxonomic diversity of mammals: the topic of our next chapter.

CHAPTER TWO

Evolution and Diversity

She has not always been as we see her today. Her form, physiology, and even her milk are the result of more than 20 million years of evolution. However, in the form we would recognize as "hyena," her ancestors have been around as members of one of two closely related genera (*Hyena* or *Crocuta*) at least since the Miocene. During the Miocene, special changes to her dentition allowed her to crush bones and become a dietary specialist. Although she may be described as looking dog-like, that resemblance is only a matter of convergent evolution. Instead, as a member of the suborder Feliformia, she is more closely related to civets, meerkats, and cats than she is to dog-like carnivores, the Caniformia (e.g., dogs, bears, raccoons). Outside of her cousin hyenas her next nearest relative is the aardwolf (*Proteles*) with whom she shares many physical traits but not the insectivorous palate. In subsequent chapters, the story of our hyena female unfolds relative to each step of her reproductive life. (Ferretti 2007; Fourvel et al. 2015; Mills 1982)

Who were the first mammals? What do we know about their reproduction? What influenced the transition from egg laying to live birth? From an evolutionary perspective, did live birth (viviparity) originate once or twice? In other words, because viviparity occurs in both eutherians and marsupials, we must consider the question: did these two lineages diverge before or after the evolution of viviparity? In this chapter, we review the evolution of mammals and then turn our focus to the origins of and changes to key reproductive features that epitomize the major modes of reproduction in mammals today. The three major groups of mammals, Prototheria, Metatheria, and Eutheria, are distinguished by differences in reproduction. We review the very early history of the lineage that led to mammals as well as the origins of mammals and lactation. Pivotal evolutionary changes in mammals relevant to females include type of placenta and invasiveness thereof; associated alterations in developmental patterns; anatomical changes of the uterus, vagina, and mammary glands; and, ultimately, changes in type of maternal care, including lactation. These key changes are reviewed in detail elsewhere in the relevant chapters (e.g., "Anatomy," chapter 4; "Lactation," chapter 9). Here we review some of the key aspects of mammalian evolution that may be important for the reader less familiar with this aspect of mammals. Detailed examples relative to these features will also be revealed in subsequent sections and thus the more seasoned reader may wish to jump ahead to these sections.

The second part of the chapter explores the phylogenetic diversity of mammals. The complexity of mammalian evolution is not completely resolved, and multiple phylogenies exist for mammals. Consequently, we cannot go into depth on specific relationships among taxa. Phylogenies are hypotheses, which are updated and changed repeatedly. Our goal here is to introduce the reader to the story of mammals as a group and provide a context for the female perspective. However, for practical purposes, we provide a basic phylogeny of some major groups of mammals with the understanding that these relationships are likely to change as the field progresses.

Mammalian Evolution: The Major Events

The story of mammalian evolution is long and complex and, despite extensive research, still contains mysteries. Fossils remain buried and the details of current relationships among extant taxa are continuously updated. Nevertheless, today we see the result: enormous diversity particularly in terms of reproductive anatomy, physiology, and behavior. In many cases, reproductive changes have been step-wise, and one sees a gradual accumulation of increasingly coordinated reproductive traits. For example, small changes in the invasiveness of the placenta resulted in changes in offspring dependence on mothers and, consequently, changes in the degree or type of maternal care. Although some ancestral features are still present, e.g., egg laying in the platypus (*Ornithorhynchus anatinus*), no mammal alive today could truly be considered primitive (box 2.1). Indeed, some ancestral reproductive modes, such as that of the opossum (*Didelphis*), have persisted for millions of years and perhaps may be better suited for a suite of environments than the derived strategies of other mammals. Where do we as mammals have our reproductive origins? What is the story of female reproduction from 350 million years ago until now?

Laying on Land: The (Really) Early Years (350 MYA to 150 MYA)

Vertebrates crawled out of the oceans to live on land roughly 350 million years ago (MYA) in the Devonian (box 2.2). These tetrapods (so called because of their four feet) had limbs rather than fins that made locomotion more efficient for an active terrestrial life. However, eggs of these earliest vertebrates remained tied to water, as are the eggs of frogs, salamanders, and caecilians (collectively, the Amphibia). Consequently, the next key innovation was to free offspring development from moist environments by developing a protective covering for the embryo and its yolk. Shells greatly reduced desiccation and allowed tetrapods to move farther inland to carry out their usual life cycles, including reproduction. The shell is (and presumably was) deposited within the reproductive tract before the egg is deposited on land. Coordinated with the evolution of a shell, the embryo also

BOX 2.1. Ancestral/Derived versus Primitive/Advanced

The phenotypes of all individuals are a result of events of both the past and the present. We are born with a set of genes from our parents and a body built from the food our mothers processed and passed on to us in utero. Before birth, our mothers filtered the world for us. Our ancestors, with their genes, provided instructions for how to use those maternally provided building blocks to create hands or hooves, skin or scales, fins or feet. Even at birth, we are a combination of the deep past and the more recent present. Each pattern we observe in extant mammals is a product of past environments and ancestral genomes. The past constrains the future in both individuals and lineages.

Just as no two individuals are identical, so too, no two lineages are the same. Thus, reproductive processes also differ. Distinctive features are often specializations for particular habitats, diets, social interactions, or against particular predators, parasites, or disease vectors. Such distinctive features are called derived traits and are shared by members of a given lineage. For instance, women have prominent breasts, a derived feature of human reproduction, whereas cows and ewes have udders, a derived feature of ruminants (figure 4.6). Both breasts and udders produce milk and work extremely well for females and their young. Neither anatomical form is more advanced, nor is either a better vehicle for milk. Similarly, the indistinct anatomy of species without prominent nipples or udders is not primitive, although it may be the ancestral condition. Each mammalian lineage is a specific combination of ancestral and derived traits that have worked well for previous generations. These suites of characters are neither primitive nor advanced, and no one way of reproducing is better than another.

changed to survive within the confines of the new maternal barrier. New tissues, the extra-embryonic membranes (amnion, chorion, and allantois) facilitated new mechanisms for gas exchange, waste collection, and protection. Collectively, vertebrates with shelled eggs are referred to as the Amniota. Forming a shell around an embryo also means conception must occur internally. Consequently, mating patterns needed to shift. Internal conception is present but rare in non-tetrapod vertebrates, but in tetrapods (except amphibians), internal conception has become the norm. Thus, this seemingly simple alteration, producing a shelled egg, changed the dynamic between mother and offspring, as well as between females and males. The shell isolated the developing embryo from the female's body. Any communication or nutrient transfer between the mother and offspring became more difficult because it now had to traverse the shell.

Which came first: the eggshell or the extra-embryonic membranes? The names for this new adaptation do not resolve the question, as the term *cleidoic egg* emphasizes the maternal shell, whereas the term *amniotic egg* emphasizes the internal membranes. Our previous discussion tacitly put the evolution of the shell first, but one might envision how the extra-embryonic membranes could be a result of maternal-offspring conflict. Perhaps the membranes were a way for the embryo to get more resources from mom or to prevent maternal influence. If so, the maternally produced eggshell may have been a way to limit transfer to or from offspring and thus a pre-adaptation to life on land. Inquiring minds want to know, but at least for now the answers are shrouded in the mists of time.

Certainly, the suite of characteristics (maternally derived, shelled eggs and the conceptus-derived amnion, chorion, and allantois) allowed an extensive radiation

BOX 2.2. Evolving Mammals in Changing Environments

What was happening during each major evolutionary transition in the ecology experienced by ancestral mammals? How did climatic or biogeographic factors influence early reproduction? Here are some relevant details for the time frame in which mammals arose and diverged.

The Paleozoic era was characterized by rapid diversification of biodiversity, as well as large-scale continental movement. Marine life dominated, but plants, insects, and vertebrates also moved onto land. At the end of the Paleozoic, the climate became hotter, with a phase of high atmospheric carbon dioxide in the early Triassic (Kidder, Worsley 2004). Biodiversity on Earth plummeted (the Permian-Triassic extinctions), but the precursors to mammals, the cynodonts, emerged. Thus, the ancestors emerged at a major transition period—the end of the Paleozoic and the beginning of the Mesozoic.

The Mesozoic is a very interesting period for mammalian evolution. The first 50 million years, the Triassic, were mostly hot and dry and the oldest-known mammals come from this time. During the first 5 million to 10 million years, the tropics may have been uninhabitable with temperatures lethal to most terrestrial vertebrate life (Sun et al. 2012). Our mammalian ancestors may have been evolving independently on either side of this lethal region. Later, much of the habitat was desert-like with periods of increased rain and humidity. The supercontinent of the Paleozoic (Pangea) was beginning to break apart from north to south, toward the end of the Triassic. In the next 50 million years (Jurassic), Pangea more or less completed its break up into the major continental regions of today. The climate was more mesic, and the dominant terrestrial life forms were dinosaurs. Undoubtedly, dinosaurs influenced the early evolution of

mammalian reproduction, but exactly how is unknown. The Jurassic is also the time when the metatherian-eutherian split may have occurred (Luo et al. 2011). The last part of the Mesozoic, the Cretaceous, was hot and humid with diverse continental landmasses. Many new niches were connected with the diversification of flowering plants and insects. A Cretaceous fossil, similar to the platypus, suggests that monotremes were well diversified by the early Cretaceous (Archer et al. 1985).

The Cenozoic brings us to modern times and the major diversification of mammals. Representatives of many major orders are present in the Eocene, of major families in the Miocene, of major genera in the Pliocene, and many species in the Pleistocene.

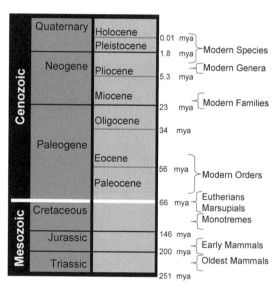

Time line of mammalian evolution. Mammals arose in the Mesozoic and ancestors to egg-laying mammals appear in that era. However, the modern therian orders first appeared early in the Cenozoic. Modern families are present about 23 million years ago (MYA), modern genera about 5 MYA, and modern species perhaps 2 MYA. Diagram by Teri Orr modified from the Burke Museum, Seattle, WA; used with permission.

of vertebrates farther inland, while anamniotes (fishes, amphibians) remained in or near water. Eventually, amniotes colonized deserts, rain forests, mountains, tundra, and even polar regions. They flew into the skies as pterosaurs, bats, and birds. A few amniotes even returned to aquatic environments, such as plesiosaurs, whales, and sea turtles.

Today, the living amniotes include mammals, birds, crocodiles, lizards, snakes, and turtles. These groups are separated into three lineages, all of which arose around the same time. One lineage (Anapsida) may have led to turtles although this is a topic of much discussion among paleontologists. A second (Diapsida) led to lizards, snakes, crocodiles, dinosaurs, and birds. The third lineage (Synapsida) led to mammals. This tripartite split means that reptiles, per se, are not ancestors to mammals but arose at nearly the same time. The term *mammal-like reptile* is often used when describing the early precursors of mammals. A more accurate term for the earliest members of the lineage would be *stem-mammals*, or *proto-mammals*. Thus, fossil groups, such as the pelycosaurs (e.g., the sail-backed *Dimetrodon* from the Permian) and therapsids, are part of the ancestral line leading to mammals.

Synapsids were abundant when they first arose in the late Paleozoic (around 260 MYA), but many became extinct between the Paleozoic and the Mesozoic, perhaps at the Permo-Triassic mass extinction. Subsequently, dinosaurs dominated the Mesozoic landscape. At some point, probably in the Triassic, one (or perhaps more than one) of the few remaining synapsids developed features we recognize today as mammalian. What were these features?

Currently, a standard suite of skeletal features assign fossils to the mammalian lineage. Among these traits are a lower jaw made of a single bone, three middle-ear bones, non-uniform teeth that differ in both location and shape (i.e., incisors, canines, premolars, molars), and the precise occlusion of teeth from the upper jaw with teeth from the lower jaw. Fossil and molecular evidence suggest an initial divergence of mammals from therapsids perhaps 200–250 MYA (Lefèvre et al. 2010). A fossil mammal, *Morganucodon*, has most of these traits.

Origin of Mammals (150 MYA to Today)

Lactation is the hallmark of mammalian reproduction, but a few other reproductive features are found in all mammals. One is embryo retention (gestation), although this is not unique to the Mammalia. Embryo retention preceded the origin of mammals and independently evolved in taxa as diverse as sharks, fish, and lizards (Blackburn 2015). Embryo retention results in an extended period of contact between developing offspring and mom, using hypertrophied extra-embryonic membranes to advance growth of offspring (Lombardi 1994). Embryo retention is central to reproduction in therians but also occurs in the egg-laying monotremes. As described in more detail in chapter 7, monotreme mothers also provide resources to developing embryos before egg laying.

Oviparity (egg laying) is probably the ancestral condition for mammals. Shelled eggs impose a barrier between mother and offspring. That barrier has advantages but also disadvantages. Removing the shell before egg laying would allow more extensive contact between developing offspring and mom. The extended contact would eventually lead to hypertrophy of the extra-embryonic membranes and apposition with maternal tissues into a variety of complex placental structures (Lombardi 1994). If an eggshell is no longer deposited around the conceptus, live birth (viviparity) occurs instead of egg laying. If numbers mean anything, viviparity was clearly an advantage as nearly all mammals are viviparous.

In viviparous mammals, the length of the placental phase of gestation varies. A limited duration of placental exchange characterizes marsupial mammals (e.g., opossums; koalas, *Phascolarctos cinereus*; Tasmanian devils, *Sarcophilus harrisii*; kangaroos, *Macropus*). Noteworthy is that not all marsupials have marsupia (pouches), and the formal name for the group is Metatheria (metatherian). Here we use *marsupial* and *metatherian* interchangeably. In this group, gestation lengths range from 2 to 5 weeks with little connection to maternal size. For example, the 35-day gestation of an 8-kg koala is slightly longer than the 33-day gestation of the 26-kg red kangaroo (*Macropus rufus*). Although gestation lengths of metatherians are often characterized as short, the 20- to 80-g *Antechinus* with a 4-week pregnancy has a longer gestation than rodents of similar size, such as mice with gestations of about 3 weeks (Hayssen et al. 1993). Thus, metatherian gestations are not always short and may even be lengthened considerably when they include delays (e.g., delayed implantation; chapter 7).

Complex and extensive maternal-fetal exchange occupies the majority of gestation in eutherian mammals. Eutherians are also the mammals with which we are most familiar and include aardvarks (*Orycteropus afer*), bats, cats (Felidae), dugongs, elephants (Elephantidae), and on through the alphabet all the way to yaks (*Bos grunniens*) and zebras (*Equus*). Their placental structures are highly diverse and the multiple variations are described in chapter 4. Note that the term *placental mammals* in place of the term *eutherians* is misleading because metatherians also have placentas (Renfree et al. 2013). Furthermore, both metatherians and eutherians are viviparous.

How often viviparity arose in mammals is unknown. The traditional view is that viviparity originated once and that the marsupial and placental reproductive modes diverged later, but some evidence suggests marsupials and egg-laying mammals (monotremes or Prototheria) are more closely related to each other. If so, viviparity must have originated twice, independently in marsupial and eutherian mammals (Sharman 1976). The jury is still out, but the traditional view of a single origin predominates current opinion (Blackburn 2015; Werneburg et al. 2016). Given that soft tissues do not fossilize well, hard data on these questions will be difficult to obtain.

The evolution of many reproductive traits is unusual in one respect. Most mammalian traits (e.g., body length, number of toes, and stomach morphology) are characteristics of individuals, and natural selection acts independently on each individual in the population. Many reproductive traits, such as gestation length or milk composition, involve two or more individuals. For instance, the length of lactation is not determined solely by the mother or by the young but rather depends on the interaction of the two. In such cases, selection acts on at least two individuals simultaneously. Reproductive characteristics, like lactation, are thus characteristics of interactions between individuals and not of a single individual. The evolutionary processes that affect reproductive characteristics have not been delineated, but trait group or interdemic (between populations) selection may be important (Wilson 1979). The study and characterization of lactation (or any reproductive trait) solely on the molecular, cellular, or individual level will miss the essential, interactive, and integrative aspects of this phenomenon. For mating, gestation, birth, and nursing, selection is operating on pairs or groups simultaneously.

Origin of Lactation

The development of lactation was probably a key feature in the origin and later success of mammals in adapting to the changing environments of the Mesozoic and Cenozoic, and was unquestionably fully functional well before the end of the Triassic.—Lillegraven 1979:260

What parental care was present in early mammals? We do not know. Snakes and lizards may lay eggs without substantial subsequent maternal care. However, if mammals are defined by lactation, then, by this very definition, maternal care, in the form of nursing was present. Today's monotremes have long periods of lactation, but we have no way of knowing how long lactation was in early mammals.

How did lactation originate? Soft tissues, such as mammary glands, do not fossilize well, and most behaviors (like nursing) do not fossilize at all. Thus, while paleontologists use skeletal characteristics (usually from the skull) to identify mammals in the fossil record, the evolutionary ecologist is often left empty handed. Teeth are key to identifying mammals and, as far as bones go, tell more of a story than one may expect from a first glance. Features associated with teeth help us interpret the evolutionary patterns of lactation.

As any breast-feeding mother knows, infants' teeth can cause pain. In fact, too much discomfort may cause a mother to stop nursing. If tooth development is delayed, then lactation can be extended. Since lactation obviates the need for biting and chewing food, delaying tooth development is not a problem for a hungry neonate. Consequently, neonatal jaws have time to grow and strengthen before teeth are necessary. This allowed teeth to develop specialized functions thereby increasing their efficiency. Unlike lizards and sharks, mammalian teeth are not continually shed and replaced, but instead mammals have a set of juvenile teeth and then a set of adult teeth—an arrangement termed *diphyodonty*. The presence of specialized dentition and deciduous teeth infers the presence of suckling (Luo et al. 2004). Thus, the presence of diphyodonty in *Morganucodon*, from the late Triassic to the early Jurassic (Sánchez-Villagra 2010), suggests lactation (or at least nursing and by reasoning lactation) was present at this time. A second derived feature of mammals, a secondary hard palate, is also associated with suckling and infers the presence of lactation (Maier et al. 1996). But what drove the origin of lactation?

The beginning of lactation required a suite of events (Hayssen, Blackburn 1985). To start, mothers needed to stay with their eggs after they were laid. Advantages to maintaining such proximity might be to reduce predation, to provide protection from the elements, or to provide warmth to facilitate more rapid growth. If selection favored some form of incubation, then it may also have favored the development of a specialized, vascularized patch on the mother's abdomen. Skin is replete with various glands, including sweat and sebaceous glands, and these would presumably be present on the hypothetical incubation patch. If so, maternal secretions could be incidentally absorbed by the eggs or ingested by hatched young. Thus, mothers could begin to provide water, nourishment, or immune protection to their young. Skin secretions have many functions, and any or all of these could be the ancestral function of proto-lacteal secretions. For instance, milk may have initially evolved to prevent the desiccation of eggs or to protect against microbial or fungal attack (Hayssen, Blackburn 1985). Examining the origins of milk constituents may shed light on this issue (Oftedal et al. 2014).

Diversity of Mammals: The Big Three

Once the first mammals were present they diversified and, in many cases, aspects of their reproduction changed. Much of the early evolution of mammals is described relative to fundamentally different reproductive modes. However, these modes are not examples of advancements but rather different strategies that function well in certain contexts. Regardless, for ease of discussion, we consider mammals in three main categories, or modes of reproduction, as demonstrated by the three taxa, Prototheria, Metatheria, and Eutheria.

Evolution within mammals is complex and has generated enormous diversity, particularly regarding reproductive physiologies and behaviors. Changes relating to reproduction have led to the gradual accumulation of increasingly unique reproductive patterns. Many facets of those patterns changed in synchrony. For example, alterations in the invasiveness of the placenta altered how offspring depended on mothers and the changes in offspring dependence altered the degree of maternal care. Natural selection did not act on these features in isolation but rather simultaneously on mother and offspring. Thus, the evolution of reproduction is the result of selection operating on at least two individuals.

No one pattern is better than another, but each suite of reproductive characteristics works tolerably well in particular surroundings, e.g., the reproductive physiology of a bat would not be suitable for a whale. As we explain in box 2.1, the term *ancestral* is not a synonym for primitive. Similarly, recent (derived) patterns are not necessarily advanced or better than their precursors. For instance, egg laying may be an ancestral form of reproduction, but thousands of species of birds and insects also lay eggs and are not disadvantaged by doing so. Today's monotremes are the successful result of millions of years of evolution as is their reproduction.

The Egg-Laying Monotremes (Monotremata or Prototheria)

The earliest mammals laid eggs, and today's echidnas (*Tachyglossus*, *Zaglossus*) and the platypus still do. In this regard, egg laying is clearly an ancestral trait (Wourms, Callard 1992). We detail the biology of egg laying in monotremes later in the book (chapter 8). Although egg laying is ancestral, the evolutionary histories of other distinctive features of the reproductive biology of the egg-laying mammals are less clear.

Collectively, the egg-laying mammals are classified as the Monotremata, from the Greek for *single hole*. The name derives from the presence of a cloaca, the single opening from which both reproductive and excretory (intestinal and urinary) products exit the body. Having a cloaca is presumably an ancestral trait, as it is shared with frogs, turtles, lizards, and birds. Other mammals, besides monotremes, also have cloacas (chapter 4). Monotremes have fully developed mammary glands and complex milk, but a second distinction of the Monotremata is the lack of nipples for the release of milk. Instead, monotreme milk is secreted through pores. Consequently, although milk and lactation are defining features of mammals, the presence of nipples, or teats, is not. Those evolved later. Epipubic bones, a pair of bones connected to the pelvis but projecting away from it, are present in monotremes (as well as marsupials) and are often associated with having an abdominal pouch. But only the echidnas have pouches. Thus, pouches may be a derived condition independent of epipubic bones.

The order Monotremata is the only surviving taxon from subclass Prototheria and is, for this reason, of extreme evolutionary interest. If monotremes represent a separate taxon from the therians (marsupials and eutherians), then that split occurred around 166–210 MYA.

The Transcendent Marsupials (Metatheria)

Some terms used to describe marsupials are based in reproductive anatomy. For example, *marsupial* refers to the *marsupium* (abdominal pouch). Despite the name, many marsupials do not have pouches. In addition, when present, pouches are highly variable. For instance, the pouch of the water opossum (*Chironectes minimus*) opens backward so that when a female swims her pouch does not fill with water. Within a single family, pouches can vary a great deal. For instance, in the Dasyuridae, pouches range from relatively flat and nonexistent to deep and extensive. Pouches are not an ancestral trait for metatherians but have evolved independently several times (Hayssen et al. 1993).

American opossums are in the family Didelphidae, and the family name, meaning two uteri, describes another feature of metatherian biology. In marsupials, nearly the entire reproductive tract is paired. All mammals have two ovaries and two oviducts, but most marsupials, in addition, have two uteri, two cervices, and two (or three) vaginas.

The most distinctive features of marsupial reproduction are (1) a gestation with a short period of placental exchange; (2) birth of neonates with a highly developed olfaction, well-developed forelimbs, and only the rudiments of other anatomical structures; and (3) a highly complex lactation often split in two phases: a teat attachment phase and an intermittent suckling phase. Milk composition varies substantially between these phases. Even more amazing, some kangaroo females can simultaneously produce milk of one composition for pouch young attached to a teat and milk of a completely different composition at an adjacent teat for a joey outside the pouch (Hayssen et al. 1985; Graves, Renfree 2013). The physiological control of lactation in marsupials is highly derived.

We have already mentioned the eutherian bias in calling ourselves placental mammals when metatherians also have placentas. We also have a tendency to think about marsupial neonates as exteriorized embryos. However, no eutherian embryo at a similar stage of development could climb unaided up its mother's abdomen, locate a teat, and attach to it. Overall, "marsupials are best regarded as alternative to, rather than primitive to, eutherian mammals" (Hayssen et al. 1985:617).

Eutheria

Eutherians are the most abundant and diverse group of mammals. The extant orders diverged from one another very rapidly between 50 and 80 MYA. Eutherians are characterized by a long phase of placental exchange during gestation and, relative to marsupials, a simple lactational physiology. The diversity of eutherian reproductive patterns forms much of the content of this book, and we will refer to many groups by their taxonomic names. Thus, we need to briefly review the taxonomy of mammals.

Diversity of Mammals: A Closer Look at Closer Relationships

Modern forms of mammals from these three modes tally to roughly 5,400 species in roughly 29 orders (figure 2.1; Wilson, Reeder 2005). The specifics of taxonomy are subject to debate by experts. For this reason, we suggest the reader find a recent overview as we only present a general phylogeny here.

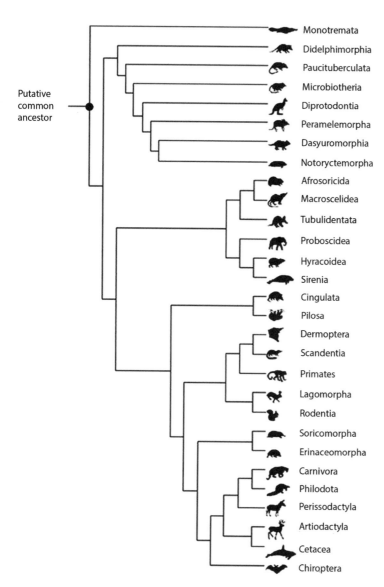

Figure 2.1. Phylogeny of 29 mammalian orders. The dot illustrates the most recent common ancestor (about 210 million years ago). The top branch denotes the egg-laying mammals (Monotremata); the next seven branches are orders of marsupials (Metatheria); the remaining branches are in the Eutheria. Phylogeny and illustrations from the Burke Museum, Seattle, WA; used with permission.

A few features of that diversity deserve mention. First, about 42% of all mammalian species are rodents (Rodentia). Rodents include laboratory rats and mice, as well as squirrels (Sciuridae), beavers (*Castor*), gophers (Geomyidae), voles (e.g., *Microtus*, *Myodes*), lemmings (e.g., *Dicrostonyx*, *Lemmus*), hamsters (e.g., *Mesocricetus*, *Phodopus*), gerbils (e.g., *Meriones*), porcupines (Erethizontidae, Hystricidae), chinchillas (*Chinchilla*), coypus/nutria (*Myocastor coypus*), capybara (*Hydrochoerus hydrochaeris*), and naked mole-rats (*Heterocephalus glaber*), among others. A second diverse group is bats (Chiroptera) with 21% of all mammalian species. For a long time, bats were divided into mega-bats (flying foxes, old-world fruit bats; Pteropodidae) and micro-bats (all the others), but current phylogenies

BOX 2.3. What's in a Name?

Animal names are important but also confusing and complicated. The current scientific naming of mammals began with Linnaeus in 1758. Since then, each mammal has been given a unique binomial name, e.g., *Homo sapiens*, comprised of its genus, e.g., *Homo*, and species, e.g., *sapiens*. Unfortunately, scientific names are not always familiar, for example, *Bos grunniens* or *Bos mutus* is not as recognizable as yak and rabbit is more obvious than *Oryctolagus cuniculus*. Unfortunately colloquial or common names can be ambiguous in several ways. First, they may refer either to a specific species, e.g., spotted hyena (*Crocuta crocuta*), or to larger group of similar mammals, e.g., hyenas. For instance, the name *rhino* refers to a family of mammals, the Rhinocerotidae that includes five species in four genera all of which are colloquially known as rhinos (white rhino, *Ceratotherium simum*; Sumatran rhino, *Dicerorhinus sumatrensis*, black rhino, *Diceros bicornus*; Javan rhino, *Rhinoceros sondiacus*; Indian rhino, *Rhinoceros unicornis*). Second, some species have more than one common name. For instance, *Rangifer tarandus* is known either as reindeer or as caribou, whereas *Cervus elaphus* is called elk or wapiti in the United States but is known as red deer in Europe and the United Kingdom. To confuse matters further, the same common name in different areas may refer to different species. For instance, in the United

Kingdom, elk refers to *Alces alces* (a moose in the United States), but in the United States, elk refers to *Cervus elaphus*. Finally, common names may be deceptive, for instance, the flying lemur (*Cynocephalus*), neither flies nor is a lemur and hence is better referred to as a colugo. Thus, common names are not as precise as scientific names, but both are useful.

Generally, in our text, we use both the common and scientific name for each taxon the first time we refer to it in each chapter. In the rest of the chapter, we then use the common name, unless ambiguity would result. Usually, we associate genus or family names with common names, e.g., elephant, Elephantidae, but if we give a species-specific common name, then we often provide the full scientific name, e.g., African elephant (*Loxodonta africana*); Asian elephant (*Elephas maximus*). On occasion, for domestic species, or when we think our meaning is clear, we will use a common name without the scientific name, e.g., cattle, women, platypus. We will also do this when the common and scientific names are identical or nearly so, e.g., dugong/*Dugong*, giraffe/*Giraffa*.

Some common names that are used frequently by mammalogists may not be familiar to those outside the field. For instance, *ungulates* refers to all mammals with hooves: the odd-toed ungulates (Perissodactyla) and the

even-toed ungulates (Artiodactyla). The perissodactyls include horses (Equidae), tapirs (Tapiridae), and rhinos. The artiodactyls include pigs (Suidae), peccaries (Tayassuidae), hippos (Hippopotamidae), camels (Camelidae), deer (Cervidae), giraffes (Giraffidae), pronghorns (Antilocapridae), bovids (Bovidae), and a few others. Recent findings have led to another term, Cetartiodactyla, giving nod to the close relationship between cetaceans (e.g., baleen whales, Mysticeti, and toothed whales, Odontoceti) and artiodactyls. Within the artiodactyls, deer, giraffes, bovids, and a few others have a specialized compartment of their stomach, the rumen, which aids in digesting the cellulose in plants. They form a taxonomic group, the Ruminantia, or ruminants. Complex stomachs to digest cellulose are also present in sloths (*Bradypus, Choloepus*), kangaroos (*Macropus*), camels, and others, and these animals may also chew their cud, or ruminate, but because their overall anatomy is not the same as that of the deer, giraffe, and bovid cud-chewers, these other species are not taxonomically ruminants. Similarly, not all dietary carnivores are taxonomic carnivores. Carnivora is the formal name for the order of mammals that includes hyenas (Hyaenidae), dogs (Canidae), cats (Felidae), weasels (Mustelidae), seals (Otariidae, Phocidae), genets (Viverridae), walruses (Odobenidae), bears (Ursidae), and raccoons (Procyonidae), among others.

suggest that the dichotomy is more complicated. Rather than fall into one camp or another in regard to bat taxonomy, we will refrain from using subdivisions (i.e., mega vs. micro or yin vs. yang chiropterans). Regardless, the relationships within bats are enormously diverse. Beyond the bats and rats, no other eutherian lineage has more than 10% of mammalian diversity. In fact, some eutherian orders include very few species, such as Tubulidentata (the aardvark, 1 species), Dermoptera (colugo [arboreal gliding mammals from southeast Asia], 2 species),

Thus, these mammals are all taxonomically carnivores (or, more properly, carnivorans), even though the giant panda (*Ailuropoda*), eats only bamboo and is taxonomically a bear. But carnivore is also the word for a dietary grade of animals, those that eat meat. In the literature for mammals, similar issues (taxonomic vs. diet ambiguities) exist for anteaters and insectivores. Generally, mammalogists refer to a group of species that share a common ancestor as a *clade*, whereas another term, *grade*, is used for group of species that share a common characteristic but are not related to each other. Thus, the taxon, Carnivora, is a clade, but the dietary group, carnivore, is a grade.

A few other names may cause confusion. First, hystricomorph rodents are suborder of Rodentia that includes porcupines, cavies, agoutis, chinchillas, naked mole-rats, nutria, viscachas, and others. The group is characterized by a distinctive (and esoteric) attachment of jaw muscles to the skull. Second, cetaceans (also discussed above relative to the Artiodactyla) come in two varieties depending on diet. The mysticetes are the huge baleen whales that sieve tons of plankton and krill from the oceans. The odontocetes (dolphins, e.g., *Tursiops*; porpoises, e.g., *Phocoena*; killer whales, *Orcinus*, sperm whales, *Physeter*) are the toothed whales whose diet includes squid, fish, and seals. As is obvious from these examples, the term *whale* refers to different taxonomic groups. The name *seal* is another term with is-

sues. It lumps together mammals with very different ancestries, the phocids (Phocidae: earless seals) and the otariids (Otariidae: eared seals, with sea lions and fur seals). Taxonomists have not decided whether the two families shared an ancestor before or after flippers evolved (Uhen 2007). Another group, walruses, is probably related to the otariids. Pinniped and Pinnipedia refer to all three fin-footed, carnivorous mammals. As a final example, the common name sloth refers to mammals of two entirely different families, the three-toed sloths (*Bradypus*, Bradypodidae) and the two-toed sloths (*Choloepus*, Megalonychidae).

Whenever we use a common name, we often provide a scientific (usually genus) name; however, listing all the scientific names may render the text difficult to parse (as you probably noticed). Consequently, we have two alphabetical taxon indexes one for common names and one for scientific names of extant (not extinct) mammalian taxa at various levels. These indexes include both grades (e.g., seals) and clades (e.g., otariids) of mammals. Associated with each entry are the chapter numbers in which the name is used. These indexes have benefits. First, readers interested in specific taxa will be able to find the topical chapters where we use that taxon as an example. Second, readers unfamiliar with specific groups will be able to identify them. Third, when we do refer to an animal without further specification, as in our spotted

hyena chapter introductions, the name will be present in an index with more detail.

Also worth mentioning is that names can be subject to bias. The names of the three subclasses of mammals, Prototheria, Metatheria, and Eutheria, are a case in point. "Eu" is from the Greek meaning good but often translated as "true" in scientific names and "theria" refers to "wild beasts." Thus, Eutheria means "good" or "true" mammals with the resultant implication that monotremes and marsupials are neither good nor true. This general eutherian bias means that the Metatheria are often viewed as transitional between the "true" mammals and the "proto" mammals. Consequently, we generally use the terms *marsupial* and *monotreme* when we refer to Metatheria or Prototheria, respectively.

In writing a book from the female perspective, we felt that using female-specific common names, e.g., queen, vixen, bitch, would be appropriate. However, these names can be distracting and thus we generally avoid their use. In some cases, however, using the female-specific name generates a more fluid and meaningful text, e.g., using *woman* instead of *human* in the title of our final chapter. Thus, we will not completely abandon words, such as *ewe, mare, sow,* or *cow,* especially for domestic species, but will use female-specific terms sparingly, when they are appropriate and when the taxonomic context is clear.

Proboscidea (elephants, 3 species), Hyracoidea (hyraxes, 4 species), and Sirenia (dugongs and manatees, 5 species; Wilson, Reeder 2005). All these names can be confusing, and we highlight that issue in box 2.3.

Evolution in Retrospect

Mammals arose about 200 MYA from synapsid ancestors, who themselves arose about 300 MYA from early tetrapods. The fossil record does not easily reveal the mysteries associated with early mammalian reproduction, but we have a basic picture. Egg-laying mammals were the earliest and survive today in the Monotremata, the two echidnas and the platypus. The viviparous therians are currently split into two groups: Eutheria and Metatheria (marsupials). Mammals today inhabit nearly all continents and a huge variety of habitats. Reproductive adaptations to these habitats are key to that success.

The story of the evolving female mammal is ongoing. Today, mammalian evolution includes adaptations to human-derived environmental changes, for example, from global warming (Chapter 14). The evolutionary history of reproduction in mammals is the basis of the rest of this book. Thus, in line with the astute comment that "nothing in biology makes sense except in the light of evolution" (Dobzhansky 1973:125), this aspect (evolution) of female reproductive biology is a key feature of our text. Evolution generated the diversity of form and function we see in mammals today as well as the diversity of interactions females have with each other, with their offspring, with mates, and with the environment. Exploring those diversities constitutes the remainder of this book.

PART ONE

THE REPRODUCING FEMALE

Reproduction is embodied by an individual female from inside to out. At her core are genes that have been subject to generations of inheritance and natural selection. These genes code for proteins that comprise the tissues that result in her structural anatomy. In turn, her anatomical structures undergo a suite of physiological processes. Each of these components, inheritance, morphology, and physiology, is key for her reproduction, and in the first part of this book, we cover each in turn.

We represent the relationships between these variables as a series of nested circles (see figure). We start our exploration of these representative circles in chapter 3 with a focus on inheritance. Nearly all the anatomy and physiology of reproduction has genetic underpinnings, but the path from gene to phenotype is incredibly complex. Chapter 3 summarizes some key genetic factors relevant to female reproduction. Genes code for the building blocks of the body, and thus next, we discuss anatomy (chapter 4)—the study of structure and how structure matches function. In the case of reproducing females, anatomical features range from the ovaries and the uterus to mammary glands and the placenta formed by the developing fetus. To conclude the inner workings of our reproductive female, we explore how anatomical structures function together through physiology. Physiology (chapter 5) operates on many levels, from mechanisms within a cell to interactions among organs and integration across organ systems. Understanding physiology, thus, is imperative to uncovering how a female interacts with her environment.

CHAPTER THREE

Inheritance

Our female hyena's anatomy and physiology are dictated by underlying genes that are turned either off (silenced) or on (expressed). The collection of all of her genetic material (her genome) is analogous to a blueprint. Her genes code for amino acids, the building blocks of proteins. Proteins are structural components of muscles and organs. As enzymes and hormones, proteins are also key to the dynamic machinery within cells. The transcription and translation of genetic material to form proteins started at her conception and will continue until she breathes her last breath. During her development, sex determination began, limbs formed, stem cells migrated, and cells eventually differentiated to become tissues and organs. All these processes depended on underlying genes and their products. After birth, development slowed but growth and maturation continued through to puberty. As an adult, her hormonal fluctuations, particularly those associated with reproduction, are driven by her perception of environmental and internal cues. However, the production and release of her hormones involves interactions between and within cells and, at the most fundamental level, the expression of genes. Our female hyena is not solely a product of her genes, because her environment shapes her as well, but that is a story for another section. (Drea et al. 1998; Frank et al. 1985; Glickman et al. 1987, 1998, 2005)

> . . . in utero development has not only provided a safe environment for fetal development, but also one in which the maternal genome plays a dominant role.
>
> —Keverne 2015:6838

The innermost circle of the diagram at the start of this section represents the core genetic material and includes sex chromosomes and sex determination as well as other core processes that affect differences between females and males. Genetic information is central not only to every cell in the body but also to the development of every individual and ultimately to the evolution of populations.

Genetic information resides on chromosomes, which are long strands of DNA (deoxyribonucleic acid). DNA has a complex structure similar to a twisted ladder. The rungs of the ladder are the nucleic acids, adenine (A), thymidine (T), guanine (G), and cytosine (C). These are the four letters of the DNA alphabet. The 64 different triplet combinations of DNA letters can be translated into amino acids, the building blocks of proteins (e.g., TAT codes for tyrosine and CAT codes for histidine). Specific triplet codons also indicate when a gene starts (e.g., ATG) or stops (e.g., TAG). The lengthy nucleic acid chains are bunched up and folded into more compact structures, chromosomes, with one long double-stranded helix of DNA per chromosome. Most chromosomes come in identical pairs with one from the mother and one from the father. The number of pairs is specific to each species (humans have 23 pairs, therefore 46 individual chromosomes). One pair of chromosomes, the sex chromosomes, differs between females and males, an observation noted by Nettie Stevens in 1905. Not all mammals have the same number of sex chromosomes, and the first section of this chapter reviews differences in sex chromosomes.

Sex chromosomes are linked to sex differentiation, that is, to the developmental processes by which an individual becomes a reproductive female or male. Other chromosomes (autosomes) also influence the wide variety of morphological and behavioral sexual differences across species. We will explore the basics of sex determination and some differences in the process across mammals.

Genes are specific segments of DNA that often code for proteins. Thus, each molecule of DNA (each chromosome) can have many different genes along its length. The entire set of genes on all the chromosomes constitutes the genome. Most of the genome is the same for both females and males, thus, genetic sex differences are small. However, most female mammals have two X-chromosomes, whereas males only have one. If X-chromosomes were small, and carried only a few genes, the small difference between females and males would not be a problem. But the X-chromosome is usually large and carries a number of genes that are not related to sex differences. In eutherians, the X-chromosome is highly conserved across taxa and usually comprises about 3% to 5% of the haploid genome (Graves 1996). X-chromosome inactivation is a process that equalizes the genome between females and males, and we describe it in some detail.

Genes interact with surrounding cellular material to orchestrate the synthesis of proteins that are, themselves, the basis for all of our tissues from our blood to our hair. Which proteins are synthesized, when they are synthesized, and how quickly they are synthesized determines development. Consequently, genes have major roles in most aspects of mammalian biology, and this is certainly true of female reproduction.

At conception nearly all cellular material comes from the mother. Thus, mothers contribute not only their ancestral history (genes) but also fats, proteins, sugars, minerals, vitamins, and other such materials. This material comes from the mother's past and current interactions with her environment such as through her diet. So mothers have both genetic and cellular input into conception and development. In addition, part of a mother's cellular contribution includes small but essential organelles, the mitochondria. Mitochondria have their own DNA, comprising 40 or so genes. These mitochondria and their DNA are present in oocytes at the time of conception. Thus, females contribute genetic material to developing offspring as both mitochondrial and nuclear DNA. Tiny bits of mitochondrial DNA may come from fathers; however, paternal contributions are rare because the mitochondria carried by the male gamete are limited to the sperm's midpiece and are not very close to the ovum. Only seldom do paternal mitochondria find their way into the ovum. When they do, mitochondrial disorders often result (Gyllensten et al. 1991).

Not only are genes the conduit of information across generations but, when expressed, they influence the phenotype and physiology of individuals. Genetic differences in specific genes (allelic variants) can also serve as a measure of evolution. For example, one definition of evolution is a change in allele frequencies in a deme (i.e., localized population) over time.

Nearly all the anatomy and physiology of reproduction has genetic underpinnings, because proteins regulate most synthesis, growth, and development. The number of genes involved in any one stage of reproduction must be staggering, but the specifics of which gene does what remain poorly understood. This will change. For instance, in 1997 more than 40 genes were known to cause reproductive failure in lab mice (Rinkenberger et al. 1997), but only 5 years later the total was more than 200 (Matzuk, Lamb 2002). A few genes have a single circumscribed effect on physiology or anatomy, but many genes have multiple effects. Thus, the function of a specific gene may vary in different parts of the body or at different times of life. In addition, many genes may need to work in concert to achieve a single result. Our understanding of how genes work is embryonic!

Besides sex chromosomes, sexual differentiation, and X-chromosome inactivation, one additional topic will be covered in this chapter: epigenetics. Like a blueprint, a set of genetic instructions can do nothing without workers and materials to build the end product. In this case, genetic information is translated within an environment (i.e., a cell) with the appropriate materials (e.g., building blocks, such as amino acids or carbohydrates) as well as organelles to implement these genetic instructions. This environment may itself influence the end product. In other words, changes in the anatomy or physiology of an organism may not be due to the genes themselves but to the environment in which those genes are located. Different cellular environments may shut down specific genes and enhance others. In addition, while genes code for proteins, those proteins are made in the cell, and proteins may be altered in shape or size by the cellular environment. These processes are broadly termed *epigenetics*. X-inactivation is but one such process, many others exist and can influence reproduction. We will end the chapter with a discussion of epigenetics and genomic imprinting.

Sex Chromosomes

A well-known feature of chromosomes is their importance for sex determination. A single pair of sex chromosomes is present in most, but not all, therian mammals. The genes on marsupial and eutherian sex chromosomes are partly homologous, and those that are similar may represent the ancestral XY pair (Graves, Renfree 2013; sidebar on page 29).

Generally, mammalian females are homogametic (i.e., all gametes have the same sex chromosome; in this case, each will have an X-chromosome), and males are heterogametic (that is, half the male gametes have a single X-chromosome and half have a small, gene-poor Y-chromosome). However, not all mammals have this familiar genetic constitution.

Monotremes (platypus and echidnas) have a complex set of sex chromosomes rather than just a single pair. Female monotremes have five pairs of sex chromosomes, while males have four to five pairs (figure 3.1; Gruetzner et al. 2006; Ferguson-Smith, Rens 2010). Therian (marsupial and eutherian) XY chromosomes are not genetically similar to monotreme sex chromosomes but are somewhat similar to a specific pair of monotreme autosomes (Veyrunes et al. 2008). In addition, for monotremes and marsupials only a subset of X-chromosome genes are conserved across taxa (Graves 1996). So therian XY chromosomes may have evolved from an autosomal pair of chromosomes after therians split from monotremes roughly 165 MYA. Therians also have variety in their sex chromosomes.

Similar to textbook examples of sex determination (as in humans and other well-studied mammals), most marsupials have the usual XX/XY sex chromosomes. However, four exceptions are well known. The long-nosed potoroo (*Potorous tridactylus*), the swamp wallaby (*Wallabia bicolor*), and the bilby (*Macrotis lagotis*) have XX/XY$_1$Y$_2$ sex chromosomes, while the spectacled hare-wallaby (*Lagorchestes conspicillatus*) has X$_1$X$_1$X$_2$X$_2$/X$_1$X$_2$Y sex chromosomes (Tyndale-Biscoe, Renfree 1987). In all these variations, females remain homogametic and do not have a Y-chromosome. The study of how these variants function and their evolutionary history could be fruitful.

A few eutherians also deviate from the usual (XX/XY) sex determination theme. One variant is the YY male. This occurs in the Eurasian shrew (*Sorex araneus*) and Indian muntjak (*Muntiacus muntjak*). This XX female/YY male variant is similar to that of the swamp wallaby: XX female/XY$_1$Y$_2$ male (Wurster, Benirschke 1970; Wójcik et al. 2003). In a second variation, found in mole voles (*Ellobius lutescens, E. talpinus, E. tancrei*), both females and males are XX; thus, the Y chromosome is lost altogether (Bakloushinskaya et al. 2013; Just et al. 2007; Veyrunes et al. 2009). In both these variants, females are homogametic as usual. However, two additional variations occur in which the female karyotype becomes heterogametic. In one variation, heterogamy occurs in both sexes (X0 females/XY males) by deleting an X chromosome. This karyotype is present in

Genome Hunter

Jennifer A. Marshall Graves (1941–)

Australian evolutionary geneticist Jennifer A. Marshall Graves is a Distinguished Professor at La Trobe University (Melbourne), Professor Emeritus at the Australian National University (Canberra), Thinker-in-Residence at the University of Canberra, Fellow of the Australian Academy of Science, and L'Oreal-UNESCO Laureate. These recognitions are for her major contributions to the field of reproductive biology, from studies on the origins of mammalian sex chromosomes and sex-determining genes to her work on the organization and evolution of the mammalian genome. Her doctoral studies at the University of California, Berkeley, were on the control of DNA synthesis in mammal cells, but she was an early convert to comparative mapping and sequencing of genes in distantly related mammals. She proposed the first sequencing of marsupial and monotreme mammal genomes and directed the Centre for Kangaroo Genomics.

For more than four decades Graves has studied the evolution of sex determination through comparative studies of sex chromosomes in distantly related mammals and reptiles. Much of her work has focused on identifying differences among eutherian, marsupial, and monotreme mammals to trace the evolution of the mammalian genome and particularly of sex chromosomes. She found that the human XY chromosome pair contains an ancient region that is conserved between eutherians

and marsupials (therian mammals), but eutherians have a region that was added more recently. This became critical when Graves discovered that the first candidate for the sex-determining gene lay in the new, rather than the old part of the Y. This sparked a search for the right gene, *SRY*, which was ultimately discovered by one of her graduate students. *SRY* on the Y chromosome acts as a switch leading to the development of testes in an embryo by activating the key testis-determining gene, *SOX9*, an autosomal gene.

Graves discovered that many regions were shared by the X and Y chromosomes, and by genome regions containing the same genes in other animals, implying that the mammalian X and Y descended from an ancestral autosome. She found that *SRY* has a partner *SOX3* on the X chromosome with similar base sequence. *SOX3*, which is conserved in all animals, is therefore the ancestor of *SRY*.

Graves has studied mammals usually considered "strange," such as kangaroos and platypuses, as well as birds and lizards, to understand how sex and sex chromosomes evolved. She found that although the genetic pathway that makes a testis is highly conserved, the switches that kick-start this process seem to have evolved independently in different lineages.

The platypus, one of the egg-laying monotreme mammals that separated very early from other mammals, has

been especially instructive. Platypus sex chromosomes (there are 10!) share no homology with the therian XY pair; instead, the conserved therian X chromosome is homologous with a platypus autosome. This confirms that the therian X and Y chromosomes evolved from an ordinary chromosome in a common mammalian ancestor and places the origin of the therian X and Y chromosomes at the divergence of therians from monotremes about 190 million years ago. Thus, human sex chromosomes are much younger than we thought.

Graves also discovered other male-specific genes on the mammal Y with partners on the X, implying that these genes were located on the original autosome but evolved a male-specific function. The few genes (45 in all) that remain on the Y are all that are left of an original 1,000 or more genes. Graves predicts that the process of gene loss from the Y is ongoing, and the human Y will self-destruct in about 5 million years.

Graves's work has illuminated many of the mysteries of mammalian sex determination, and she is noted for identifying the male-determining Y chromosome as a degraded and mutated version of the X chromosome.

(Photo of J.A.M. Graves and a joey by Micheline Pelletier/GAMMA used with permission.)

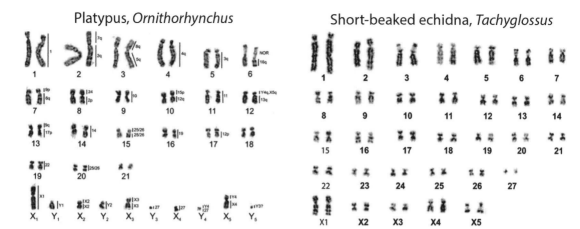

Figure 3.1. Monotreme karyotypes. *Left,* male platypus (*Ornithorhynchus anatinus*); *right,* female short-beaked echidna (*Tachyglossus aculeatus*). Both species have 5 pairs of sex chromosomes, but the platypus has 21 autosomal pairs, whereas the echidna has 27. The banding patterns result from ex-situ treatment with Giemsa, a stain that makes DNA rich in adenine and thymine appear darker than DNA rich in guanine and cytosine. From Rens et al. 2007 (open access).

the creeping vole, *Microtus oregoni,* and in 43% of female mandarin voles, *M. mandarinus* (Charlesworth, Dempsey 2001; Zhu et al. 2003). X0 females also occur in two species of spiny rats (*Tokudaia osimensis, T. tokunoshimensis*), which discard the Y chromosome, as well as an X chromosome, leaving both sexes with a single X (X0 females/X0 males). A related species, *T. muenninki,* has normal XX/XY sex chromosomes (Kobayashi et al. 2007). Alternatively, in some lemmings (*Dicrostonyx torquatus, Myopus schisticolor*), South American field mice (*Akodon*), and African pygmy mice (*Mus minutoides*) females may replace an X chromosome with a Y chromosome leaving both sexes XY (XX or XY females/XY males; Bull, Bulmer 1981; Fredga 1994; Veyrunes et al. 2009).

These are the classic examples of sex-chromosome diversity. However, chromosomes themselves do not determine sex but, rather, the genes they carry provide the appropriate instructions to create a female or a male. Thus, any sex-determining genes could be translocated (transferred) to the autosomes and still perform the same functions. Consequently, genetic females cannot be absolutely identified by the lack of a Y-chromosome.

Sex Determination

The means by which sex chromosomes function to determine sex are diverse. Most mechanisms lie along a genetic versus environmental continuum. An example of an environmental mechanism occurs in some turtles where sex is determined by the incubation temperature of the egg. However, mammals exemplify the other end of the continuum and exhibit genetic sex determination

(Beukeboom, Perrin 2014). In all but a few unusual cases, genes on the sex chromosomes determine whether an individual will be a reproductive female or male. In fact, whether the genital ridge of the early embryo develops into an ovary or a testis is influenced by genes on the sex chromosomes. Gonads, once formed, influence the sex of the individual further by producing hormones to organize sexual development. Hormonal exchange among siblings may also influence sexual development (see chapter 7, "Gestation"). Some genetic and molecular pathways that orchestrate sex differentiation have been elucidated (Beukeboom, Perrin 2014) but much remains unknown.

Research on sex determination has, to a large extent, focused on males, specifically, the testis-determining factor (aka SRY the sex-determining region of the Y chromosome). This work rests on the model that, in the absence of testicular development, the ovary passively forms as the "default" gonad. This model highlights the classical stereotype of active males and passive females. However, "the XX gonad is not an innocent bystander in sex determination" (Edson et al. 2009:632). In fact, the XX gonad produces a factor "Z" that actively stimulates the ovarian differentiation cascade. In males, SRY or some downstream target of SRY inhibits the ovarian cascade. ß-catenin is a major pro-ovary and anti-testis factor and may be factor Z (Edson et al. 2009). If so, ß-catenin actively stimulates the production of an ovary, but in males, SRY inhibits this cascade. Thus, the differentiation of an ovary is an active and not a passive process.

Usually in mammals, SRY is a single-copy, male-specific gene located on the Y chromosome (Fernandez et al. 2002). The gene encodes a protein with a highly conserved, 79 amino-acid region flanked by highly variable, species-specific regions (Fernandez et al. 2002). These variable regions offer the potential for pre-zygotic reproductive isolating mechanisms, but this idea remains open to debate (Graves 1996). Several mammalian species lack the SRY gene and thus must use a different testis-determining mechanism. For instance, male monotremes have Y-chromosomes but no SRY gene (Ferguson-Smith, Rens 2010). In addition, the two eutherian spiny rats (*Tokudaia osimensis, T. tokunoshimensis*) lack both a Y-chromosome and the SRY gene (Kuroiwa et al. 2011). Meanwhile, females in at least one species, *Microtus cabrerae*, have multiple copies of SRY, although these copies appear inactive (Fernandez et al. 2002).

Once an ovary develops, ovarian hormones are available for continued differentiation of the female reproductive tract, mammary glands, and other female tissues. In marsupials, SRY determines the testes but then testicular hormones control the development of other male features (as they do in eutherians; Graves 1996).

Renfree et al. (2014) reviewed hormone-independent sex differentiation focusing on marsupials. In this case, gonadal development occurs after birth. In marsupials, the urogenital opening is identical in females and males. The opening is the terminus for both the renal and genital ducts, each with its own sphincter.

In addition to gonads, marsupials have other sexually dimorphic structures, the mammary glands and scrotum (Graves, Renfree 2013). These structures differentiate independently of endogenous hormones or treatment with estrogen or testosterone, suggesting direct genetic control of their development and differentiation. The current hypothesis is that a mammary gland versus scrotum switch is located on the X-chromosome. Because XO females possess ovaries and a scrotum but no pouch, the switch may operate by dosage (i.e., two copies make a pouch and mammary glands, one copy makes a scrotum) or by imprinting (a paternal X is required for a pouch; Graves, Renfree 2013). The fact that XXY males possess testes and a pouch but no scrotum suggests a dosage effect (Graves 1996).

Overall, the development and evolution of sexually dimorphic structures differ in monotremes, marsupials, and eutherians. Monotremes lack a scrotum and only echidna females have pouches. In most marsupials, the pouch-mammary gland versus scrotum development is mutually exclusive and independent of the Y-chromosome. Scrota are variable in eutherians (absent in many species, prepenile in some, post-penile in others), and scrotal anatomy probably evolved in response to species-specific natural selection. Thus, the eutherian and marsupial scrota may be the result of convergent evolution with no connection to female anatomy (Graves 1996).

X-Inactivation

X-chromosomes are often large and carry a number of genes with important, non-reproductive functions. In humans, the X-chromosome has 1,000 to 2,000 genes, whereas the Y-chromosome has fewer than 100. Females have two X-chromosomes, one from their mother and one from their paternal grandmother (via their father). The presence of two X-chromosomes in females, but only one in males, presents a challenge to the operation of the rest of the genome. With twice the possible production of X-chromosome genes in females over that in males, the integration with the rest of the genome is difficult (Lyon 1961). To compensate for this dosage difference, one of the X-chromosomes in females is inactivated early in development. The dosage-compensation hypothesis (termed *the Lyon hypothesis*, or *Lyonization*) for X-inactivation was first developed by Mary Lyon (Lyon 1961; sidebar on page 33).

In eutherians, X-inactivation occurs twice early in development. The first X-inactivation begins at roughly the 2–4 cell stage (Huynh, Lee 2003). At this point, the paternal X is inactivated and stays inactive from early cleavage until the mid-blastocyst developmental stage (Huynh Lee 2003; Chuva et al. 2008). At the mid-blastocyst stage, the eutherian conceptus is a hollow ball of cells with a clump of cells (inner cell mass) inside one edge. The outer ball of cells will become the placenta (extra-embryonic membranes), whereas the inner cell mass will become the embryo proper. Cells in the inner cell mass reactivate the paternal X-chromosome so that, briefly, both Xs are functional (Mak et al. 2004). An-

The Woman Who Silenced Genes—Lyonization
Mary Lyon (1925–2014)

Mary Lyon was educated in Britain during World War II, not an easy time for women to study. Fewer educational positions were open to women, and women only received titular degrees despite attending the same classes as men. In this environment, Mary Lyon attended both Girton College and Cambridge University and became a geneticist by way of embryology.

Her interest in embryology was sparked by the work of C.H. Waddington at a time when genetics was not even offered as a course! She decided to pursue a PhD and joined the lab of R.A. Fisher but before finishing moved to Edinburgh and joined D.S. Falconer's lab where she completed her degree. Thus, her academic pedigree is a who's who of genetic titans.

In Edinburgh, she joined the MRC radiobiology unit in Harwell where she was active from 1962 to 1987. Lyon's research made a massive contribution to biology through her discovery of X-chromosome inactivation.

Using mutant mottled mice, she carefully conducted breeding experiments and described what is now known often as Lyonization. Lyon's "Gene Action in the X-Chromosome of the Mouse (*Mus musculus* L.)," published in *Nature* in 1961, demonstrated that in the somatic cells of the body only one sex chromosome was (usually) expressed and coded for proteins, the other X chromosome was silent. Lyon explored the implications of her discovery (X-inactivation) for various aspects of mammalian biology, including development. Her discovery led her to a lifelong path of research trying to understand the dynamics of X-inactivation and the evolution thereof.

However, her work was by no means on this topic alone. Lyon also studied the biology of sperm and how reproductive organs differentiate between male and female embryos, as well as other aspects of chromosome expression, including abnormalities in various strains of mice. Much of her work was with radiation-induced mutant mice. Even after retirement she still worked in her lab.

Mary Lyon became a Fellow of the Royal Society in 1973 and has an award named for her by the British Genetics Society.

(Photo from Jane Gitschier 2010; open access.)

other 24 hours later in lab mice, these embryonic cells again inactivate one of the X-chromosomes, but this second inactivation is random, that is, either the maternal or paternal X is silenced (Chuva et al. 2008 Okamoto et al. 2004). All cells have the same genetic make-up, but the expression in each cell differs with maternal X-chromosome genes expressed in some cells and paternal grandmother X-chromosome genes expressed in others. In lab mice, the timing of this second inactivation is about 5 days post-conception, which is also approximately the time of implantation. Extra-embryonic tissues retain the early inactivation, keeping only the maternal X active (Cheng, Disteche 2004). Consequently, the paternal X is inactive while the zygote is loose in the female tract before implantation. Pre-implantation X-inactivation is a dynamic process and may lead to sex differences in gene expression in the early blastocyst (Bermejo-Alvarez et al. 2010). Such early sex differences provide a means for mothers to identify the sex of their offspring before implantation and thus alter the sex ratio at birth. After implantation, the paternal X is inactive in the tissues (placenta) that have the most intimate connection with the maternal oviducts and uterus. Even this small inactivation of possible paternal antigenic tissue may have beneficial consequences for females.

Much later in development, X-inactivation is reversed again but this time only in the primordial germ cells (PGC) as they migrate to the genital ridge (chapter 6, "Oogenesis to Conception"). PGCs are cells in the embryo destined to become the next generation of gametes when the embryo becomes an adult. The limited reactivation occurs about 9–11 days post-conception in lab mice (Chuva et al. 2008). The reactivation is a response to signals from cells in the genital ridge. Thus, somatic cells and not the PGCs induce the reactivation. Presumably both X-chromosomes remain active henceforth in the PGCs so that all gametes have active X-chromosomes.

X-inactivation is not limited to eutherians and occurs in both marsupials and monotremes (Graves 1996), but with a different biochemical mechanism. In marsupials, the paternal X is inactivated, but not completely and not to the same degree in different tissues (Graves, Renfree 2013). Research on X-inactivation in additional mammal species, including diverse marsupials and especially monotremes would be valuable. The early paternal X-inactivation in eutherians, as well as the maintenance of the paternal X-inactivation in extra-embryonic tissue, suggests that paternal X-inactivation is an ancestral condition and random X-inactivation is the derived state (Graves 1996).

Epigenetics: Genetics beyond Genes

The phenotype of an organism reflects, in part, the genetic information inherited from its parents. Many differences among species and among individuals are due to differences in their genes. However, for a gene to have an effect (i.e., result in a phenotype) it must be expressed, that is, activated. Genes are turned on and off at different times, for variable lengths of time, and to different degrees. Some variation among organisms is due to differential gene expression, not just different genes. However, even that is an oversimplification. Even after expression, the cellular environment can alter the product of the gene.

Although genes provide a blueprint for a potential phenotype, the environment surrounding those genes influences which parts of the blueprint will be realized and to some extent how structures are built. If these environmental changes (silencing part of the genome rather than changing nucleotide sequences) can be passed to the next generation, then evolution will occur outside the traditional genetic mechanisms. Broadly, epigenetics encompasses both the effects of the environment on genes within an individual and the possible cross-generational effects of the environment on genetic lineages. In either case, epigenetic phenomena provide a set of molecular mechanisms for how the environment can change the reproduction and behavior of individuals as well as lineages (Champagne, Curley 2009; Dickins, Rahman 2012).

Parent-of-Origin Gene Expression—Genomic Imprinting

For most genes, alleles from both parents are expressed as needed during the life of the individual, and, when not needed, both alleles are inactive. However, for a small proportion of genes (about 0.005% in humans), only one allele is ever expressed while the other is always inactive. Because one allele stems from each parent, the genetic information from the silenced parental allele is unavailable for use by cells unless "unpackaged." In other words, in these few cases, gene expression is related to the parent of origin and only the variant from one parent is functional. Single alleles can be blocked or clusters of alleles on a single chromosome can be silenced by blocking their regulatory region. Either maternal or paternal alleles may be silenced, and silencing may occur on autosomes or sex chromosomes. Rarely, a maternal allele may be blocked in one tissue and a paternal allele in a different tissue (Keverne 2014). The clearest example of parent-of-origin gene expression is the paternal X-inactivation discussed earlier. Other examples of an unequal parental genetic contribution to the very early conceptus include (a) genes on the X chromosome that are not inactivated and thus present in a double dose (only in an XX conceptus); (b) genes present on the Y chromosome that are not present on the X (only in an XY conceptus); (c) material from the sperm that passes into the zygote during conception; and (d) maternal cytoplasm and organelles present in the ovum.

The number of genes with parent-of-origin expression differs across the major mammalian reproductive lineages: about 100–150 genes in eutherians, fewer in marsupials, and none (to date) in monotremes (Renfree et al. 2013; Keverne 2014; Kusinski et al. 2014). The blocked loci may be highly conserved, and the sex of the blocked parent may also be evolutionarily conserved. For instance, in eutherians, blocked loci are the same in lab mice and in humans, suggesting a long (at least 90-million-year) history (Tycko, Morison 2002).

The functions of these blocked genes are diverse, ranging "from protein ubiquitination to growth factor clearance to transcriptional modulation, and also include the production of several classes of non-translated RNAs" (Tycko, Morison 2002:253). Functional identification is done by comparison with known genes and thus is biased in favor of genes with importance in biomedical science, such as those with connections to cancer. Therefore, the function of many imprinted genes is unknown and speculation regarding the selective benefits of imprinting based on functional significance is subject to the same biomedical bias. In addition, in eutherians, imprinted genes with expression in the placenta, brain, and early embryo have received the most attention, while in marsupials, imprinted genes in the placenta and mammary glands are often singled out (Graves, Renfree 2013; Renfree et al. 2013). Again, the tissues chosen can bias the interpretation.

Why did parent-of-origin expression evolve? Several possible benefits have been hypothesized, such as prevention of parthenogenesis, promotion of the evolution of viviparity, an outcome of conflict between maternal and paternal genomes, a facilitation of cooperation between maternal and paternal genomes, or a consequence of co-adaptation of mothers and offspring (Kevern, Curley 2008). Of course, not all imprinted genes need be selected by the same adaptive process. Thus, each of these hypotheses may be valid for different genes, or sets of genes, or different lineages.

The same molecular mechanisms that prevent any allele from being expressed (such as patterns of DNA methylation and modification of histones) also occur on a parent-of-origin basis. Such blocking of expression is maintained for the life of the cell and is transferred to any daughter cells. The blocking is erased during gametogenesis, and then, sometime after conception, the parent-of-origin blocks are created anew. Maternal alleles will be blocked in XY conceptuses; paternal alleles will be blocked in XX conceptuses. To the extent that the imprinting (blocking) occurs very early in development, in utero, and uses maternally inherited cytoplasmic enzymes and organelles, the process will be biased in favor of, and regulated by, the maternal lineage. Thus, the major players in the regulation of these genes are the mother and the offspring.

Genetics: The Central Core

Genes are key for the successful differentiation of the sexes, as well as for subsequent development, maturation, and eventual reproduction. Only in the past 30 years has genomic imprinting (epigenetics) been described and increasingly, albeit gradually, understood. This is an active area of study and as we learn more about the mechanisms (e.g., what makes one gene dominant and another recessive) we anticipate much more will be determined, especially in currently poorly studied, non-model taxa. In the meantime, much is known about the anatomy and physiology that results from the transcription and translation of the genetic core. The next chapters turn to those structures and functions of a female's body related to reproduction.

CHAPTER FOUR

Anatomy

Much research on female hyena biology has been focused below the belt, and a slew of male-centric terms are used to describe her anatomy: peniform clitoris, pseudo-scrotum, pseudo-penis. She is considered masculinized, virilized, or sex reversed. From a human perspective, these terms make sense and generate intense interest, but from the hyena's perspective, her robust genitalia epitomize the female form of her species. Nonetheless, her genitalia are exceptionally derived. Not only is her clitoris greatly enlarged but it also functions as a birth canal. Consequently, birth itself is not a straight shot but involves a sharp turn. The anatomy of hyenas sets a context for examining other female structures, especially some, such as the clitoris, which have largely been ignored in many other taxa. We can also use the hyena example to ask about our basis for comparing female versus male structures. (Cunha et al. 2003; Frank et al. 1990; Harrison Matthews 1939; Neaves 1980)

It is the first organ any mammal makes.—Power, Schulkin 2012:1

In this chapter, we turn our attention to the physical arena within which off-spring grow and develop as well as other tissues and organs essential to those processes. Ovaries and the uterus are obvious components. Here, ova are formed, ova and sperm interact, conception occurs, as does implantation and embryo development. Also obvious are mammary glands, the namesake of mammals. Other features of female anatomy with essential roles in reproduction are the vagina, cervix, and clitoris. Less obvious, but equally key, are oocytes and placentas, structures associated with the very start of female development. Of course, other anatomical structures, such as the pelvis (Tague 2016), differ between the sexes but are not specific organs of reproduction.

One bias in the usual treatment of anatomical structures is to view them as static. Many images and photographs from dissections promote this perception of stasis, but reproductive structures change in shape, orientation, and size not only developmentally and over a life span but also on much shorter time frames. Muscle fibers are an important component of reproductive organs and allow for short-term flexions, extensions, contractions, and spiraling at specific times during reproduction. They can squeeze off one section from another, direct the transport of gametes, flush out debris, expel eggs or offspring, or pull teats from the mouths of suckling young. Unlike these short-term dynamics due to muscles, longer-term change results from swelling and growth. The dimensions of reproductive structures are not stable, and lest we focus solely on size, color changes also occur, for instance, to the sex-skin of baboons. Dynamic changes are constant. For instance, ovarian cells are often in flux because as follicles enlarge or involute nearby cells must also change positions. This activity is not normally considered part of ovarian dynamics but is certainly a component of ovarian physiology. Our reliance on two-dimensional photographs as well as our historical explorations of structure in non-living specimens has led to the static bias in our traditional view of anatomy. The advent of ultrasound recordings, laparoscopic illumination, and video, as well as the electronic means to share these dynamic records of anatomy in action, should lead to a more complete understanding of the four-dimensional nature of structure. However, this exploration is embryonic. Thus, our chapter relies heavily on the traditional static view of anatomy, supplemented by what little is understood of its dynamic aspects.

A second bias in texts that describe reproductive anatomy (e.g., Flowerdew 1987; Hafez, Hafez 2000) is the focus on structures that have developmental parallels in males. Thus, the focus is usually on gonads as well as the sex differentiation of a pair of embryonic ducts (the Müllerian and Wolffian ducts). Müllerian ducts form the female reproductive tract and atrophy in males, whereas Wolffian ducts form the male reproductive tract and atrophy in females (figure 4.1). The general scholarly tendency is to describe male structures first and then to explore female structures in contrast to males. Thus, the usual coverage concerns gonads and the reproductive tract and either ignores, or relegates to minor importance, structures that are well developed in females, such as mammary glands, placentas, and oocytes. The omission of the nominate feature of mammalian reproduction, mammary glands, as a component of female reproductive anatomy, is especially curious.

What about oocytes and placentas? The omission of these features reflects an emphasis on adult structures rather than on developmental ones. However, this explanation is unsatisfactory because the comparative emphasis on female and male anatomy is based on development. One could understand that oocytes, like sperm, are incomplete, having only half a genome and thus are excluded, but texts often discuss sperm structure and taxonomic differences but not oocyte structure and differences. For instance, the third edition of Knobil and Neill's seminal *Physiology of Reproduction* (Neill 2006) begins with a chapter on spermatozoa but has no chapter on the oocyte. The fourth edition (Plant, Zeleznik 2015), however, includes an excellent chapter on the oocyte.

The placenta is the first structure constructed by the conceptus. During its existence, the placenta is the largest organ of the conceptus and is the only organ

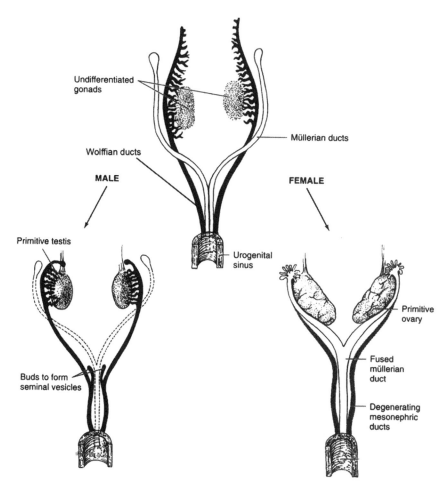

Figure 4.1. Differentiation of reproductive ducts in eutherians. Generally, Müllerian ducts form the female reproductive tract and atrophy in males, whereas Wolffian ducts form the male reproductive tract and atrophy in females. From Tanagho and McAninch 2008; used with permission.

ever discarded (Power, Schulkin 2012). Its anatomic description is usually included in discussions of gestation but not reproductive anatomy. This placement reinforces the concept that the placenta consists of both maternal and fetal tissues, whereas in reality the placenta is a fetal, not maternal tissue. The placenta coordinates the exchange of materials between the mother and the developing embryos and acts as a filter between them. The placenta is also the object of our enigmatic initial quotation.

Our anatomical exploration is organized more or less chronologically. We start with the female gamete and then examine the placenta. We next move to structures of adults: ovaries, the reproductive tract, external genitalia, and, finally to mammary glands, the quintessential feature of mammals.

As a field of study, functional morphology tries to match the specifics of structure with adaptive advantages. Explaining the adaptive function of external structures, such as limbs or wings, is commonplace. This approach is less common in studies of reproductive anatomy. Where possible, we will explore these avenues.

Tangentially, detailed anatomical descriptions of female reproductive tracts from various mammals are not currently in vogue; xenarthrans are an exception (Favoretto et al. 2015; Rossi et al. 2011). As a result, we must rely on much older literature for information. For instance, the most recent anatomical description of the female tract for hippos, *Hippopotamus amphibius*, is a paper by H.C. Chapman in 1881. In addition, although the classic literature is often of high quality, finding these sources is difficult because current search engines are not geared for finding older sources that do not have embedded key words. A second issue is that much of the early anatomical literature is in German or French while most current search engines use English key words. In addition, in science, emphasis is usually finding the most cutting-edge and recent source rather than the most extensive and accurate one, which may be much older. The issue is one of specificity. Individual papers in the older literature are more comprehensive in scope, whereas today's searches are specific in focus. Thus, a search for "hippo cervix" generates articles on a biochemical signaling pathway but not anatomical descriptions. When possible we use both new and older papers to highlight the value of earlier detailed descriptions of reproductive anatomy.

Oocyte/Ovum

The oocyte is a highly differentiated, molecularly complex product of gametogenesis, despite its outwardly simple morphological appearance.

—Mtango et al. 2008:224

Reproduction for females starts with the formation, development, and maintenance of female gametes. Female gametes are highly specialized both in structure and in function. Not only do they contain half the genetic information needed to form a new individual, but they also contain all of the nutrients, and most of the cytoplasmic components, necessary to initiate embryonic development (Lombardi 1998). Although the earliest stages of oocyte formation occur while a female is developing in her mother's uterus (chapter 6, "Oogenesis to Conception"), the final phases leading either to ovulation or atresia (degeneration of follicles that do not ovulate) occur in the adult ovaries. Thus, adult ovaries contain gametes in various stages of development and disintegration.

Female gametes (box 4.1) can be roughly grouped into one of four age-categories: (a) immature, diploid, primary oocytes; (b) secondary oocytes embarking on meiotic division, (c) mature haploid ova ready for ovulation, and (d) ova in the process of disintegration (atresia). The transitions between categories are gradual and the distinctions between them can be hazy.

BOX 4.1. What Is an Egg? Egg, Ovum, Embryo

In laboratory parlance, and even in print, the oocyte . . . , ovum, zygote, morula and blastocyst are frequently referred to indiscriminately as the "egg."—Perry 1981:321

Oogonia, oogonium, oocyte, ova, ovum, egg: which is which? Female gametes have many names, and their distinctions are related to the age or stage of the gamete. We describe how female gametes are created (oogenesis) in chapter 6. Here, we will name names. Oogonia (oogonium is singular) are the earliest stage en route to mature female gametes. Oogonia are formed in the fetal ovary very early in development from cells called *primordial germ cells* (PGC). PGC arise by normal cell division (mitosis). In contrast, oocytes arise from oogonia by a special cell division, meiosis. Meiosis (see figure) is a series of two cell divisions that results in four daughter cells. During the two-division process, a female's DNA (ignoring genetic material in the mitochondria) first doubles and then halves so that, in the end, each of the daughter cells has only half the DNA of the original oogonium. Oocytes can be further classified as primary or secondary oocytes, depending on the stage of meiosis (see figure). Oocytes are early-stage female gametes, whereas ova are later-stage gametes. Although oogenesis results in four daughter cells with equivalent amounts DNA, the rest of the cell contents are unequally divided. One cell gets almost all the contents and the other three get very little. The depauperate three cells are called polar bodies. The cell with the majority of the original contents is called an ovum (ova is plural). Both ova and polar bodies can fuse with sperm, but as the cell contents are important to the development of the conceptus, polar body-sperm fusions seldom result in viable embryos.

What about egg? As the epigraph indicates, the word *egg* is ambiguous. It can refer either to a female gamete by itself, to the female gamete plus maternally derived tissues (e.g., a chicken egg), or to an early conceptus, as when we say an egg implants in the uterus. In other words, the definition depends completely on context. In addition, the phrase *fertilized egg* invokes a female passive-male active tone as discussed in chapter 1. Thus, we refrain from using the ambiguous *egg* and instead generally use *oocyte* to refer to female gametes and *zygote* to refer to the earliest conceptus.

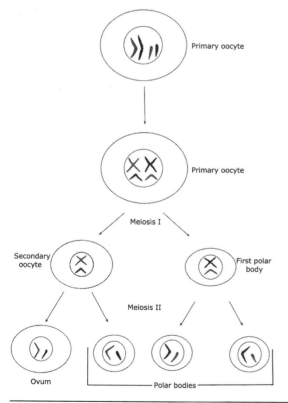

Meiosis and the creation of polar bodies. The creation of gametes requires two cell divisions for the eventual production of four gametes (an oocyte and three polar bodies). Chromosomes come in pairs (one from each parent), thus the mother cell is diploid (two pairs of chromosomes are illustrated). Each gamete only has a single chromosome from each original pair and thus is haploid. To get the same amount of DNA in each of the daughter cells, the genetic material must first double to generate four copies of each chromosome and then divide twice. Although genetically each of the four daughter cells has the same amount of genetic material, one daughter cell receives nearly all of the original cellular contents. Thus, cell division is equal for the genetic material but very unequal for the cytoplasmic contents. Illustration by Abigail Michelson.

Primary oocyte

Primary oocyte

Meiosis I

Secondary oocyte

First polar body

Meiosis II

Ovum

Polar bodies

Oocytes enlarge as they develop. For example, immature lab mouse oocytes grow from about 12 μm to about 72 μm (6-fold increase), whereas human oocytes grow from 36 μm to about 120 μm (3-fold increase; Griffin et al. 2006). Ova are large. In comparison most body (somatic) cells are about 10–20 μm in diameter and male gametes (oblong sperm) are ~10 × 6 μm. Most oocytes degenerate and never reach ovulation. Thus, the ovary contains degenerating oocytes as well as remnants of prior stages of unequal meiotic divisions, the polar bodies (box 4.1). Regardless of age, female gametes have the same general structure.

Female gametes are single cells with the usual cellular components. They have extensive cytoplasm with multiple organelles that increase in number and size as the ova mature. The cytoplasm and the constituent organelles (mitochondria, ribosomes, Golgi apparatus, endoplasmic reticulum) are all maternal in origin and provide a continuity of matrilineal descent from mother to daughter to granddaughter. As oocytes develop into mature ova, they accumulate the molecules and organelles that provide both the first chemical building blocks as well as the energy reserves required for the initial development of the conceptus. The oocyte cytoplasm is absolutely necessary for normal development (Krisher 2004). Thus, the maternal contribution to the next generation is much more than just DNA. The non-genetic components of the oocyte are critical and constitute a wholly matrilineal form of inheritance. Identifying differences among species in the composition of the cytoplasm of oocytes and connecting those differences to their functional outcomes is an open field.

The cellular components of an oocyte are important to survival as is the membrane that encloses those components. In general, "cells" (from Latin for small room) are identified as such because they have a membrane that separates them from other cells. Oocytes are no different, although their cell surface has its own name, the *oolemma*. Like other cell membranes, the oolemma (a) regulates the transportation of materials into and out of the cell and (b) is the intermediary between chemical signals from other cells (in this case, other ovarian cells). Specific to the oolemma is a multifaceted role in conception, including regulation of sperm entry, incorporation of paternal genetic material into the ovum once a sperm is selected, and restriction of other paternal materials from entering the ovum. Again, membrane structure varies across mammals and connecting membrane differences to their functions is an open field. One especially important function is species recognition, so that ova only fuse with sperm of the appropriate species. This function may be assisted by material external to the oolemma, such as the ovarian cells that accompany the oocyte after ovulation. However, for this step to be necessary, females must have initially mated with the wrong species, an unlikely event given extensive mechanisms for pre-copulatory mate choice.

Like plant and bacterial cells, the oocyte has an external coating outside the oolemma. Called the zona pellucida, this specialized extra-cellular matrix is primarily composed of cross-linked, sulphated glycoproteins (Menkhorst, Selwood 2008). The zona serves multiple functions. One function is mechanical, whereby the zona keeps the oocyte intact as the oocyte enlarges. Second, the zona filters the transfer of material from the ovary to the oocyte and thus is part of the regulation of oogenesis. Third, zona proteins are involved in choosing and activating (capacitating) sperm to allow conception and prevent polyspermy. Fourth, the zona plays a role in speciation (reproductive isolation) by preventing access to the ovum from sperm of other species. The zona persists after ovulation and is augmented or modified by oviductal or uterine secretions both before and after conception. These additions include regulatory molecules, inhibitors, and growth factors. These extra components have a variety of cellular functions in both conception and early development (Menkhorst, Selwood 2008).

The extent and development of the post-conception coatings (egg coats) in mammals varies across the major reproductive groups (monotremes, marsupials, eutherians). Monotremes lay eggs with complex multilayered shells. As with other mammals, the monotreme zona pellucida immediately surrounds the conceptus. Outside the zona is a cushioning protein layer of albumin. While still in utero, the monotreme egg enlarges greatly in size, and the permeable shell allows transfer of materials from mother to offspring until the egg is laid (Menkhorst, Selwood 2008). This permeability could be a pre-adaptation for lactation or viviparity.

External additions to the zona also occur in therians. In marsupials, post-conception coatings are prominent but not as extensive or as complex as those in monotremes. Among eutherians, rabbits (*Oryctolagus cuniculus*) provide their offspring with egg coats as elaborate as marsupials, but most eutherians, e.g., rodents, primates, ungulates, carnivorans, have limited additions to the zona. Conceptus- or oocyte-derived additions inside the zona may also be present, e.g., in marsupials, European moles (*Talpa europaea*), rabbits, lab mice, baboons (*Papio*), humans, cats, dogs, European badgers (*Meles meles*), martens (*Martes*), spotted skunks (*Spilogale*), fur seals (*Callorhinus*), and horses (Denker 2000; Enders 1971; Enders et al. 1989; Menkhorst, Selwood 2008). Thus, although key components may be structurally similar and share an evolutionary history, modifications of the zona are diverse.

The modified zona continues its protective role by mechanically preventing the conceptus from either disintegrating or breaking apart, as the zygote undergoes early cell divisions. These actions also protect against mono-zygotic twinning (box 4.2). In addition, the zona isolates the conceptus from any post-mating detritus as well as any immune cells that are cleaning up that detritus. A second immune function may be to protect the "foreign" conceptus from the maternal immune system. Of course, the egg shell of monotremes protects the embryo from the external environment after egg laying. Apart from protective functions, the zona may have an important role in orienting the development of the early embryo and trophoblast (the cells of the early conceptus that become the extra-embryonic

BOX. 4.2. Litter Size and Twinning

Litter size is the number of young in a single pregnancy and ranges from one in many species to perhaps 30 in naked mole-rats (see figure). Measures of litter size have included counts of ovarian corpora lutea, full-term fetuses, number born, or number of offspring first seen after birth, such as when juveniles initially emerge from a den or a cavity. Many mammalian species have only one young at a time, and singleton births are characteristic of many lineages: kangaroos, phalangers, sirenians, cetaceans, pinnipeds, pangolins, elephants, aardvarks, anteaters (Myrmecophagidae), and sloths, as well as most bats, primates, and ruminants. A number of rodents, such as porcupines, also have singleton births. For other species, litters are more common, e.g., most carnivorans, squirrels, mice, suids, shrews, lagomorphs, hyraxes, and elephant shrews.

Ovaries contain many follicles, but only a select number proceed through to ovulation at any one time. Ovulating only a single ovum (monovular) is a complicated process. To do so requires stopping the development of oocytes in the opposite ovary (contralateral inhibition) as well as preventing development of all but one oocyte within the ovary in question (ipsilateral inhibi-

tion). The physiological control for accomplishing both ipsilateral and contralateral inhibition requires both local (within an ovary) and systemic (between the ovaries via the circulation) regulation. Once these control mechanisms are in place, they allow other anatomical changes to evolve, such as the shape of the uterus or the number of mammary glands.

Monovular species are at a disadvantage if the environment changes, and producing multiple offspring rather than just one would increase reproductive success. However, some armadillos (*Dasypus*) have managed to remain monovular yet produce litters of 4 to 8 young (perhaps 12 in *Dasypus kappleri* or *D. septemcinctus*; Hayssen et al. 1993). They do so by producing identical quadruplets or octuplets. That is, they generate multiple offspring from a single conception, essentially creating litters of clones. This polyembryony allows females to escape any physiological or anatomical constraints that might prevent ovulating more than a single ovum (Craig et al. 1997; Galbraith 1985).

Of course, most species with litters release more than one gamete from each ovary at each reproductive episode. This condition is referred to as

polyovular. Unfortunately, the term *polyovular* can be confusing, as it also refers to having multiple ova in a single follicle. Whereas armadillos were an example of a monovular species that produce litters, camels are one of many species that ovulate many ova but produce singletons. In camels (*Camelus*), 12–19% of females ovulate multiple ova, but twin births are extremely rare. In addition, the single pregnancy is almost always in the left uterine horn (ElWishy 1987). As both ovaries are functional, transuterine migration must occur in addition to some form of litter-size reduction.

Overall, across mammals, the eventual litter size of an individual is a function of the number of ova ovulated, the number of conceptions that occur, the number of blastocysts that implant, the number of embryos that are not rejected, and the number of offspring that survive birth.

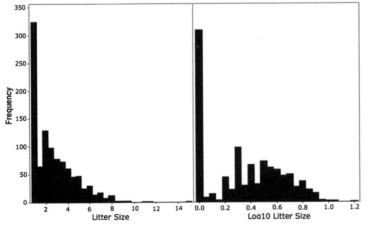

Litter size distribution. Litter size has a skewed distribution with many mammals producing singleton litters. The $\log_{10}$ transformation suggests that litter size may be bimodal; that is, species produce either litters of one or litters of more than one. Producing singletons requires a more complicated physiology than producing litters (see text). 1,037 species from Hayssen 1985, 2008a.

membranes). The coatings facilitate proper implantation and spacing of blasto-cysts in utero (Menkhorst, Selwood 2008). For therians, the blastocyst eventually "hatches" from the zona and implants in the uterus at which point the zona disin-tegrates. Zona disintegration also occurs within the ovary for oocytes that will not be released.

A Word about Polar Bodies

Polar bodies are formed during the meiotic division of oocytes (box 4.1). Instead of meiosis yielding four cells of equal size and constitution (as in spermatogene-sis), meiosis in oogenesis is unequal and yields one large cell with most of the cytoplasm as well as three smaller cells termed *polar bodies*. Polar bodies have a nucleus, but little cytoplasm and, consequently, only a few cytoplasmic organ-elles. Unequal cell division may also lead to genetic differences between oocytes and polar bodies and, in a complicated process, lead to changes in chromosome structure (De Villena, Sapienza 2001).

In plants and some animals, polar bodies have a significant role in reproduc-tion, but, in most female mammals, polar bodies degenerate and the fragments are trapped in the zona pellucida (Schmerler, Wessel 2011). Occasionally, a polar body will not degenerate and conception is possible, engendering fraternal twins. Tangentially, human fertility specialists have exploited the fact that because at least one of the polar bodies has the identical genetic complement as the ovum, polar bodies can be removed and tested for genetic abnormalities rather than us-ing mature ova.

Placenta

The placenta is probably the most variable of all mammalian organs.

—Power, Schulkin 2012:16

The placenta is a complex organ that bears a genetic make-up identical to the fetal genome and is associated with the developing fetus until birth or hatching, at which point it is discarded. The ancestral origin of the placenta began with the amniotic egg of terrestrial vertebrates. The innovation of the amniotic egg was the presence of three extra-embryonic membranes: amnion, chorion, and allan-tois, all within a protective shell. These membranes provided the developing em-bryo a means for gas exchange (chorion), protection (amnion and chorion), and collection of waste products (allantois). A fourth membrane has an older origin. This fourth layer, the yolk sac (vitelline membrane), encloses the maternally derived yolk that surrounds the embryo and is present in lineages that pre-date mammals, e.g., lamprey, fishes, sharks, and frogs. Yolk provides resources (energy and nutrients) for the early growth of the embryo. In mammals, the four extra-embryonic membranes fuse in various combinations to form the placenta. For example, a chorio-vitelline placenta, the predominant placenta of monotremes and marsupials, is a fusion of the chorion and the yolk sac, whereas a chorio-allantoic placenta, the predominant placenta of eutherians, is a fusion of the chorion and the allantois.

Not only is the placenta an old structure (evolutionarily speaking), but it also arises very early in development. By the 64-cell stage of conceptus development, the 13 or so cells destined to become the embryo (the inner cell mass in eutherians) and the precursor placental cells (the trophoblast) are distinct layers that do not exchange cells upon further divisions. This distinction is the first differentiation of cell types in mammalian development (Gilbert 2000).

Placental anatomy is replete with a bewildering array of technical terms. Much of the terminology depends on the tools used for observation. With the naked eye, the shape and location of the placental attachment to the uterus can be described as diffuse, cotyledonary, zonary, or discoid (figure 4.2). With microscopes and histological analysis the number of tissue layers between the fetal and maternal circulation (figure 4.3) can be resolved, resulting in the use of terms such as *epitheliochorial, syndesmochorial, endotheliochorial,* or *hemochorial.* This microscopic perspective is the basis for the oldest classification of placentas (Grosser 1909, 1927). More recent observations use tools with even greater resolution, especially for exploring the type of maternal-fetal interdigitation. These new observations provide additional categories for placental morphology, including labyrinth, trabecular, folded, villous, or cups. These morphologies are even more complex when specific cell types are considered. For instance, specialized cells (the decidua) from the uterus may become closely associated with the placenta in some animals and be discarded with it. All these levels of analysis have great value as descriptive tools for veterinary pathologists, zoo veterinarians, and phylogeneticists (Carter 2012; Carter et al. 2013), but their functional relevance has not always been clear (Capellini 2012).

One difficulty is that placentas are not static structures, they change over the course of gestation. Early classifications of placental types relied on examining placentas near term, and the dynamic nature of placental structure was ignored. A second issue is that these categories are vague and do not apply to all placentas. Intermediate conditions exist, which do not adhere to established conventions. Finally, the basic function of the placenta, the transfer of materials, occurs at the molecular, biochemical level, a level that does not always have visual equivalents. Our understanding of the biochemistry and the genetic regulation of that biochemistry is still in its infancy. Thus, species-specific characterizations of placentas are not always possible or functionally relevant (Power, Schulkin 2012).

The placenta is the conduit for maternal-offspring exchange of chemical signals (e.g., hormones), nutrients (e.g., glucose, oxygen), and waste products (e.g., urea, carbon dioxide). How, and perhaps if, placental structure is correlated with fetal growth rates, fetal development, litter size, or gestation length is not well understood, in part because placentas are physiologically dynamic and structur-

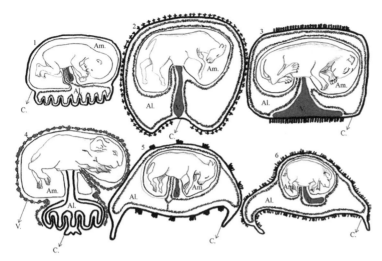

Figure 4.2. Diagram of placental types in six species that show the contributions of different extra-embryonic membranes. Key to structures: *Al*, allantois (*fuzzy line*, all but guinea pig; *bold line*, guinea pig); *Am*, amnion (*fine line*); *C*, chorion (*bold line*); *V*, vitelline (yolk sac, *shaded*). Key to species: *1*, human (discoid); *2*, horse (diffuse); *3*, cat (zonary); *4*, guinea pig (discoid with inverted yolk sac); *5*, sheep (cotyle-donary); *6*, pig (diffuse). From Fernandes et al. 2012, citing Leiser, Kaufmann 1994 (open access).

ally varied. Also, some regulation of transfer between mother and offspring may be more a function of molecular or cellular processes (e.g., fetal hemoglobin; Carter 2012) rather than placental structure.

As humans, we often forget that for many mammals, singleton births are not the rule. Multiple births, with multiple placentas, make gestation a more complicated process. With several offspring simultaneously in utero, siblings may compete for maternal resources. Thus, offspring may have mechanisms to obtain more nutrients or to control the growth of their siblings. Placental structures are likely involved in these processes. If the multiple offspring are identical, e.g., in nine-banded armadillos (*Dasypus novemcinctus*), which give birth to identical quadruplets that share a single placenta (see box 4.2), mechanisms could evolve for sibling cooperation rather than conflict. Tangentially, non-identical siblings may also share a single fused placenta, as do the triplet offspring of marmosets, *Callithrix* (Stevenson 1976).

Two specialized references on the placenta are exceptional. First is a traditional book, *The Evolution of the Human Placenta* (Power, Schulkin 2012). This monograph covers all mammalian placental structures. The second is an electronic resource with hundreds of images and descriptions of placentas from more than 150 species in about 15 eutherian orders. This repository is from K. Benirschke's examination of placentas available to him over decades of research at the San Diego Zoo (placentation.ucsd.edu/).

The placenta performs many functions. These include nutrient and immune transfer, protection, coordination of timing between mother and offspring, and in utero communication (see also chapter 7, "Gestation"). From the maternal perspective, the placenta may be a site for embryo rejection, such as rejection of offspring due to the presence of foreign (i.e., paternal) DNA. The major function of a therian placenta is to allow in utero exchange of materials between a mother

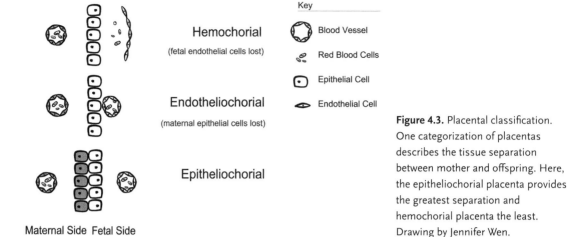

Key

Hemochorial

(fetal endothelial cells lost)

Endotheliochorial

(maternal epithelial cells lost)

Epitheliochorial

Maternal Side Fetal Side

Blood Vessel

Red Blood Cells

Epithelial Cell

Endothelial Cell

Figure 4.3. Placental classification. One categorization of placentas describes the tissue separation between mother and offspring. Here, the epitheliochorial placenta provides the greatest separation and hemochorial placenta the least. Drawing by Jennifer Wen.

and her offspring. This includes nutrients, wastes, gases, hormones, and other signaling factors. For materials to get from mother to fetus, they may pass through multiple tissues, maternal blood and associated vasculature, maternal structural tissues of the uterus, placental tissues and vasculature, and fetal structural tissues and vasculature, before finally reaching the fetal blood supply. Not all of these separate layers are present all the time or in all species.

An early assumption was that the fewer the layers separating the maternal and fetal circulation, the more advanced the placenta. The underlying reasoning (and a second assumption) was that the mother and the fetus were a cooperative unit. Thus, any structure or physiological process that increased the efficiency of exchange was advantageous. Given that the species making these assumptions was human, the human hemochorial placenta with the fewest layers was considered the most advanced. However, neither of the fundamental assumptions turns out to be completely valid. First, the evolution of placental types suggests that the invasive (endotheliochorial or hemochorial) placenta is actually ancestral for eutherians (Ferner et al. 2014). Thus, humans have an ancestral, not a derived, placenta. Second, we now understand that the maternal-fetal unit may not always be a cooperative enterprise. The mother and offspring may be under different selection pressures. Over the course of her life, a mother's fitness is increased by successfully raising as many offspring as possible. Her current offspring will most often be a positive contribution to that count, but, at times, aborting a current reproduction and starting again may in the long run provide her with more grandchildren. In those cases, the fetal-maternal unit is not a cooperative one.

Overall, the ideal placental structure will allow for some maternal control over the speed and duration of gestation, the transfer of nutrients, and the exchange of wastes. Evolutionarily, mothers will want to control offspring growth and to maintain their ability to abort a reproductive bout. Clearly, embryo rejection is not in the best interests of any offspring. Thus, on the fetal side, mechanisms may be in place to promote fetal growth and development. Fetal efforts are capped by maternal physiology, because if fetal demands push a mother to early mortality, the fetus also perishes.

We cannot leave any discussion of the placenta without recalling another eutherian bias. The name, placental mammal, gives the impression that only eutherians have placentas, but as discussed earlier, marsupials have a fully functional placenta. In marsupials, the yolk sac forms the definitive chorio-vitelline placenta and a few marsupials, e.g., bandicoots (*Isoodon macrourus, Perameles nasuta*), have a chorio-allantoic placenta. All marsupials have an additional non-cellular, permeable coat between the trophoblast and the uterus for the first half to two-thirds of gestation. The marsupial placenta has a shorter life span than the eutherian placenta but has a significant physiological role regardless. For instance, the tammar, *Macropus eugenii*, placenta produces hormones and growth factors (Renfree 2010). Thus, all therians, marsupials and eutherians, are placental mammals.

Ovaries

Now that we have explored the earliest reproductive structures we will examine those of adult females. The multi-functional ovary and the female gametes it nurtures are essential for reproduction. As endocrine organs, the ovaries produce both reproductive steroids, such as progestogens, estrogens, and androgens, as well as non-steroidal regulatory compounds, such as growth factors. Ovaries are also the site of oogenesis, the growth of female gametes. Oogenesis occurs within specialized balls of ovarian tissue, called *follicles*, that change in character as female gametes mature. Thus, ovaries regulate folliculogenesis as well as ovulation (chapter 6).

Ovaries vary in location. They are paired organs usually found in the lumbar region of the abdominal cavity. They can be more anterior (e.g., near the anterior portion of the kidneys as in *Cavia*, guinea pigs) or posterior (e.g., in the pelvic region as in women; Mossman, Duke 1973). Their exact location across species varies with posture (upright postures lead to more posterior organs) and with the length and shape of the uterus (the longer and straighter the uterus, the more anterior the ovaries; Mossman, Duke 1973). Overall, ovaries are positioned near the uterus, but sufficiently distant so as to be away from harm as the offspring get larger and the uterus stretches. Ovaries maintain their position via ligaments and other tissues (the mesovarium) connecting them to the body wall. The smooth muscle in the mesovarium provides a mechanism either for some movement of the ovaries relative to other organs or for maintaining the ovaries in one place, for example, during locomotion.

Neither ovaries nor oviducts are directly connected to one another; instead, both are open to the abdominal cavity. Interesting consequences of this openness are (a) ova can migrate across the peritoneal cavity from one ovary to the opposite oviduct, (b) ectopic pregnancies can occur (i.e., a pregnancy where the embryo implants outside of the uterus), (c) sperm or other debris can contaminate the abdominal cavity, and (d) peritoneal fluid can drain into the genital tract. Two evolutionary solutions reduce these potentially hazardous outcomes. First, the oviduct is very close to one end of the ovary. This part of the oviduct is shaped like a funnel with a fringe of tentacles (called fimbria). This funnel may enclose much of the ovary, and thus trap any released ova. Second, at the other end of the ovary, the tissues that attach the ovaries to the rest of the body may form a pocket (bursa) around each ovary. This pocket may have a wide or very narrow opening and the fimbria of the oviduct are generally located near this opening or may project into it (Bedford et al. 1999). Although different species use different combinations and elaborations of fimbriae and bursas to enclose the ovary, the functional result is the same: reducing exchange of material between the peritoneal cavity and the reproductive tract. In turn, these seemingly simple structures are key for facilitating a successful conception.

Ovaries come in various sizes and shapes. As mentioned in chapter 1, the metaphor of ripening ova is a poor and male-biased analogy with fruit. Similarly, ovarian shapes are often compared with foods, such as mulberries in sows, bunches of grapes in mysticete whales, almonds in ewes, and kidneys in mares. Ovaries are not particularly large, even in whales. Human ovaries of reproductive age women are only 3–10 g each (Perven et al. 2014) or about the weight of a U.S. nickel (5 g). The ovary of a laboratory mouse is a thousand times smaller at 0.005–0.007 g (Bhattacharya 2013). For comparison, ovarian weights of other well-known mammals are rabbit 0.4 g, mares 80–120 g, pony mares or African elephants (*Loxodonta africana*) ~40 g, sows 7–20 g, cows 3–18 g, and blue whales (*Balaenoptera musculus*) 30 kg (Aurich 2011; McEntee 1990; Miller et al. 2007; Zoubida et al. 2009). Often an ovary is roughly 0.01%–0.03% of an adult female's mass, but given that cow ovaries are about the same size as the ovaries of human women and those of pony mares are about the same size as those of the African elephant, body size is not the only factor determining the total size of the ovaries.

Ovarian shape and size are not static. Ovaries change in both parameters with age, season, and reproductive state. The ovarian weight of a pregnant blue whale is 30 kg, but ovaries of non-pregnant females are only 7 kg. Ovary size changes over gestation. Although whale ovaries are huge during pregnancy, ovaries of mares in the last third of gestation have regressed to the point at which they appear inactive. In addition, females that reproduce seasonally may reduce the size of their ovaries during the non-breeding season (McCue 1998), and ovaries may also shrink at the end of a female's reproductive life. In contrast, fetal ovaries at later stages of gestation may be as large as or even larger than those of adults (e.g., African elephants, *Loxodonta*, 60 g vs. 40 g; gray seals, *Halichoerus*, 18–29 g vs.

10–14 g; harbor seals, *Phoca*, 32–36 g vs. 25–28 g; Allen et al. 2005; Amoroso et al. 1951; Glickman et al. 2005; Hobson, Boyd 1984). Given that fetuses are smaller than adults, the large size of these fetal ovaries is even more striking.

The taxa with large fetal ovaries are not closely related but all have singleton litters and long gestation lengths. Furthermore, the fetal testes of males of these same taxa are similarly enlarged. The size increase of the fetal gonads is primarily due to hypertrophied tissues that generally produce steroids. Thus, the fetus, rather than the mother, may be generating hormones to maintain pregnancy in the latter part of gestation. Consistent with this hypothesis is that these fetal gonads generally regress before birth (e.g., equid foals: 25–50 g each at 250 days of gestation vs. 5–10 g each at birth; Hay, Allen 1975). Whatever the causes or consequences, the change in fetal ovarian size throughout gestation is another example of the dynamic nature of the mammalian ovary.

Although size and shape vary, the general structure of the ovary is relatively stable across species. The bulk of the ovary is an aggregation of balls of cells (follicles) often set off from the rest of the body by an outer surface (capsule). Ovaries of small mammals, e.g., woodland jumping mice, *Napaeozapus insignis*, may lack an outer capsule. When present, the capsule has two layers. The most external is a thin layer of epithelial cells, usually misleadingly called the germinal epithelium although it does not give rise to primordial germ cells. This outer layer lies over a thicker layer of fibrous connective tissue, the tunica albuginea. Within the external capsule lies the bulk of the ovary and contains developing and regressing follicles (each with an oocyte) and supporting tissue. For many species, ovulation can occur over most of the surface of the ovary; however, mares have a distinct ovarian region, the ovulation fossa, from which ova are released (Kimura et al. 2005). This fossa is not present in the related tapirs and rhinos (Lilia et al. 2010; Zahari et al. 2002).

Ovarian organization is often described from a primate bias. In primates, the follicular tissue (area of the ovary where the follicles occur) surrounds a relatively small, vascular, central core with blood and lymph vessels, ligaments, and other support tissues. This primate condition led to the terms *cortex* (outer) for the follicular region of the ovary and *medulla* (inner) for the support region. Across mammals, these terms are only useful in a general sense or not at all. In addition, not only is separation between the regions usually indistinct, but the ovarian organization is radically different across mammals. Three examples will illustrate the issue. First, in horses and nine-banded armadillos, the regions are reversed; the support tissues lie in the outer regions of the ovary, and follicles comprise the central core (Kimura et al. 2005; Mossman, Duke 1973; McCue 1998). Second, in the plains viscacha (*Lagostomus maximus*), the support tissue is greatly reduced and the ovaries have a highly convoluted anatomy; this anatomy may support the ovulation of 400–800 ova per reproductive cycle (Espinosa et al. 2011). Third, in some moles (e.g., *Condylura cristata*, *Talpa europaea*), follicular tissue occupies

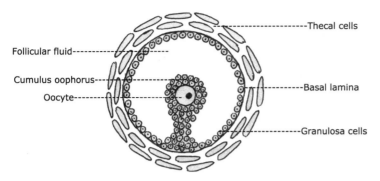

Figure 4.4. Anatomy of a follicle. The major cellular (granulosa, theca, cumulus oophorus) and acellular (basil laminar, follicular fluid) components of an antral (pre-ovulatory) follicle enclosing an oocyte. Modified by Abigail Michelson from Edson et al. 2009.

one end of the ovary, and the other end is devoted to steroidogenic cells, producing primarily testosterone (Bedford et al. 2004). Tangentially, testosterone production led to the misleading categorization of the ovary as a hermaphroditic ovotestis. However, (a) the ovary does not produce sperm, and (b) testosterone and other androgens are normal products of ovaries (and adrenals) of most female mammals. Regardless of the problems with biased terminology, the major structural and functional unit of the adult ovary is the follicle.

Anatomy of the Follicle

At its simplest, a follicle comprises one to many layers of maternal cells surrounding an oocyte. The number of concentric layers, the shape of the cells in those layers, and the characteristics of sandwiched non-cellular layers between the concentric layers all vary with the age and state of the follicle. These layers of maternal cells are named, depending on their proximity to the oocyte and the stage of follicular growth (figure 4.4). Granulosa cells are closest to the oocyte. As granulosa cells become more numerous, the follicle grows. Eventually, those closest to the oocyte are called cumulus cells and the collection of all cumulus cells is referred to as the cumulus oophorus. As the follicle matures, a basal, non-cellular layer (basal lamina) surrounds the granulosa cells. Outside the basal lamina are thecal cells that, as they become more numerous, differentiate into theca interna (nearest the basal lamina) and theca externa. Within a mature follicle, the cumulus oophorus and the oocyte float in a fluid antrum with only a small stack of granulosa cells connecting them to the rest of the follicle. Numerous follicles are present in an ovary, and they are usually in many different stages of development or atresia (regression). However, individual follicles follow an established pattern of growth (folliculogenesis) as described in chapter 6.

The textbook view of a mature follicle is of a capsule with a central gamete that floats in a fluid-filled antrum attached to the rest of the follicle by a stalk of maternal cells. However, not all mammals conform to this textbook anatomy. First, follicles may have more than one gamete. Such polyovular follicles occur occasionally in all domestic animals but are more common in cats and dogs

(McEntee 1990). Although unusual for domestic animals, such follicles are abundant in tenrecs (*Tenrec*), with up to five oocytes in a single follicle (Nicoll, Racey 1985). Tenrecs have a second distinction, the follicles of *Setifer, Tenrec,* and others lack an antrum or central cavity; instead, the gamete remains nestled in a solid ball of maternal cells (Enders et al. 2005). Tenrecs are also unusual in that conception occurs within the follicle after the follicle as a whole is ovulated, rather than ovulation of just the oocyte and cumulus (Nicoll, Racey 1985).

The Reproductive Tract: From the Ovaries to the Outside

Infinite diversity in extraordinary combinations describes the variety of reproductive tracts across mammals. Although not quite as variable as the placenta, the female reproductive tract has been modified in numerous ways to accommodate the diversity of maternal-offspring interactions across species and exhibits large variation during the reproductive lives of individual females. After reviewing the basic pattern, we will explore some of this diversity.

The ancestral structure of the female reproductive tract is a pair of hollow tubes each with a ciliated funnel at the anterior end to capture ova released from the ovary. This funnel is followed by a muscular segment with regional modifications. The pair of tubes remains separate near the ovaries but often fuses to form a larger space for offspring development. The reproductive ducts may terminate directly to the outside of the body or may connect with the urinary ducts to form a urogenital sinus (e.g., elephant shrews, Macroscelididae). In some mammals (e.g., echidnas, tenrecs), the urogenital sinus may also fuse with the terminal end of the digestive tract to form a cloaca. The entire reproductive tract is secretory, but what substances are secreted varies by segment and species.

Several regions of the reproductive tract are normally identified. These are the oviducts, uterus (uteri), vagina, and vulva as well as transitional zones between these, such as the utero-tubal junction and cervix. Each region is modified to perform a variety of functions, including sperm storage; conception; shell deposition in the case of monotremes; blastocyst implantation; pregnancy recognition; nutrient provisioning; water and gas exchange; production of regulatory hormones and factors; removal of debris; providing a barrier to outside influences; embryo rejection; spacing of embryos; ameliorating development of embryos; release of the egg, embryo, or fetus; and repair or refurbishment of uterine tissues after birth (Wagner, Lynch 2005). Not all functions are equally important for all species, and as a result, not all regions or transitions are present in all species. We briefly review the anatomy of the reproductive tract and explore its taxonomic diversity (figure 4.5).

Oviduct

The oviduct (aka Fallopian tube) is a fluid-filled, partially ciliated, muscular tube that extends from the ovary to the uterus and is usually the site of conception and

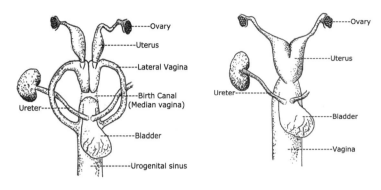

Figure 4.5. Relative location of ureters and reproductive ducts in a marsupial (*left*) and a eutherian (*right*). In marsupials, the developmental route the ureters take to reach the bladder is between the reproductive ducts, thus preventing fusion of the uteri. Modified by Abigail Michelson from Tyndale-Biscoe 1973.

very early development. This muscular component is important as it allows the oviduct to move relative to the ovary. These muscles allow flexion, shortening, and squeezing. The orientation of muscle fibers (Hafez, Hafez 2000) is consistent with the oviduct mechanically influencing gamete and zygote transportation. Although usually a paired structure, in some species one oviduct atrophies or is highly reduced (e.g., some bats).

Key is that the oviduct is not a passive environment. Ova are transported from the ovary to the site of conception and sperm are transported in the opposite direction from the uterus to the oviduct. At mating, sperm are not physiologically able to fuse with ova, nor are they able to reach the site of conception unaided. Oviductal fluids not only transport sperm but also capacitate them, enabling conception. In addition, sperm may be stored in the oviduct or actively destroyed there. Finally, once conception has occurred, oviductal secretions are a major source of early nutrition for the conceptus.

From ovary to uterus, four oviductal regions are named: (1) infundibulum, (2) ampulla, (3) isthmus, (4) utero-tubal junction. The fringed, funnel-shaped ovarian end of the oviduct (the infundibulum) partially encloses the ovary, whereas at the uterine end (utero-tubal junction), the oviduct connects continuously with the uterus. In between, the ampulla and isthmus may or may not be distinct. Length is not closely related to body size, e.g., both the 500-g, four-toed hedgehog (*Atelerix albiventris*) and the 75-g Asian musk shrew (*Suncus murinus*) have oviducts ~7.5 mm long; yet, the oviduct of the 4- to 5-g least shrew (*Cryptotis parva*) is 4–5 mm, and the oviduct of the 70-g eastern mole (*Scalopus aquaticus*) is 20–30 mm in length (Bedford et al. 1997a, 1997b, 1999, 2004).

Internally, the oviduct has a mucous lining (mucosa), which is rich with folds (rugae) that increase surface area. In the Asian musk shrew, the oviductal folds nearest the uterus function as crypts for sperm storage (Bedford et al. 1997). Similar crypts in moles (*Condylura cristata, Scalopus aquaticus*) are ciliated, have leukocytes (Bedford et al. 1999), and may also be a site of sperm storage. Leukocytes here may remove debris or sperm.

The oviduct varies across species in (a) the shape and extent of the infundibulum and (b) the nature and degree of coiling along the tube (Lesse 1988). Variations in the infundibular funnel were presented above with the ovarian bursa. The rest of the oviduct is equally varied and topographically complex. The oviduct may be looped (e.g., bears, *Ursus*; squirrels, *Sciurus*), coiled (e.g., pigs, *Sus*), kinked (e.g., kangaroo rats, *Dipodomys*; mink, *Neovison*; pocket gophers, *Geomys*), or relatively straight (e.g., rabbits, *Oryctolagus*; women; Mossman, Duke 1973). Spirals and loops increase the length of the oviduct and may affect transit time for ova and sperm. Thus, one function of oviductal length may be to regulate the timing of conception relative to ovulation and mating.

In addition to species-level variation, the oviduct changes character along its length from the ovary to the uterus as well as over time. The mucosa thickness and its folds are greatest near the ovary and decrease near the uterus, whereas the muscular layer is a larger proportion of the oviduct near the utero-tubal junction (Lesse 1988). Ciliated cells are more prominent near the ovary, whereas secretory cells are more abundant near the uterus (Lesse 1988). Temporal variation also occurs. The hypertrophy and atrophy of various oviductal cells vary with reproductive stage (e.g., pregnancy, peri-ovulation). As a result, the volume and composition (electrolytes, water, oxygen, hormones, sugars, amino acids, immunoglobulin, etc.) of the oviductal fluid also changes (Lesse 1988; Lesse et al. 2001). In sum, across species the basic tubular structure of the oviduct is similar, but the morphology, histology, and biochemistry of oviducts vary not only across species but also within an individual at different ages or in different reproductive stages.

Uterus

The uterus is yet another multi-purpose organ. Functions of the uterus include maintaining embryo health before implantation, communication with ovary, pregnancy recognition, placenta formation and attachment, sperm transport and storage, gestation and eventual expulsion of offspring and placentas, clean up, and microbial health. Structurally, the conventional categorization of uterine architecture (simplex, bipartite, bicornuate, duplex; figure 4.6) belies the variety and complexity of uterine anatomy across mammals.

Anatomically, the uterus is a continuous structure starting from the oviducts and terminating at the vagina or vaginas if present. At the oviductal end, the utero-tubal junction separates oviduct from uterus whereas the cervix (or cervices) is at the vaginal end. Although the oviducts are never fused, the luminal space of the remaining reproductive tract is often fused to varying degrees. In some taxa, the left and right lumens never fuse and the resultant track has two of everything, including cervices. This duplex morphology occurs in a variety of taxa including; dermopterans (colugo), some pteropodid bats, all lagomorphs, and some rodents (e.g., *Cavia*, *Cricetomys*, *Cuniculus*, *Dasyprocta*, *Rattus*), as well

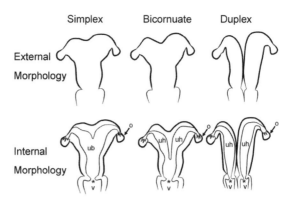

Figure 4.6. Uterine shape is not always obvious at first glance. The top row illustrates the external morphology of three major uterine morphologies (simplex, bicornuate, duplex) when the peritoneal cavity is exposed, whereas the bottom row illustrates the internal condition when the uterus is dissected. Simplex and bicornuate uteri may not be distinguishable by external morphology. Uterine morphology is determined by the degree to which the left and right uterine cavities coalesce not by the external appearance. Illustrated are *ub*, uterine body; *uh*, uterine horn; *asterisk*, cervix; *v*, vagina; *o*, ovary. Drawing by Jennifer Wen.

as marsupials and monotremes (Akinloye, Oke 2010; Favoretto et al. 2015; Hood 1989; Mayor et al. 2011, 2013). At the opposite extreme, the uterine lumens can be completely fused along their entire length, from the utero-tubal junctions to a single cervix and vagina. This fused (simplex) uterus is found in primates and nine-banded armadillos (Enders et al. 1958; Haig 1999). Partial fusion is very common and termed *bicornuate* or *bipartite*. Shrews (Soricidae), artiodactyls, perissodactyls, cetaceans, carnivorans, and some bats have uteri with some degree of luminal fusion. For example, mares have a large uterine body with small horns near the oviducts, whereas female dogs and cats have a small uterine body with long horns.

Fusion refers to the internal lumen and not to external morphology (figure 4.6). For instance, the fused portion of an externally Y-shaped uterus may have two cervical canals leading to two cervices. Thus, the uterus externally appears to be bicornuate but is actually duplex.

Other arrangements are not easily accommodated by these categories. For instance, the uterus of the brown-throated, three-toed sloth (*Bradypus variegatus*) has no uterine horns and thus would be termed *simplex*, but it also has two cervices opening into a single urogenital sinus (Favoretto et al. 2015). Wislocki (1928) described a single uterine cavity in sloths terminating in a vaginal canal that divides into two parts farther away from the uterus. The opposite occurs in some antelope (e.g., *Addax*, *Hippotragus*, *Oryx*), as well as in wildebeest (*Connochaetes*), all of which have a single cervix that bifurcates into two cervical canals, each leading to a separate uterine horn (Hradecky 1982). Thus, luminal fusion may occur either from the top down (cranial to caudal) or the bottom up (caudal to cranial;

figure 4.6). One extreme condition is a reproductive tract that is completely doubled from the oviducts down to a pair of vaginas, as in marsupials. The opposite
extreme also includes paired oviducts, but these open into a single lumen in the
uterus connected to a single cervix at the caudal end of the uterus, and opening
into a single vagina with a single external orifice (e.g., three-banded armadillo,
Tolypeutes; Cetica et al. 2005).

Other features of the uterus are more homogeneous across mammals. The
uterus comprises three layers. The outermost layer, the serosa, is in contact with
the peritoneal space of the abdomen. It is a thin layer of cells that prevents material
from the abdominal cavity from crossing into the uterus. In addition, ligaments
that support the uterus in the abdominal cavity attach to the serosa, especially
near the cervix. The middle layer of the uterus is the muscular myometrium with
inner circular and outer longitudinal smooth muscles. These muscular layers are
key for contractions during parturition as well as general movements of the uterus.
Finally, the innermost layer, the endometrium, is the most dynamic. The endometrium changes with reproductive state and in response to ovarian hormones. In
particular, it is reduced when not reproductive and becomes thick during gestation or even earlier in preparation for supporting the conceptus. The thickened
part of the internal layer (aka decidua) may be sloughed off periodically (menstruation) or with the placenta at birth. The uterus also enlarges and stretches
during gestation. After parturition, the uterus shrinks to a non-pregnant state.
This process can be rapid; for example, the rabbit uterus weighs 40–50 g immediately post-partum and two days later is down to 20–30 g (Zoubida 2009).

External Genitalia: Cloaca, Urogenital Sinus, Cervix,
Vagina, Clitoris, Prostate

[Female] external genitalia are rarely mentioned in the recent literature.
—Plön, Bernard 2007:147

[The clitoris] has been considered a rudimentary organ without apparent function.
—Toesca et al. 1996:514

Although plenty of emphasis is put on external male genitalia, the distal/cau-
dal part of the female reproductive tract receives short shrift in anatomical descriptions. Yet the lower part of the female reproductive tract is essential and
highly variable. From our human perspective, the uterus ends at the cervix,
which opens to the vagina, which opens to the outside world. However, not all
species have a cervix, or even a vagina, and a few have two of each (or three in the
case of kangaroos). But let us start with the familiar.

In humans, the terminal end of the uterus is the cervix, a circular neck of soft
tissue-encased cartilage with a (usually) small opening to the vagina. The cervix
is a dynamic structure with both short- and long-term changes over a female's

life. Early in development, the cervix is much larger than the uterus proper, but during puberty, the uterus grows much faster and enlarges to adult size, dwarfing the cervix (Ellis 2011). Mucus glands in the cervix are responsive to repeated ovarian hormone changes, as is the collagen composing the stiff but flexible ring. The tightly closed ring provides one level of protection against infectious agents but must dilate extensively during parturition and slightly during mating. The cervix is also the region most tightly connected to the rest of the body. Therefore, the cervix is where the uterus is able to flex forward or backward. The uterus can even turn itself inside out if a full uterine prolapse occurs.

In humans, the vagina is a muscular, elastic conduit that connects the reproductive tract with the outside world. Like the oviducts and the uterus, the vaginal canal has concentric, cylindrical layers of diverse tissues. The outermost is a sheath of epithelial tissue with some structural connective tissue. The middle layer has smooth muscle fibers in various orientations, and the innermost, luminal layer may be folded into rugae. The vagina is infused with a complete vasculature and extensive innervation. In addition, the vaginal lumen supports an extensive and dynamic microbiome, microbes that help maintain the health of the reproductive tract. The vaginal opening and the enclosed microbiome receive some protection via concentric folds of skin (the labia). Also, a vaginal membrane (hymen) may be present. Within the labial folds lies the clitoris, an erectile organ that is highly vascular even when not aroused (Şenayli 2011). From a human perspective, those (cervix, vagina, vulvar folds, and clitoris) are the basics, but these structures vary widely across mammals.

Rather than start at the uterus, we will begin an exploration of variability at the other end. Human females have separate external openings for their digestive tract (rectum and anus), their bladder (urethra), and their reproductive tract (vagina and vulva). This arrangement exists for females of many other species, but some females and all males combine the latter two openings (urinary and reproductive) into a urogenital sinus. In addition, some females pocket all three openings into a single vestibule (cloaca). Developmentally, all embryonic mammals have a cloaca. In many mammals, a septum (perineum) develops, separating the digestive tract from the urogenital tract. Subsequently, in many females, the urethra and reproductive tract are further separated. Some mammals retain a cloaca (e.g., monotremes, *Tachyglossus, Zaglossus, Ornithorhynchus*; tenrecs, *Echinops, Hemicentetes*; porcupines, *Erethizon*; and pikas, *Ochotona*; Duke 1951; Marshall, Eisenberg 1986; Mossman, Judas 1949; Riedelsheimer et al. 2007).

Cloacal and urogenital sinuses can be quite complex. For instance, inside the cloaca of the lesser hedgehog tenrec (*Echinops telfairi*), a sphincter controls the opening and closing of the intestinal opening, and a second opening leads to the urogenital sinus (Riedelsheimer et al. 2007). Conversely, inside the urogenital sinus of sloths (*Bradypus, Choloepus*), the opening to the reproductive part of the sinus is sealed during pregnancy but otherwise open (Wislocki 1928).

As with sloths, a vaginal closure (hymen) may separate the distal end of the vagina from the outside, the urogenital sinus, or the cloaca. Often such a membrane seals the vaginal canal before puberty and the vagina subsequently remains open. However, the position and composition of the membrane, as well as the duration and extent of opening, vary across species. European moles have yet another anatomical twist. Their vaginal opening forms spontaneously during the breeding season but is covered by unbroken skin between the clitoris and the anal papilla the rest of the year (Harrison, Matthews 1935). The patency of the vagina in small mammals, e.g., moles, voles (*Microtus*), is used to characterize their reproductive status as either perforate (open) or imperforate/not perforate (sealed). As the definition of perforate is to pierce, penetrate, puncture, etc., use of the term carries with it the connotation that some action (e.g., mating) has occurred to open the vagina. However, the physiology of the process is unknown and patency may be hormonally regulated.

To return to urogenital sinuses, other therian females, besides sloths, retain a urogenital pouch, but that pouch can vary in size. For instance, in hippos, the vaginal canal and the urethra open into a "genito-urinary vestibule" with a large sinus (Chapman 1881:141), but the urogenital sinus of elephant shrews (*Elephantulus, Macroscelides, Petrodromus*) is shallow (Tripp 1971). For xenarthrans, a urogenital sinus is present in genera from four different families, e.g., lesser anteater (*Tamandua*, Myrmecophagidae), sloths (*Bradypus*, Bradypodidae; *Choloepus*, Megalonychidae), armadillo (*Dasypus*, Dasypodidae), suggesting it may be a feature of that group (Enders et al. 1958; Favoretto et al. 2015; Rossi et al. 2011; Wislocki 1928). Not so for lagomorphs, in which one family, the Leporidae (cottontails, *Sylvilagus*, and hares, *Lepus*), has a urogenital sinus (Elchlepp 1952; Hewson 1976), but the related family, Ochotonidae, encases its urogenital sinus in a shallow cloaca (Duke 1951). Marsupials have a urogenital sinus (Kirsch 1977), as do females from some eutherian lineages, for example, tree shrews (*Tupaia*), golden moles (*Amblysomus*), armadillos (e.g., *Chaetophractus*), binturongs (*Arctictus*), elephants (*Elephas, Loxodonta*), hippos, and spotted hyenas (figure 4.7; Balke et al. 1988a, 1988b; Cetica et al. 2005; Chapman 1881; Fuchs, Corbach-Söhle 2010; Kuyper 1985; Perry 1964; Story 1945). Therefore, the range of taxa with urogenital sinuses includes some that are usually considered ancestral/basal, e.g., marsupials, armadillos, golden moles, and tree shrews, as well as others with more recent origins, such as carnivorans and artiodactyls.

Although a cloaca or a urogenital sinus might be considered ancestral features, their presence across such diverse taxa suggests they may have adaptive value and could be derived. But what is the value of the various external openings? A positive consequence of a single cloacal opening is the reduction in the number of access points for pathogens or debris. A negative is that a single opening allows for easier contamination from the digestive tract into the reproductive or urinary tracts. A second negative is that the control of the cloacal opening

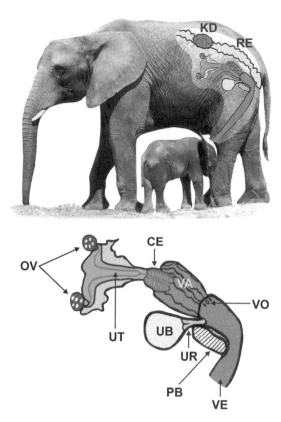

Figure 4.7. Position of reproductive tract in an African elephant (*Loxodonta africana*). *Top*: *KD*, kidney; *RE*, rectum. *Bottom*: *CE*, cervix; *OV*, ovaries; *PB*, pelvic bone; *UB*, urinary bladder; *UR*, urethra; *UT*, uterus; *VA*, vagina; *VE*, vestibule/urogenital canal; *VO*, vaginal os, with two blind pouches. From Hildebrandt et al. 2006; used with permission.

must synchronize the requirements of three different organ systems with exceptionally different outputs: feces, urine, and offspring. Separation of the openings allows specialized regulation to match the character and timing of the various outputs. This advantage might be especially important when giving birth to large young. A final consideration is that the reproductive tract must allow input during mating but neither the urethra nor the rectum need to accommodate that function. These thoughts are all speculative, and this is an open area for study.

Vagina

For the most part, the vagina is simply an extension of the uterus with the same tubular shape and histological layers (external protection, middle muscle, internal glandular serosa). The uterine terminus of the vagina is often marked by the muscular constriction of the cervix. But (1) a cervix is not always present, e.g., narwhals (*Monodon*; Plön Bernard 2007); (2) the cervix may not be muscular, e.g.,

Scientist, Adventurer, Editor
Barbara J. Weir (1942–1993)

In her early career, Barbara J. Weir worked at the Wellcome Institute of Comparative Physiology, part of the Zoological Society of London, where she pioneered studies of the reproductive biology of hystricomorph rodents. Her doctoral work on chinchillas and their relatives (Cambridge University, 1968) sparked an intense fascination with this group. At a time when women were not expected to be field biologists, she made several expeditions to Argentina, Bolivia, and Peru to study these animals and became a recognized authority on their biology.

In a series of papers, Weir carefully described female breeding habits of these little-known species. She carried out a comparative study of the breeding biology of wild guinea pigs (*Cavia aperea*) and two related species (*Galea musteloides* and *Microcavia australis*). At this point, almost nothing was known about wild guinea pigs compared to their domestic relatives.

Weir's discovery regarding what cues females use to time estrus set the stage for her next article, "The Evocation of Oestrus in the Cuis, *Galea musteloides*,"

published in 1971. Species related to *Galea musteloides* (which is also known as the common yellow-toothed cavy) experience estrus cycles in the absence of males. Weir explored why this is not the case in *G. musteloides*. Weir experimented by exposing female cavies to varying levels of male hormones and scent. Some females were kept completely isolated from males, some were able to see and smell a male in a neighboring cage, and others were able to see and smell the male and his urine and feces. Females did not respond to these cues, except for a few rare instances of spontaneous periods of estrus. In general, estrus began only once females had direct contact with males. Weir's study strongly suggested that female cavies may need to be in physical contact with males in order to begin an estrus cycle.

In her 1973 article, "The Induction of Ovulation and Oestrus in the Chinchilla," Weir expanded her studies of induction of estrus to the long-tailed chinchilla (*Chinchilla lanigera*). Weir explored the dosages of exogenous gonadotropins needed to trigger estrus

and ovulation in female chinchillas. This research allowed chinchilla breeders to breed chinchillas year-round, as the correct dosages of gonadotropins could induce ovulation and make female chinchillas receptive to mating outside of the normal breeding season.

After her extensive work on hystricomorphs, Weir turned to the editorial side of science. She was the editor of the *Journal of Reproduction and Fertility* (now *Reproduction*) from 1973 to 1991, producing more than 50 volumes, including several significant ones on equine reproduction.

An international symposium in her honor was convened in June 1973 and resulted in the classic "Biology of Hystricomorph Rodents" (Zoological Society of London Symposium No. 34, 1974). In spite of decades with significant ill health, she maintained a dedication to science and a productive career.

(Photo courtesy of the Journal of Reproduction and Fertility, used with permission.)

plains viscacha (Weir 1971a, 1971b; sidebar on this page); and (3) a female may have two cervices if the uterus is fully duplex, e.g., rabbits (Bensley 1910).

Like the uterus, the vagina can be fused throughout its length (as in humans) or remain a paired structure, as in monotremes and marsupials. In marsupials, a third vaginal canal forms between the two lateral canals to function during parturition (figure 4.5). The vagina can be quite long. For instance, that of the European mole extends to about 40% its body length during the breeding season (Bedford et al. 2004). Our knowledge of vaginal structure has a size bias, with features of vaginal anatomy primarily known for large mammals.

The internal vaginal architecture may have folds, rings, or funnels. For instance, longitudinal folds are present in African and Asian elephants; circular folds are present in *Tenrec*; and transverse interlocking fibrous ridges that generate an internal spiral are present in hippos (Balke et al. 1988a, 1988b; Bedford et al. 2004; Chapman 1881). In cetaceans, these folds "look like a chain of successive funnels with the mouth pointing towards the cervix" (Plön, Bernard 2007:148). The vagina can have diverticula or pouches, as are present in the plains viscacha and hippos (Chapman 1881, Weir 1971a, 1971b). In dugongs, the capacious vagina has a keratinized vaulted ceiling (Nishiwaki, Marsh 1985). In seals, the vagina has longitudinal folds invested with muscle fibers that can constrict the vaginal canal (Atkinson 1997). Speculations on the functions of the vaginal architecture include protection of the uterus from pathogens, providing a barrier to the elements; facilitating orgasm, mating, conception, sexual conflict, birth, or species recognition; and eliminating debris. These hypothetical functions lack firm support.

Clitoris

Before leaving external genitalia, two other features deserve mention: the clitoris and the prostate. The clitoris is rarely illustrated or included in descriptions of the female reproductive tract; yet, it is present in all mammals and presumably provides positive reinforcement for mating via stimulation of the limbic system. As the clitoris and penis have the same embryonic origins, they can also share characteristics such as the presence of a prepuce, an erectile ability, a complement of glands, and the presence of a bony core (os clitoridis aka baubellum). The clitoris of many rodents superficially resembles the male urinary papilla (or perhaps the male urinary papilla resembles the clitoris). Females of some species (e.g., hedgehogs, *Erinaceus*; moles, *Talpa*; spider monkeys, *Ateles*; and, of course, spotted hyenas, *Crocuta*) have a prominent, pendulous, and erectile clitoris.

Descriptions of clitoral size can be ambiguous. For instance, the clitoris of an elephant (*Elephas*) can be 37 cm long (Eisenberg et al. 1971), but that of the hippo (*Hippopotamus*), described as "large," is only 6.5 cm in length (Eltringham 1995:29). Similarly, *Erinaceus* and *Orycteropus* (aardvark) are said to have large clitorises and *Daubentonia* (aye-aye) a short one, but without measurements, these descriptions are unhelpful (Hayssen et al. 1993; Pocock 1924). On the other hand, the clitoris of the European mole at 0.5 cm (Harrison, Matthews 1935) would seem to be small, but as it is 70% the size of the male's penis, the clitoris is comparatively large. The equivalent human-female clitoris would be 3.5 inches (8.89 cm) long.

In general, anatomists appear to use the human clitoris-penis size-differential as a yardstick. If a female of a different species has a larger clitoris than expected for humans, her clitoris is described as enlarged, masculinized, penile, or peniform. A second bias is that although the clitoris has erectile tissue, when it becomes turgid the term *clitoral enlargement* is used rather than *clitoral erection* (Maurus et al. 1965). Consequently, an "enlarged" clitoris ambiguously refers either to a clitoral erection or to the flaccid size of the organ.

Clitoral size may change with age. For instance, the clitoris of a juvenile female fossa (*Cryptoprocta*) is larger than that of adults, has a bigger os clitoridis (an internal bone), and more ornamentation in the form of keratinized spines (Hawkins et al. 2002). Overall, clitoral structure is highly varied.

No matter the yardstick, spotted hyenas have an impressive clitoris, one that as previously mentioned performs many functions. Females urinate, mate, and give birth via the clitoris. Thus, the whole structure is effectively a urogenital sinus (Cunha et al. 2005). The tissue is erectile and used in social displays. It can also expand to accommodate the birth of 1- to 1.5-kg pups. This adaptation, and the social dynamics surrounding it, are clearly beneficial as spotted hyenas are the dominant predators of the African savannah, with populations surpassing those of lions (*Panthera leo*), leopards (*P. pardus*), and cheetahs (*Acinonyx jubatus*).

The "Female" Prostate

Another component of the female reproductive system that receives scant mention is a gland situated between the vagina and the urethra. This paraurethral gland is present in both sexes but, for unknown reasons, is given different names: Skene's gland in females, prostate in males. In both sexes, the gland has the same structure. In females, the gland is smaller, but well vascularized and has abundant secretory cells (Santos et al. 2003). It cannot be considered either rudimentary or immature (Santos et al. 2003). The gland is present in women and female cottontails, as well as some rodents (e.g., *Arvicanthis, Lagostomus, Meriones, Praomys, Rattus*), bats (e.g., *Coleura, Nycteris, Taphozous*), hedgehogs (*Erinaceus*), tenrecs (*Hemicentetes*), and moles (*Talpa*; Elchlepp 1952; Flamini et al. 2002; Santos et al. 2006). It may be much more common than this list suggests because its existence is poorly known and therefore often ignored in anatomical studies. In women, it may also be the location of the G-spot (Eichel, Ablin 2013).

As for function, the composition of the gland suggests both exocrine and endocrine roles. The gland releases a thick, whitish ejaculate in women (Zaviačič 1987; Rubio-Casillas, Jannini 2011), with a composition similar to that in males. The timing of the release indicates a role in mating, but the specific function is unclear. The gland also secretes material in smaller amounts to regulate the activities of nearby organs. Again, too little is known about the specific compounds released, the timing of the secretions, or the cells on which the secretions might act to speculate about specific functions. Clearly, the prostate gland could use more study with respect to female reproduction.

Vasculature, Portal Systems, Innervation

How does the reproductive system receive oxygen and nourishment? The anatomy for this transfer includes the complex vasculature of the ovary and reproductive tract as well as communication with accessory reproductive structures such as mammary glands. Vasculature and innervation coordinate various anatomical parts during reproduction. For instance, lactation usually inhibits ovulation, also, gestation, in some mammals, inhibits milk production. Some of these actions involve local control, i.e., communication that occurs directly between organs rather than via the brain.

The local transfer of substances often involves a specialized vascular arrangement, the rete mirabile. This specialized system of vessels, in which arteries and veins effectively run parallel to one another but in opposite directions, achieves a process called counter-current exchange. This allows materials, e.g., steroids, to be localized in specific regions rather than become part of the general circulation (Lesse 1988). The architecture of the rete also allows signals to be delivered directly to an adjacent organ. Vascular counter-current exchange structures allow communication from ovary to ovary, from oviduct to ovary, from oviduct to uterus, from uterus to ovary, and from vagina to uterus (Einer-Jensen, Hunter 2005). The exchange can involve venous blood, arterial blood, lymph, or the fluid between cells. For instance, embryonic signals for the maternal recognition of pregnancy can be delivered from the uterus directly to the ovary to maintain pregnancy or signals in the ejaculated seminal fluid can be delivered from the vagina to the uterus to increase sperm transport. At the mammary gland, local control can help match supply and demand (Knight et al. 1998). Thus, the vasculature allows chemical communication between mother and offspring, between mating partners, and between one side of the female reproductive system and the other.

The reproductive system also has extensive innervation and a lymphatic vasculature. In mammary glands, lymphatic vessels filter materials before they are synthesized into milk. In the oviducts and uterus, the lymphatic system handles post-mating debris. The oviduct also has afferent (lower thoracic nerves) and efferent sympathetic and parasympathetic nerves (Lesse 1988; Mossman, Duke 1973). This innervation can coordinate mating and ovulation. The anatomical details of the reproductive system offer clues to the avenues by which the system controls and regulates diverse processes from fluid currents, to muscular contraction, to litter size, to immunological or steroid-based control of ova, sperm, conception, and early embryo migration and development. The basic anatomy of most species has yet to be described, and such investigations, although not currently in vogue, are likely to generate more fruitful understandings of reproductive processes than simply sequencing the genome of each species.

Mammary Glands

For successful reproduction, milk must be produced and provided to offspring. The fundamental milk-producing cells are the same across mammals, but offspring do not come in only one size or with uniform needs. Consequently, the organization of these lactocytes and surrounding mammary tissue differs across mammals. The anatomy of mammary glands is as diverse and derived as are offspring themselves.

Numerous small hollow cavities (alveoli) lined with milk-secreting cells make up the functional unit of the mammary gland. Adjacent alveoli form lobules, and lobules have collecting ducts that open at a nipple, as in primates, or into a storage cistern connected to a teat as in cetaceans or ruminants (figure 4.8). The number and arrangement of alveoli and their ductwork, as well as the final external outlet, vary across mammals. For instance, in cetaceans, the ducts coalesce and empty into a larger cistern where milk is stored until called on for release via the teats (Plön, Bernard 2007), whereas in monotremes, the lobule ducts open independently onto an areolar region rather than into a teat (Griffiths et al. 1973). While this monotreme morphology seems well suited to the snouts (pointed in the echidna or flattened and paddle-like in the platypus), these distinctive cranial features are reduced in the neonates. Thus, the areolar region is unlikely to be an adaptation to the anatomy of the monotreme neonate. Women also have an areolar region with 15–20 pores encircling the tip of the nipple. The innervation of the mammary gland influences the control of milk release, a process that involves communication between mother and offspring.

Contrary to popular assumptions, milk is not pulled from the mammary gland by the neonate. Instead, stimulation applied to the nipple or teat, either by neonates or by milking machines, is sent to the brain, which releases a hormone, oxytocin, from the posterior pituitary. Oxytocin then moves through the circulation and reaches the mammary gland where the hormone elicits contraction of muscle fibers surrounding the alveoli and ducts in the mammary gland (milk

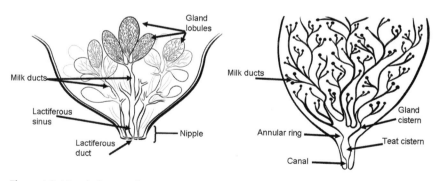

Figure 4.8. Morphologies of mammary glands with nipples (*left*) or teats (*right*). Note variation in number of cisterns and lobules. Drawing by Jennifer Wen.

letdown) and, thus, pushes milk into the neonate. Milk release is a result of this neuro-endocrine milk-ejection reflex. Neonatal sucking is only the stimulus for the release of milk and, at least technically, sucking, itself, does not pull milk out of the mammary gland.

The nature of this milk-release reflex means that besides alveoli and ducts, other anatomical aspects of the mammary gland are important. For instance, the arrangement of muscle fibers determines how alveoli will be squeezed, and the vascular network will influence the order in which different regions of the gland will receive the hormonal signal. Of course, the pattern of innervation determines which parts of the gland will be responsive to the neonatal signal. Overall, the location and extent of the muscular, vascular, and neuronal networks within the gland will influence milk release.

Other components of mammary glands have functional relevance, for instance, connective tissues, such as tendons and ligaments. These tissues give the mammary gland its external shape but more importantly hold the gland to the body. In primates, the relatively light mammary tissue is attached to pectoral muscles, but in dairy cows, with empty udders weighing perhaps 25 kg and full ones about 50 kg, strong ligaments attach the udder directly to the pelvic bones (Cowie 1974).

The number and position of mammary glands vary across mammals, but they are usually present in even numbers. For instance, like humans, the little brown bat (*Myotis lucifugus*) has a pair of thoracic mammae, but the capybara (*Hydrochoreus hydrochoerus*) has seven pairs of mammae, distributed along the sides of her body: two thoracic pairs, four abdominal pairs, and one inguinal pair (Husson 1978). Litter size and the number of mammary glands would be expected to be related. Not surprisingly, then, for many mammals, especially rodents, the number of mammary glands is often twice the average litter size (Gilbert 1986), but this is a very rough estimate. For instance, capybara have many more mammae (14) than their usual litter size (3–4), whereas average litter size (11) for naked mole-rats (*Heterocephalus*), equals the usual number of mammae, and the maximum litter size (28) is twice the number of mammary glands (Chapman 1999; Hayssen et al. 1993; Sherman et al. 1999). Like naked mole-rats, for a number of dasyurid and didelphid marsupials, litter size at birth exceeds the number of mammary glands. However, in these species, neonates who do not latch onto a teat die, whereas naked mole-rat neonates can share access and most are weaned. Clearly, litter size is more variable within a species, and the number of mammary glands generally is not. The number of mammae is often a species-specific, if not genus-specific, character (Hayssen et al. 1993). In this regard, the Neotropical singing mouse (*Scotinomys teguina*) is unusual. These females have different numbers of mammae in different geographic regions. Northern females have three pairs of mammae, whereas southern females have only two pairs (Hooper 1972). Surprisingly, litter size is the same across the range (Hooper, Carleton 1976).

For most mammals, teats and nipples are on the surface of the body (figure 4.8), although they may be buried in the fur when not in use. However, several groups of mammals keep their teats in pouches. The most well known of the pouched mammals are kangaroos and their relatives. Many, but not all, marsupials have teats within a pouch. The egg-laying echidnas are a second group of females who protect their mammary areolar region within a pouch. The third major group of pouched mammals is the cetaceans. Whales and dolphins have two teats, each one located in a small pouch on either side of the genital vent. The teats protrude during lactation but otherwise remain within the mammary slit (Plön, Bernard 2007).

Before we leave mammary glands, we note that some mammals have nonfunctional teats or papillae. These look like teats and may be derived from them. For instance, megadermatid and rhinolophid bats have a pair of inguinal or pubic papillae that look like nipples (Hayssen et al. 1993; Simmons 1993). Young can grip these papillae when being carried rather than holding on to and perhaps damaging the mammary nipples.

Anatomy: Unanswered Questions

Although female anatomy has been studied for thousands of years, our investigations have been limited to species we are drawn to or find accessible. Our technological skills have allowed us to see below the surface, and current advancements will soon let us observe and measure how structures change over time. For each structure, many questions remain. For the oocyte, how is structure involved in species recognition or embryonic development? For the placenta, does structure dictate the kinds of exchange that can occur between mother and offspring or among offspring?

For the ovary, how does structure relate to ovulation? Is the nearness to vasculature important? Does the extent of steroid-producing, non-follicular tissue have a functional significance and temporal variability? What does the location of connective tissue and muscle fibers tell us about the development of follicular waves or the choice of oocytes for ovulation? Does the morphology of the external capsule (bursa) dictate ovulation rate or location?

For the oviduct, how does the internal anatomy relate to gamete transport and storage? Does the regional amount of coiling (or the diameter or the arrangement of cilia) alter the timing of transport? As in the ovary, does the location and extent of glandular tissue have a functional significance? How do the location and orientation of muscle fibers determine the type, degree, and direction of oviductal motility? How does the oviduct collect the ova?

For the uterus, how does shape relate to litter size? How does uterine anatomy relate to uterine spacing and migration of embryos? How does uterine shape affect interactions among embryos?

For the external genitalia, how does the morphology of the internal and external genitalia alter the microbiome, mating, protection of embryos, birth, or social behavior? What is the function of the prostate? What is the function of diverse clitorises? What do differences in the musculature and the vasculature of genitalia suggest about differences in function and action?

As these questions illustrate, distinctions between anatomy and physiology are blurred. Not only do we need to look at how anatomical structures change as they execute their functions, but we also need to understand how physiological processes regulate those changes. Consequently, our next chapter provides a brief overview of some major components of physiological regulation as they affect female reproduction.

CHAPTER FIVE

Physiology

Cells, Systems, Populations, and Ecology

Although she reaches independence from her mother later relative to both bears and canids, in almost all other regards she resembles a medium-sized canid (to which she is only distantly related). Indeed, despite being a highly carnivorous mammal, her life history associated with reproduction more closely resembles that of omnivorous canids or bears. She has a long gestation with large offspring and a long period of lactation, longer than expected for her body size. Her distinctive genitalia are formed by a unique endocrine (hormonal) profile. Development of her robust clitoris occurs well before the capacity for steroid synthesis has been detected in fetal ovaries. Not until external genital differentiation is completed do fetal ovaries develop substantial androgen-synthesizing capacities. Androgens are likely to influence critical aspects of sexual development in both our female spotted hyena as well as her male counterpart. Both clitoral and penile development are advanced by day 30 of gestation, and sex-specific differences in internal genital organization are evident at day 45 of gestation.

Physiological changes due to hormones are not just key for morphological development but also for organizing neural biology. Hormone exposure in utero may result in variable levels of aggressive behaviors in pups. During pregnancy, females appear to exert negative feedback control of gonadotropins and androgens, a physiology deemed "modified" at least relative to other mammals. Also their prolactin levels are relatively low during lactation. (Licht et al. 1992, 1998; Lindeque et al. 1986; Conley et al. 2007)

> It is unfortunate that the hyena is not a convenient laboratory animal, for experimental work on this aspect of its physiology would likely be of considerable interest.
> —Harrison Matthews 1939:74

The genetic blueprint (chapter 1) provides instructions for creating the molecules and anatomical structures (chapter 2) that make up a female. Physiology focuses on the dynamic interaction of those parts, not only internally, organ to organ and tissue to tissue but also the mechanisms by which a female's body reacts to stimuli from her environment. Thus, the study of physiology occurs on multiple scales, including processes between cells, between organ systems, between individuals, and between individuals and the environment. Although physiology integrates of all these levels, most physiologists constrain their focus to just one or two stages.

Cell physiology and systems physiology include all of veterinary and human medicine as well as the study of cell and molecular processes. Researchers who call themselves "cell or systems physiologists" focus on commonalities across species as well as abnormalities among individuals. Often they are looking for practical applications, such as curing disease or improving production. As such, they study a limited number of species. These include (a) domestic mammals: horses, pigs, cows, sheep, goats; (b) companion mammals: cats, dogs, guinea pigs (*Cavia porcellus*), golden hamsters (*Mesocricetus auratus*), gerbils (*Meriones unguiculatus*); (c) laboratory mammals: mice, rats, rhesus macaques (*Macaca mulatta*); (d) farmed mammals: rabbits, fox (*Vulpes vulpes, V. lagopus*), mink (*Neovison vison*), red deer (*Cervus elaphus*), alpaca (*Vicugna pacos*), llama (*Lama glama*); (e) mammals killed commercially: whales, dolphins (Cetacea); and (f) ourselves and our nearest relatives (great apes and monkeys). The study of cells and systems provides the foundation of reproductive physiology, and we will review some of the basic principles in this chapter.

A second set of physiologists, ecological physiologists, explores the physiologies of mammals facing challenging environments or lifestyles, such as adaptations to flight, to a diet of fruit, or to desert environments. Historically, their focus is not on reproduction but on specific processes—hibernation, water balance, or fasting—directly related to a specific external challenge. Physiological ecologists often study ensembles of mammals with similar ecologies, such as arboreal folivores, desert rodents, or volant mammals (i.e., bats); alternatively, they study individual species with specific adaptations, such as naked mole-rats (*Heterocephalus glaber*). Throughout this book, we explore the discoveries these physiologists have made regarding reproduction.

Finally, a third set of biologists, life-history theorists, are not usually perceived as physiologists but do, in fact, study physiological systems. They explore how mammals allocate the fundamental materials they take from the environment (energy, nutrients, water, and air) and apportion those materials into sometimes competing functions, such as growth, survival, storage, or reproduction. These physiologists are primarily interested in patterns across species rather than in processes within individual bodies. Often they take data from as many mammals as possible to look for such patterns. The influence of body size (allometry) is especially important, as is the influence of ancestry. We will cover some of their work in the later part of this chapter.

Given the breadth of study that falls under the umbrella of *physiology*, we clearly cannot cover the entire field, nor even just those parts directly connected with reproduction. However, some basic physiological mechanisms are important for understanding how reproduction works. Thus, part of this chapter explores fundamentals of reproductive physiology connected with integration and coordination, in other words, hormones and hormonal mechanisms. More detail on the cellular and systems side of reproductive physiology can be found in exhaustive reviews, such as Knobil and Neill's *Physiology of Reproduction* (Plant, Zeleznik 2015). In addition, specific aspects of physiology are integrated into chapters in the next section of the book that covers the reproductive cycle.

We hope to illustrate the diversity of issues physiologists address concerning reproductive physiology. To this end we explore one topic and three concepts in some detail. The topic we chose is maternal recognition of pregnancy and the concepts are allometry, energetics, and trade-offs. We focus on pregnancy recognition, because it is one of the few topics for which we have considerable cross-species understanding of the physiological and biochemical diversity of mechanisms that allow a female to "know" that offspring are present in her uterus. We chose allometry, energetics, and trade-offs because these concepts elucidate physiological constraints on reproduction from a life-history perspective. Thus, the topic, maternal recognition of pregnancy, takes a molecular-cellular approach to physiology, whereas the concepts exemplify an ecological-evolutionary approach.

Before we start, we need to explore four difficulties that pervade the study of physiology at many levels. These pitfalls include (1) degree of categorization and the reductionist approach, (2) neglect of reproduction in the study of physiology, (3) lack of the female perspective in the study of physiology, and (4) problems associated with relying on the human perspective to study reproduction.

Four Pitfalls

Traditionally, physiology is the study of function whereas anatomy is the study of structure. Assigning functions in the context of physiology comes from a human predilection to look for purpose in life and living systems. Categorizing physiology also has a pragmatic component in health sciences. For both these reasons, physiology is often broken into functional categories, such as respiration, circulation, or reproduction. However, these categories are not (and cannot be) discrete or exclusive. Take birth as an example. Normally, the diaphragm is used continuously for respiration, but it is also a crucial, bearing down muscle during labor. The circulatory system is also key, as it is disrupted when maternal contact with the placenta is broken. Birth thus directly involves the respiratory, nervous, circulatory, and muscular systems, not just the anatomical reproductive system. In short, reproduction alters the physiological processes of the entire body, not just the ovaries, uterus, and mammary glands.

This brings up a second pitfall. Most scientists and physicians treat reproduction as secondary to other bodily processes. In other words, the effects of reproduction on, for example, circulation or metabolism are not considered part of normal function. Phrased another way: normal is what happens when a female is not reproductive. However, from sexual maturity until death, most females are in a reproductive state. Furthermore, our physiological systems have evolved with successful reproduction as the priority. Reproduction is the mainstay of natural selection. Females with the most progeny shape the next generation. Survival as a metric of natural selection is only important if reproduction ensues. Consequently, reproduction is primary and other facets of physiology are secondary in natural selection. Physiological function must be molded to the needs of reproduction. Selection on physiological function is strongest on reproductive females, less so on offspring and males. Ignoring the reproductive state of a female when studying respiration, digestion, metabolism, or circulation severely limits understanding of how these systems are integrated into female lives. Reproduction is not peripheral to how the body functions; reproduction is central to physiological integration.

A third bias in physiological research is a consequence of studying non-reproductive animals. Female physiology occurs under variable levels of different hormones. In comparison, male physiology occurs under relatively constant, high levels of testosterone. Unfortunately, most physiological research is conducted on males, ostensibly to avoid the confounding effects of variable hormones. As a consequence, male physiological systems (constant, high testosterone) have become the established baseline for normal physiological function (Beery, Zucker 2011). This is a clear male bias, as male physiology is treated as the normal baseline state. Females, with their variable hormone levels, are the deviant individuals. However, a fluctuating hormonal milieu is the basis of a normal female physiology. To understand the physiology of females, scientists need to study females.

In addition to the hormonal milieu, female physiology may be have a different activity schedule. Both sexes need to find food, avoid predators, and cope with the abiotic environment. However, the constraints on these basic activities differ for females and males. Simplistically, a female's day is spent near her offspring or carrying them, whereas males typically pass the day roaming a larger spatial area, looking for females, and perhaps fighting rival males. Female activity often occupies a smaller spatial area. Her activity is constrained by the weight of her pregnancy or the need to stay near her young. Her scope for foraging, predator avoidance, and accommodating abiotic stress occurs within a smaller context, but one with which she may be very familiar.

In this basic case, the need for endurance and strength may differ between females and males. Consequently, the function of exercise or muscular activity may vary between the sexes. Recent arguments suggest that research on performance traits, such as maximal sprint speed or endurance, would benefit from examining how the demands of female reproduction, such as pregnancy or lactation, may set limits on physiological parameters, such as weight-bearing or nutrient turnover. Thus, the demands of reproduction in females may select for traits that enable increased performance in both sexes. Also, reproductive traits such as lactation (chapter 9) are themselves performance traits but have not been studied in this context (Orr, Garland 2017). If females are tied to a nest area to nurse, then they are unlikely to travel great distances (in fact, doing so may be problematic). In addition, females must be keenly aware of the resources and dangers within this area and must have the mental capacity to analyze and retain this information. In contrast, males roam, find food and resources in new areas, and can run far from predators, all without having to return to a nest site and hungry mouths. The metabolic, muscular, energetic, and neural consequences of these very different lifestyles have ramifications on the trade-offs between different physiological systems and different behavioral choices. For instance, females could be sit-and-wait predators, whereas males could widely forage for a meal. This is one example of how focusing on female physiology leads to new questions. One might even posit that males can carry the increased mass of sexually selected traits, such as antlers, because female physiology was already adapted to carrying the extra weight of heavy offspring. Perhaps an initial selection for larger and heavier neonates subsequently promoted a physiology that was able to carry that extra weight; this is a female physiology that males could subsequently adapt for carrying antlers rather than progeny. Of course, the muscles used for carrying babies versus antlers would differ considerably. Regardless, by focusing on the demands of female reproduction, we are able to ask questions about the evolution of mammals in a different context.

Finally, humans have a fourth bias when examining reproductive physiology. Most human females in the developed world are not constantly pregnant nor nursing. Instead, they undergo repeated ovulatory cycles. This cultural pattern contributes to the notion that non-reproductive physiology is the norm. However, repeated ovulatory cycles without gestation is an aberrant condition for female mammals. Thus, the estrous cycle is a human concept with unclear biological relevance. Note that the estrous cycle, as commonly measured, differs from the larger reproductive cycle. The reproductive cycle moves from ovulation to conception to implantation to birth to weaning and back to ovulation. The estrous cycle is measured from ovulation to ovulation or menstruation to menstruation with no conception in between. In nature, females have as few estrous cycles and as many reproductive cycles as possible. Non-human females are either pregnant, lactating, or non-reproductive. Females in the wild do not repeatedly cycle from ovulation to ovulation without an intervening conception, and if they are repeatedly cycling something is wrong. The purpose of reproduction is to produce offspring and a female with continuously repeating estrous cycles is not doing that, she is just wasting energy. If the time or conditions are not appropriate for raising young successfully, a female would do better to shut down her reproductive system and bring it back up when conditions improve. In sum, continuous estrous cycles, per se, are not a feature of mammalian reproduction, except in zoos and other captive or domesticated environments, but they are easy to measure and, thus, a regular feature of reproductive physiology.

A similar issue concerns the extensive use of domestic, companion, or laboratory mammals for the study of reproductive physiology. These species have been released from the usual biotic and abiotic selection pressures for many generations. Thus, the physiological and endocrinological patterns we document are those that exist without major constraints. For instance, these species are not subject to predation, to starvation, or to competition. To the extent that these selection pressures modify physiology, their effects are not known. Smale et al. (2005) make a strong argument that traditional models from data on laboratory animals can be misleading. Overall, reproductive physiology in mammals probably bears some resemblance to what is known from laboratory studies, but just how comparable laboratory results are to reproduction in the wild is unclear. With these caveats in mind, we can now examine some of the fundamentals of the physiology associated with reproduction.

A Brief Overview of Reproductive Physiology

What must a female's physiology achieve for successful reproduction? To start, she must create viable gametes; find and choose a mate; find a suitable location for mating; mate; choose appropriate sperm; get rid of, degrade, or make use of possibly harmful substances transferred by the male, including pathogens; and facilitate conception. She must time mating or implantation so her needs later in reproduction are coincident with available resources or she must turn off her reproductive system if environmental conditions are poor and start it again when conditions improve. She must reject defective or excess embryos without compromising healthy embryos or future offspring. She must obtain food, water, and other resources both for herself and her developing embryos. She must find or build an appropriate place for birth of her offspring after which she will need to remove post-partum debris or leave the area. She must synthesize milk in appropriate amounts and composition and synchronize its availability with the needs of her young. She must wean her offspring. She may also provide other maternal care, such as teaching her offspring what to eat or how to find food or helping her offspring obtain the best rank possible if in a social group. All these tasks are functions of reproductive physiology and most of them fall under the rubric of coordination: coordinating female reproduction with the environment, with mates, with offspring, and with the other demands of living.

If coordination of processes is the goal, then communication, the sending and receiving of signals, is the mechanism for achieving that goal. Two signal pathways are in place: neuronal and chemical. On the neuronal side, the peripheral and central nervous systems transfer information from the environment to the brain to coordinate subsequent activities. For instance, sensory information from the environment, for example, day length, is used to coordinate timing of reproduction. Sensory information is also used to identify mates and one's own off-spring as well as to coordinate the social aspects of reproduction, such as mating and nursing. The external sensory information is transferred via nerves to the brain where it is integrated and fed to specific control regions of the hypothalamus to regulate circadian rhythms, thermoregulation, activity, hunger and thirst, reproduction, parenting, and social bonding. Nervous signals control muscular actions, such as the contractions needed for birth. The nervous system can be involved in both the receipt of information and the reaction to that information.

Chemical signals are most often used internally. Hormones coordinate the actions of cells and organs within a mother's body. For instance, ovarian hormones coordinate the activity of the uterus in advance of ovulation, conception, and gestation. Of course, a female's specific actions involve both nerves and hormones, but long-term internal coordination is primarily via chemical signals. The importance of chemical signals for reproduction cannot be overstated, and we will explore what they are and how they work in some detail.

Chemical signals have various names, such as pheromones, hormones, neurohormones, neuromodulators, and neurotransmitters, but all function to integrate bodily activities. Often the ensuing integration is indirect. In other words, hormones may regulate or prime physiological systems to facilitate appropriate outcomes at appropriate times rather than by directly triggering a specific action (Adkins-Regan 2005). Thus, chemical signals may also influence the durations of specific reproductive stages, such as gestation or lactation and may have several different targets. For instance, when oxytocin triggers birth, it also ends gestation and initiates lactation via milk letdown.

Chemical signals may have local or systemic effects. That is, they operate along a continuum of spatial scales. Different names may be applied to those different scales. At the largest scale, pheromones are chemical messengers released from the body of one organism and carried through the environment (via air or water currents) to alter the behavior of another organism. Within an organism, hormones are synthesized in one area of the body, secreted into the bloodstream, and circulated throughout the body until they reach their target organ far removed from the site of synthesis. At a smaller scale, paracrine hormones are released from the site of synthesis into the areas between cells to operate locally on nearby cells. Neurotransmitters are released into the synaptic cleft between one neuron and an adjacent neuron or muscle cell. Modifications of chemical signals within cells facilitate or inhibit the actions of the signals. For instance, neurosteroids (steroids synthesized within the brain) may act within the same cell in which synthesis occurs (Adkins-Regan 2005). Clearly, neurotransmitters and neurosteroids blur the line between neuronal and chemical (hormonal) signaling.

The association of androgens with masculine traits and estrogens with feminine traits is also a poor fit with nature's ways. (Adkins-Regan 2005:6)

The conversion of androgens to estrogens refutes the general view of estrogens as "female" hormones and androgens as "male" hormones. (Note the name *androgen* comes from the Greek "andro" for "a male human," whereas *estrogen* is from the Greek "oestrus," meaning "frenzy" or "gadfly.") Both hormones are present in both sexes and both are necessary for normal reproductive function. Intracellular conversion of androgen to estrogen is commonplace and occurs within the brains of both sexes. Testosterone is converted to estrogen in the brains of male mammals. Thus, estrogen controls many aspects of male behavior. Androgens are also important to female reproduction. In short, androgens are not exclusively male hormones, nor are estrogens exclusively female hormones.

Not only do hormones work at different spatial scales (local vs. systemic), they also work at different temporal scales. That is, many hormones have age-specific or season-specific effects. In the mid-1960s, an extremely productive dichotomy was framed to understand hormonal effects on behavior. This classic dichotomy proposed that hormones work in two ways: first, they organize behavior developmentally and, second, they activate behaviors in the mature adult (Phoenix et al. 1959; Young et al. 1964). The early organizational effects alter either neurological or anatomical systems. In this context, hormones are key instruments of sexual differentiation with lasting effects on adult behavior. In the adult, hormones activate neural and anatomical systems at sexual maturity or when coming out of a non-reproductive phase. Hormones can have both short and long-lasting effects. Although this classic dichotomy has been an exceptionally valuable paradigm for framing the diverse effects hormones have at different times or stages of development, the plasticity and flexibility of behavior suggests that the embryonic organization of neural substrates can be altered during adulthood and thus may not be set in stone (Adkins-Regan 2005). Nonetheless, thinking about the organizational and activational effects of hormones is often a useful way to frame questions or make predictions about a new system or species.

Chemically, most hormones are either steroids or proteins. The major reproductive steroids, estrogens, progestogens, and androgens, are derivatives of cholesterol (as are the gluco- and mineralo-corticoids, such as cortisol or aldosterone). Although corticoids are produced in the adrenal glands, steroid production is widespread. For instance, androgens and estrogens are synthesized not only in ovaries and testes but also in the adrenal glands, in the liver, in fat cells, and in brain tissue. In addition, progesterone, which is produced in the corpus luteum within ovaries as well as the adrenals, is a precursor to other steroid hormones and has its own physiological actions.

Although many hormones are associated with particular endocrine organs (estrogen with ovaries, prolactin with the pituitary, progesterone with the corpus luteum), secondary sources of particular hormones may be key. At times, we may attribute the actions of a hormone to the major source of that hormone rather than to a secondary source. For instance, systemic oxytocin generally arises from the pituitary. However, the corpus luteum also produces oxytocin and releases it into the ovarian-uterine circulatory system. This local secretion of oxytocin is involved both in the early stages of pregnancy recognition and in the later stages of uterine contractions for birth. In this case, the corpus luteum, and not the hypothalamus and pituitary, is regulating the effects of oxytocin on pregnancy. The lesson here is that the major source of a particular hormone may not be the specific source for the actions under investigation. Both the source and the chemistry of hormones affect their actions. The chemical characteristics of steroid versus protein hormones are related to their mechanisms of action.

Steroids are fat soluble, relatively small, and can pass through the complexly adorned lipid bilayers that constitute the membranes of the cell, the nucleus, or other vesicles. Steroids also permeate the blood-brain barrier. The structure of specific steroids shows few changes across a broad array of vertebrates even though the functions of those steroids vary a great deal. For instance, "testosterone . . . is the exact same molecule in anchovies, axolotls, adders, antbirds, aardvarks, and apes" (Adkins-Regan 2005:6). Other small chemical signals, such as dopamine, epinephrine, norepinephrine, prostaglandins, melatonin, and thyroxine are also similar across broad groups of species.

Size and mobility make small hormones effective messengers. Classically, the mechanism of their action involves the hypothalamus and the pituitary as well as the "target" organ, either the adrenal (for glucocorticoids) or the gonads (for reproductive steroids). Deconstructed, the process works as follows. Various regions of the brain receive sensory information from the body via chemical and electrical processes and, after integration, send that information on to the hypothalamus. The hypothalamus integrates that information and, if warranted, sends a signal to the pituitary. The pituitary releases the appropriate hormones into the circulation. Eventually, those hormones reach their target organs: gonads or adrenals. Those target organs then release steroids into the circulation, which are dispersed to other organs or tissues. Thus, the circulation is used to send signals from the pituitary to the gonad or adrenal glands and again to send a subsequent specific steroid to other target cells. An additional complication is that, in the blood, steroids may be attached to binding proteins, and these proteins may influence the amount of hormone available at a target cell.

Once at a target cell, steroid hormones pass through the cell membrane and may attach to steroid-specific intracellular receptors. The steroid-receptor binds to DNA, inducing the DNA to initiate the synthesis of proteins. The proteins do the work, building the uterine endometrium or degrading it, stimulating ovulation or repressing it, and creating other proteins for synthesis of other hormones. Steroids directly turn genes on or off. The multiple steps involved with steroid action mean that the time frame over which steroids work is usually on the order of days, not hours. Complicated positive and negative feedback loops between the central nervous system, hypothalamus, pituitary, and target organs allow very sensitive regulation of hormone release. Overall, the hypothalamic-pituitary-adrenal (HPA) axis and the hypothalamic-pituitary-gonad (HPG) axis are the classic signal pathways for steroid action. The HPA axis primarily coordinates responses to stress, whereas the HPG axis is involved in reproduction. However, as we cautioned earlier, secondary sources of steroids may exert more influence in specific cases than the primary HPA or HPG axes. In addition, changes to the HPA axis may influence the HPG axis and vice versa (Toufexis et al. 2014).

Protein hormones differ from steroids in chemistry, synthesis, and mechanism of action. Chemically, protein hormones are, as may be obvious, made from amino acids linked together often into long chains. The instructions for the exact sequence of these amino acids come directly from DNA; that is, protein-sequence structure is genetically determined. Protein-hormone synthesis follows the classic central dogma of molecular biology: nuclear transcription of DNA to RNA then ribosomal translation of RNA to protein. In contrast, steroid synthesis is indirectly connected to genes. This indirect connection is via proteins such as those that are necessary for the transportation of steroids, for the receptors that recognize steroids, and for the enzymes that convert steroids from one form to another. The more direct connection of genes to protein hormones also means that the variability in protein hormones is more directly influenced by evolutionary processes.

Amino acid chains that comprise protein hormones are sometimes short and in such cases may be referred to as peptides rather than as proteins. Peptide hormones can be quite short, nine amino acids in the case of oxytocin, 16 for α-endorphin. But most protein hormones have long amino acid chains subject to elaborate folding. As a result, such proteins are bulky, large, and do not pass unaided through cell membranes. Protein hormones relevant to the female mammal include adrenocorticotropic hormone (ACTH), gonadotropin-releasing hormone (GnRH), prolactin, luteinizing hormone (LH), and follicle stimulating hormone (FSH).

Protein action can be more direct and faster than steroid action. Oxytocin or prolactin, when released from the brain, travel directly to their target organs. Once at a target cell, the bulky protein hormones do not easily cross cell membranes; instead, they bind receptors (also proteins) on cell surfaces. This binding causes the receptor to change shape, and that shape change can rapidly initiate a cascade of other biochemical alterations within the cell leading to an effect. Thus, protein hormones can bypass any direct interaction with genes and more quickly cause the desired action.

Cell-surface receptors are highly variable in their affinity for, and sensitivity to, peptide and protein hormones. For example, affinity can vary across tissues within the same individual. In addition, the number and visibility of the receptors can fluctuate over time and in different tissues. Also, the synthesis of receptors within a cell can be altered by steroid hormones or other peptide hormones. As a result, the interactions of different hormones can be exceedingly complex. All this is happening on a biochemical level and differences across mammals are not well studied on this level.

Prolactin is a particularly good example of the variability in hormone structure, regulation, and function. Prolactin in mammals is composed of a single chain of about 200 amino acids; however, the sequence of the 20 different amino acids, which make up the protein, varies greatly across species. For instance, the human sequence differs from that in pigs, sheep, and horses at about 50 positions and from that in rats at about 80 places. In other words, human and horse prolactins are 80% identical (Lehrmann et al. 1988); human and pig, 77% the same; human and sheep, 73% identical; and human and rat, 60% the same (Shome, Parlow 1977). In addition to the variation in sequence, the protein can be altered after it is made, for instance, by adding sugars to it or by splicing it into smaller pieces. With such great differences, why are all these variants called prolactin? First, no matter what the species or sequence, mammalian prolactins have similar effects and thus, as might be expected, serve similar functions. Second, they are all called prolactin because changes in the variants can be traced back to ancestral forms of the protein.

To say that prolactin has similar functions, no matter what the variant, is somewhat disingenuous (or at least a simplistic generalization). This is because prolactin has more than 300 known roles and is synthesized from multiple sources in the body, e.g., anterior pituitary, various regions in the brain, placenta, uterus, mammary gland, milk, and lymphocytes (Freeman et al. 2000). Some of prolactin's functions are reproductive. In the mammary gland, prolactin influences mammary development and growth, milk synthesis, and milk secretion. In the ovary, prolactin influences the function of the corpus luteum and can be either stimulatory (luteotrophic) or inhibitory (luteolytic). Behaviorally, prolactin may alter the frequency of sexual activity and has effects on maternal care from nest building to nursing. Prolactin has non-reproductive functions as well, such as in homeostasis, immune function, osmotic balance, and angiogenesis (blood vessel development; Freeman et al. 2000). How is one hormone able to carry out so many roles?

The variability in function is partly due to variations in the hormone itself but also due to numerous regulatory molecules. For prolactin, the long list of regulatory hormones includes dopamine, norepinephrine, epinephrine, serotonin, histamine, acetylcholine, oxytocin, vasopressin, opioids, thyrotropin-releasing hormone, vasoactive intestinal polypeptide, somatostatin, gamma-aminobutyric acid, atrial natriuretic hormone, calcitonin, neuropeptide Y, angiotensin II, galanin, substance P, bombesin-like peptides, neurotensin, cholecystokinin, glutamate, aspartate, and nitric oxide (Freeman et al. 2000). These molecules influence the secretion of prolactin. Once released, prolactin binds to cells via a prolactin-receptor protein, and this receptor also has variants. In addition, soluble prolactin-binding proteins exist that can influence prolactin's actions (Freeman et al. 2000). The end result is an extremely complicated biochemical system that regulates the actions of this protein hormone. Although prolactin may be an extreme example, other hormones also have significant regulatory complexity.

This complexity means that hormones do not have only one effect on physiology or behavior. They operate in hugely variable combinations to achieve diverse physiological outcomes. The same combination of hormones may have synergistic or inhibitory effects in different tissues, at different times, in different sexes, or in different species. In many ways, the versatility of hormones is much greater than appreciated, and the total number of hormones a female possesses greatly underestimates the total number of effects those hormones precipitate.

Understanding the diversity of hormones and their actions is but one small part of reproductive physiology. The larger context is understanding how chemical signals interact with the body to solve some physiological problem. We will explore the maternal recognition of pregnancy as an example of this larger context.

Maternal Recognition of Pregnancy

Just as any given hormone has multiple functions, so too, any given physiological problem has multiple solutions. As an example of the multiple and complicated mechanisms used by females to achieve a reproductive goal, we ask the question: How does a mother know she is pregnant? In other words, what switches the maternal physiology from a non-pregnant to a pregnant state? Immediately after conception a female does not physiologically recognize that she is pregnant. The only way in which that information can be conveyed is if the conceptus sends a signal of its presence. This chemical signal for pregnancy must come from the conceptus and is the earliest documented communication between mother and offspring. The communication is necessary, not only to preserve a viable fetus but also to ensure that a female can restart reproduction if an embryo is defective or if too few healthy embryos have implanted. The signal may have other consequences. In the case of red deer, the signal is sexually dimorphic, providing females a mechanism by which to "choose" a son or a daughter by only accepting blastocysts of the appropriate sex (Flint et al. 1997). Thus, the maternal recognition of pregnancy can have far-reaching effects.

For mothers, the very earliest stages of pregnancy are indistinguishable from the non-pregnant state. After ovulation, for many females, the now-empty follicle reconstitutes itself as an endocrine organ (the corpus luteum, or CL) and produces progesterone. Progesterone is transferred to the uterus by a special ovarian-uterine vasculature (rete) where it promotes and maintains the uterine wall to facilitate implantation, early embryogenesis, and the remainder of gestation. The corpus luteum is necessary for pregnancy in eutherian mammals (Flint et al. 1990). After a period of time, specific to each species, the CL degrades, the uterine lining sloughs off or is adsorbed (a process called decidualization), and the female prepares new follicles for ovulation. Maintaining the CL (and thereby progesterone levels) is usually important for the continuance of gestation. Degrading the CL (decreasing the production of progesterone) is important if conception does not occur. Additional signals from outside the ovary can be either luteolytic (causing degradation of the CL) or luteotrophic (causing growth and maintenance of the CL). Over the course of gestation, the CL is not always key, but the production of progesterone usually is, and while the source of the progesterone could be the CL, progesterone can also be provided by the placenta or the embryo. No matter what the source, the recognition of pregnancy occurs early in gestation, usually around the time of implantation (Flint et al. 1990). Thus, *maternal recognition of pregnancy* refers to the physiological mechanisms by which luteal progesterone is maintained.

What influences the longevity of the CL? Species differ greatly in this, as for most aspects of comparative physiology. The species for which we have data are generally those with some clear connection to humans. We know quite a bit about humans and domestic ungulates (cows and sheep), so we will start with them. Then, we will explore taxonomic differences: marsupials, with their short pregnancies; canids, with their so-called pseudo-pregnancies; and then the half-dozen or so other taxa for which we have information.

In humans, the conceptus secretes a luteotrophic signal, which enters the ovarian-uterine circulation and maintains the CL. The signal is chorionic gonadotropin (CG; Bazar 2013), the same substance used by many pregnancy detections tests. Usually, CG is released by the conceptus about 6 days after conception. CG belongs to a family of protein hormones (gonadotropins), including luteinizing hormone and follicle-stimulating hormone, which are produced by the anterior pituitary and regulate many aspects of reproductive function. Luteinizing hormone, in particular, is also luteotrophic. Effectively, the human conceptus takes over the role of the mother's pituitary to maintain the CL.

Cows and ewes have a different system (figure 5.1). Instead of sending a luteotrophic signal, the new conceptus prevents the maternal secretion of a luteolytic hormone. Without a conceptus, the female's uterus sends this luteolytic hormone to the ovaries a set time after ovulation to degrade the CL. The conceptus, if present, prevents the secretion of that uterine signal; thus, the CL is maintained.

As may be obvious, the details are quite complicated and involve hormones from the pituitary, ovary, uterus, and conceptus. In cows and ewes, pulsatile uterine secretions of prostaglandin ($PGF2\alpha$) are luteolytic. These secretions are also oxytocin dependent. Thus the luteolytic signal is an oxytocin-dependent release of pulses of $PGF2\alpha$. The synthesis of $PGF2\alpha$ also relies on estrogen. When a cow or a ewe is pregnant with a viable embryo, the conceptus secretes interferon, which inhibits oxytocin and estrogen receptors, preventing the synthesis and pulsatile release of $PGF2\alpha$ and the degradation of the CL. Tangentially,

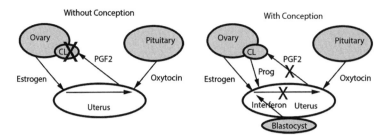

Figure 5.1. Maternal recognition of pregnancy in domestic cows (simplified). Progesterone is required for the maintenance of pregnancy. The initial progesterone comes from the corpus luteum (CL) in the ovary. *Left,* Without conception, signals from the pituitary (oxytocin) and ovary (estrogen) to the uterus prompt the uterus to secrete prostaglandin, which triggers the involution of the CL (luteolysis) and, consequently, stops the production of progesterone. *Right,* When conception occurs, the conceptus (blastocyst) produces a signal (interferon) that disrupts the internal signaling of the uterus. Consequently, prostaglandin is not released, and the CL is maintained and continues to produce progesterone. Thus, interferon is the pregnancy recognition signal in cows. Diagram by Virginia Hayssen.

in vitro female blastocysts produce twice the amount of interferon than male blastocysts (Larson et al. 2001). This difference provides a mechanism for mothers to select the sex of their offspring. For cows and ewes, four chemical signals are involved in pregnancy recognition, each from a different source: uterine PGF2α, luteal oxytocin, ovarian estrogen, and conceptus interferon (Bazar 2013). Does this pattern hold in distantly related taxa? Probably not. Although few have been studied, some examples are available for closer scrutiny and exemplify the diversity of molecular mechanisms species use to achieve the same physiological goal.

What about marsupials? For marsupials, pregnancy recognition might seem to be a moot point as the natural duration of the CL matches the interval between ovulation and birth. However, marsupials also have pregnancy recognition (Renfree 2010), and, at least in some, the feto-placental unit, not the CL, influences the physiology of the uterus. The yolk sac, not the embryo, is the probable origin of the recognition signal (Renfree 2010; sidebar on page 83).

Fascinatingly for some eutherians, pregnancy recognition may be unnecessary. For instance, in domestic dogs, the CL is maintained for the length of a normal pregnancy, even if no mating has occurred and no offspring are present. In the wild, only rarely would a female be unable to find a willing mate, but many domestic bitches are kept away from males, making ovulation without mating common. These females may display behaviors associated with pregnancy, including maternal behavior and mammary gland development. The term *pseudopregnancy* was coined to explain this relatively uncommon condition; females appear to be pregnant but are not. An open question is, What eventually causes the pseudo-pregnant CL to degrade in the absence of birth?

The Opossum Lady from Oz
Marilyn Renfree (1947–)

A native Australian, Marilyn Renfree graduated from the Australian National University, where she completed her PhD with Hugh Tyndale-Biscoe. From her initial work on the maternal-fetal interactions in mammals, Renfree's research expanded to understanding reproduction in general with a focus on her fellow Australian mammals: marsupials. This work led to investigating reproductive delays such as embryonic diapause. Renfree's research was part of an Australian renaissance exploring the extensive physiological diversity of marsupials.

After completing her PhD, Renfree did a Fulbright Fellowship at the University of Tennessee and investigated uterine proteins in the North American opossum. She also spent time in Edinburgh for an extended period as a Ford Fellow, where she worked with another of our showcased scientists, Anne McLaren (see sidebar on page 98). In

fact, they coauthored an article in *Nature*. Renfree has also documented the control of reproduction in marsupials, the evolution of sex chromosomes, and other aspects of reproduction, in particular embryonic diapause.

Much of her work focused on Tammar wallabies (*Macropus eugenii*) because, as she says, "the only day they are not pregnant is when they give birth." Her papers explore both proximate causes as well as evolutionary adaptations of reproduction in mammals. Menstruation is another topic Renfree has researched extensively.

Her most recent articles use modern molecular methods to explore the biology of marsupials, e.g., "Marsupials in the age of Genomics" with yet another of our scientists, Jennifer A. Marshall Graves (see sidebar on page 29). But her focus remains on reproductive biology: the origins of milk, gene expression associated with implantation and

placentation and in the mammary gland, differentiation of male and female reproductive tracts, as well as gonad development and gametogenesis. Her recent work investigates the evolution of genomic imprinting, using the fact that some genes are not imprinted in marsupials. Renfree's research continues to be both interesting and current. A recent publication "Everything about Kangaroos" also attests to her breadth.

From her academic home at the University of Melbourne, Renfree is internationally recognized as a leading expert on marsupial biology and reproductive biology. She has received a series of awards, including election as a Fellow of the Australian Academy of Science in 1997.

(Photo courtesy of Marilyn Renfree, used with permission.)

Pseudo-pregnancy may also occur if a female mates with a sterile male. In this case, the mechanical stimulation of mating may be sufficient to trigger the maintenance of the CL (as in domestic cats or laboratory mice). In natural situations, this type of pseudo-pregnancy would be extremely rare because wild females would most likely mate with fertile males.

Unfortunately, the term *pseudo-pregnant* has also been applied to other situations in which the CL persists post-implantation. In other words, pseudo-pregnant is used when a female is clearly (not falsely) pregnant, such as in rabbits

(Nowak, Bahr 1983). Here, the female retains control of the CL after offspring have implanted. Thus, rabbit mothers have more control over pregnancy with less embryonic influence. The increased maternal control may allow for reduction in litter size during pregnancy (as happens frequently in rabbits).

Returning to canids, ovulation and mating occur once annually and are often correlated with a CL that persists nearly the same duration as that of pregnancy. Therefore, canids have neither a luteolytic signal from the uterus nor a luteotrophic signal from the conceptus (Songsasen et al. 2006). Effectively, they do not recognize pregnancy. Long CL duration has been documented in many canids: *Canis latrans, C. lupus, C. simensis, Chrysocyon brachyurus, Lycaon pictus, Nyctereutes procyonoides, Speothos venaticus, Vulpes lagopus, V. vulpes, V. zerda* (van Kesteren 2011). Cooperative breeding occurs for at least two of these, the African wild dog (*L. pictus*) and the Ethiopian wolf (*C. simensis*).

In cooperative breeding units, a dominant female produces all or most of the offspring, and other subordinate females help provide maternal care. In these cases, the hormonal consequence of long-duration CLs in non-reproductive, subordinate females may facilitate allo-parental care, including allo-nursing as occurs in the cooperatively breeding Ethiopian wolf (van Kesteren et al. 2013). In other words, females who are not pregnant may be hormonally ready to participate in maternal care of infants in their social group.

Although canids may not recognize pregnancy, other eutherians do. However, the molecules that support or terminate the CL differ greatly across species. The following short examples (with italicized headings) may be of interest as they confirm that molecular mechanisms are not uniform in fulfilling specific reproductive objectives.

Red deer. The system in red deer (*Cervus elaphus*) is similar to cows and sheep in most respects. The conceptus uses interferon to stop the oxytocin induction of prostaglandin. However, in red deer (called elk in the United States), the source of maternal oxytocin may be the pituitary rather than the CL (Bainbridge, Jabbour 1999). This provides a mechanism for behavior, or other aspects of the central nervous system, to influence pregnancy recognition. We earlier mentioned the ability of female red deer to select the sex of their offspring by using the physiology of pregnancy recognition. The mechanism is by way of a sexual dimorphism in the production of interferon by the trophoblast. This leads to differential loss of male or female blastocysts based on the dominance status of the mother (Flint et al. 1997). As a result, more sons are born to dominant than to subordinate hinds.

As an aside, interferons are generally considered part of the body's response to pathogens, such as viruses or tumor cells. However, in the context of pregnancy, the conceptus is using interferon to prevent a signal from being sent from the uterus to the ovary. The conceptus is not treating its mother as a pathogen, nor is the conceptus preventing a maternal immune reaction against the conceptus. Biochemically, the conceptus is simply preventing oxytocin from stimulating prostaglandin release. Thus, interferon exemplifies how our general understanding of the physiological function of particular molecules may not be consistent with their function in a reproductive context.

Roe deer. Roe deer (*Capreolus capreolus*) differ reproductively from other cervids in many ways. (1) They ovulate only once per year. (2) They mate in the summer, whereas other cervids mate in the fall. (3) Implantation of the blastocyst is delayed for up to 5 months. (4) The conceptus does not release interferon. (5) The uterus does not release pulses of PGF2α even though the blastocyst has not released interferon. (6) At the end of the gestational delay, the blastocyst secretes a unique protein that initiates a cascade of physiological events to allow implantation and embryonic development (Lambert 2005). Thus, pregnancy recognition in roe deer does not occur until months after conception.

Pigs. For sows (*Sus*), uterine PGF2α is also luteolytic; however, the conceptus secretes estradiol, not interferon, and the secretion is only effective if at least two embryos are present in each uterine horn. Also, although estradiol prevents PGF2α from getting into uterine-ovarian circulation, it maintains PGF2α within the uterine lumen (Bazar 2013).

Horses. For mares (*Equus*), as with cows, pulsatile uterine production of PGF2α is luteolytic. Although the signal that prevents PGF2α secretion is unclear, mobility of the conceptus is essential (Klein, Troedsson 2011). If the conceptus is stationary before implantation, embryo rejection will occur (see chapter 7, "Gestation").

Mice and rabbits. In laboratory mice (*Mus*), neither the conceptus nor the uterus is involved in maintaining the CL. Instead, cervical stimulation (via mating) induces surges of pituitary prolactin, which are luteotrophic (Osada et al. 2001). Like mice and sows, rabbits (*Oryctolagus*) give birth to more than one offspring at a time. Unlike mice and sows, pregnancy recognition in rabbits has not received scientific attention since the early to mid-1980s. Rabbits are unusual in that pregnancy recognition occurs well after implantation, but the biochemistry of the process is unknown (Browning et al. 1980; Nowak, Bahr 1983).

Armadillos. In armadillos (*Dasypus novemcinctus*), the placenta produces progesterone to maintain pregnancy (Buchanan et al. 1956).

The specific details of these examples serve to make one point. Just as bridges can be built of many types of materials yet in all cases span rivers, so, too, the molecular processes used by different species can vary widely but fulfill the same reproductive goals. The specific goal for females is the maternal recognition of pregnancy, but the more general consequence of reproduction is the production of offspring that carry on the genetic complement of their ancestors. Thus, the molecular mechanisms vary, but the goal is the same.

Life-History Physiology: The Top-Down Perspective

Up to this point, we have focused on topics that are central to the work of veterinarians, molecular biologists, and animal scientists. Ecologists and evolutionary biologists are also interested in reproductive physiology but often from a different perspective. Classically, this aspect of physiology is concerned with traits that alter the life history (box 5.1) of the population. Evolutionary ecologists do not usually ask how a female knows that she is pregnant but, rather, how body size influences the length of gestation; they do not ask what causes pre-natal mortality but, rather, how embryo rejection alters population growth; they do not ask how ovulation produces either singletons or litters but, rather, what is the consequence of litter size on neonatal size. What is the difference in these perspectives? Simplistically, the focus of life-history theorists is on how the physiology of individuals alters their reproductive potential and how these patterns result in population-level patterns. In this context, we review several major topics that are important to understanding differences in reproductive potential across species: allometry, metabolism or energetics, and trade-offs or resource allocation.

The study of allometry describes the influence of body size on aspects of an animal's biology, in this case physiology (Schmidt-Neilson 1984). For example, larger mammals have more cells, and if cells of different animals are roughly the same size and grow at similar rates, then larger animals would take longer to reach adult size than smaller ones. If so, larger animals could be expected to have longer pregnancies and periods of nursing. Thus, all else being equal, growth to an equivalent stage of maturation should take longer for larger animals. If both variables change at the same rate (1:1) the relationship is termed *isometric* (*iso* = same, *metric* = measure), meaning that one variable, such as gestation, scales in a constant fashion with another variable, such as body size. However, things get interesting when this is not the case, as is often true in biological systems. For instance, gestation lengths in blue whales (*Balaenoptera musculus*, 11 months) are half the duration of those of elephants (*Elephas maximus*, 22 months), but blue whales weigh about 100 times more than female elephants. Whales time their gestation to a yearly environmental pattern and hence have a shorter than expected gestation length. In this case, gestation scales allometrically (*allo* = different, *metric* = measure); in other words, a set increase in mass does not relate to a similar increase in duration of gestation. Why elephants have such a long gestation remains a mystery but may be related to selection on their longevity (Lee et al. 2016).

Body size also relates to simple geometry. As an object becomes larger, both its surface area (the outside skin) and its volume (the body itself) increase, but they do so at different rates: the volume increases faster than the surface area. Thus, a large mammal has less skin relative to its body contents than a small mammal.

BOX 5.1. Personal History versus Life History versus Evolutionary History

Reproduction can be understood on several scales: individual, species, and lineage. As individuals, each of us has a personal reproductive history (first menses, first sexual partner, length of our menstrual cycle, etc.). As humans, our collective personal histories determine the reproductive (life history) strategies of our population and, if considered globally, our species. For instance, components of our life history as humans include a gestation length of 40 weeks, a usual litter size of one, sexual maturity (puberty, menarche) at 10–14 years of age, and reproductive senescence (menopause) at 45–55 years. Some of these components are more or less specific to humans, e.g., menarche, menopause. Other life-history variables, for instance, litter size or gestation length, are not. In any

case, the commonly assessed, species-specific, life-history parameters for mammals are litter size, neonatal mass, gestation length, lactation length, weaning mass, litters per year, and age at sexual maturity. Besides our personal and life histories, we share an evolutionary history with other mammals. Some of our reproductive traits reflect this shared history. For instance, as therian mammals, we give birth rather than lay eggs. Going even farther back, we share with all mammals the distinctive trait of nourishing our young with milk.

Depending on the questions of interest, various scales may be more or less relevant. Your physician wants to know your personal history. Ecologists and conservation biologists find species-specific life histories of interest,

whereas scientists trying to understand ancestral patterns wish to understand the evolutionary history of reproduction. In this book, we are primarily interested in the latter two options, species-specific life histories and how those histories evolved. We explore differences in reproduction across mammals and how those differences may be adaptations to particular environmental challenges these mammals experienced or whether the differences are holdovers from the past. Natural selection is the major process by which the different life histories (reproductive patterns) in mammals arose, and thus, it is of particular importance to our dealings with the female mammal as she reproduces.

Similarly, a baby, because of its smaller size and therefore larger surface area to volume ratio, loses heat (i.e., cools down) much more quickly than its mother and becomes hypothermic long before she will. Reducing surface area may be one advantage of the shorter and fatter limbs of human neonates. Surface area to volume relationships are critical to thermoregulation and resulting energy budgets. Thus, thermoregulation influences neonatal development and neonatal development influences the length of gestation. Thermoregulation also influences litter size, as a larger litter that huddles effectively has a more favorable surface area to volume relationship. Overall, body size either directly or indirectly has large influences on reproduction.

In some cases, body size accounts for nearly all the variation in a specific reproductive trait (Charnov 1991). For instance, more than 90% of the variation in neonatal mass can be attributed to female body mass across a wide variety of taxa, such as squirrels (Sciuridae), deer (Cervidae), and bats (Chiroptera; Hayssen 2008b, 2008c; Hayssen, Kunz 1996; Jabbour et al. 1997). However, female mass accounts for only about 50% to 60% of the variation in gestation or lactation lengths in these same groups. While maternal mass is a strong determinant of the energetic input into individual offspring (i.e., neonatal mass), other factors are at least as important in explaining the temporal aspects (i.e., length of reproduction). We will explore these other factors in the second section of this book on specific parts of the reproductive cycle.

Related to body size, metabolism (energetics) is an aspect of physiology with fundamental reproductive consequences. Reproduction takes energy, often a lot of energy. Pregnant and nursing females must digest the food they eat, reconfigure the digested food into either milk or building blocks for tissues, and then deliver these products to their offspring. Thus, the processes, and the products, require energy (Cretegny, Genoud 2006).

Scientists interested in reproduction often evaluate the energetic input into reproduction by measuring metabolic rate, but measuring the appropriate rate is not straightforward. The concept of basal metabolic rate (BMR) is a standard metric for comparing metabolism across species. Experimentally, BMR has technical specifications: it is the oxygen consumption of a healthy adult (i.e., is not growing), at rest (but not asleep), alert, fasted (i.e., not digesting food), not stressed, *not reproductive*, at a comfortable temperature (i.e., within its thermoneutral zone), and within the quiescent phase of the species' daily cycle (Speakman 2013). Scientists understand that wild mammals never exist in this state, but the ability to reliably and repeatedly perform this measure across many different mammals has led to widespread use. The key issue with using BMR for estimating reproductive effort is that reproduction changes metabolism (Stephenson, Racey 1995). Metabolism normally increases during reproduction. For instance, in a subterranean rodent, *Ctenomys talarum*, metabolic rate in lactating females was 151% higher than in non-reproductive females (Zenuto et al. 2002).

Other factors influence metabolism and may alter the energetics of reproduction. First, metabolism increases with body temperature, which is partly why mammals need more energy to live than do crocodiles or turtles. Sloths have lower body temperatures than monkeys, thus reproduction may be slower (or take less energy) in sloths. Second, metabolic rate is related to body size. On a per-gram basis, bigger animals have lower metabolism than smaller mammals. But overall, bigger mammals need more energy than small mammals. However, bigger mammals retain heat better (the surface area to volume geometry discussed earlier) and have different proportions of their bodies devoted to less metabolically active tissues (e.g., bone), which brings us to a third point. Different tissues have different metabolic rates. Consequently, the relative amounts of different tissues will affect overall metabolism. For instance, animals with large brains very often have higher metabolism, and animals that maintain large amounts of blubber or bone may have lower metabolism. To complicate matters further, diet often influences metabolic rate. For instance, anteaters and arboreal folivores generally have lower metabolisms (McNab 1986). In addition, metabolism may have a phylogenetic component. Overall, the effects of different metabolisms on reproduction are not always straightforward. For instance, in shrews (Soricidae) increased metabolism is associated with shorter gestation, larger litters, and faster fetal growth rates, but in a related family, tenrecs (Tenrecidae), metabolism does not correlate with any of eight reproductive variables explored. One explanation for this unusual observation is that in tenrecs, elevated metabolic rates improve homeo-thermy (maintenance of a stable body temperature) rather than increase repro-ductive output (Stephenson, Racey 1995).

The previous example illustrates the concept of trade-offs. A key tenet of life-history theory is that resources are limited, and each unit of a limited resource can only be apportioned into one of four general areas: (1) growth, (2) survival/tissue maintenance (including homeothermy), (3) storage, or (4) reproduction (Charnov 1993; Stearns 1993; Roff 2002). The assumption that these categories are mutually exclusive is embodied in the idea of trade-offs. Females make deci-sions to put resources into either one category or another. For example, resources devoted to growth cannot also be used for reproduction. Or, in the previous ex-ample, shrews put their additional metabolic output into reproduction, whereas tenrecs put it into maintenance via homeothermy. Another concept often used as a heuristic tool by life-history theorists is optimality. To the field of life-history theory, optimality is a concept involving individuals finding the "best" solution for allocating resources be they energy, nutrients, or time. These main concepts of limited resources, resource allocation, and optimality have generated much cogent, but complicated, theory. For additional information, avid readers may wish to read the golden triad of Charnov (1993), Stearns (1993), and Roff (2002).

Life-history theoreticians have uncovered many exciting rules of thumb that organisms may follow, but these scientists would also admit that no organism acts optimally and that resources may not always be limiting. Furthermore, trade-offs may not occur among competing functions. For example, females might choose to reproduce only when resources are not limiting. Admittedly, such females are likely at a luxury as such conditions may be rare. Nevertheless cases may exist. For instance, after a winter die-off, deer mice (*Peromyscus*) in the spring might not face limited resources or at least have decreased competition for those resources. Winter mortality may reduce competition for the new crop of insects and plants. Also, the numbers of parasites and predators may be fewer but perhaps hungrier in spring. In such a scenario, overwintered females could have sufficient resources to put energy into both survival and reproduction. This example is putative but illustrates that while trade-offs and resource allocation may be important, times when resources are not limiting may also be a construc-tive area of investigation. Studies into lactation in mice, for example, have found that females provided an excess of food during lactation cannot produce an ex-cess of pups either because their digestive tract is unable to harvest this excess or because they suffer from an excess of heat production, which they are unable to dissipate (chapter 9; Hammond, Diamond 1992; Speakman, Krol 2011).

Trade-offs do not apply only to immediate allocation of resources to growth or reproduction but may balance now versus later. That is, females can choose to reproduce immediately with available resources, or they may defer reproduc-tion and use current resources for survival (maintenance), storage, or growth

(Charnov et al. 2007). This concept was initially coined *reproductive value* (Fisher 1930) and considers the trade-off between reproducing now versus later. Following this tenet, females allocating resources to reproduction necessarily devote less to growth and perhaps, most important, to survival. In calculating population growth rates, usually only the number of female offspring per female at a given age is used, in part because paternity is uncertain. Thus, early life-history theorists focused on females, daughters, and granddaughters in their mathematical models.

Returning to the concept of reproductive value one might ask, for instance, if food is abundant, should a female increase her current litter size, or store resources as fat so she can have a second smaller litter later in the season? If a female "knows" that she has a low probability of surviving until the next opportunity to reproduce, should she increase her litter size now at the expense of shortening her life even further? These are questions falling under the scope of reproductive value. The assumption here is that a female's contribution to her population is the sum of her current reproductive output plus her future reproductive output. A female's total reproductive value is the sum of her past, current, and future contributions. As she ages, her future output is minimized and becomes less predictable, because she may not survive to the next reproductive season. Thus, females, as they age, should place more resources into their current reproductive attempt, whereas younger females should place less, because the young female has a larger potential future output. This concept explains the general sense that females have peak or prime reproductive years. It also explains why older females may nurse longer than younger ones and why late-season females may have longer lactations than early-season females. These ideas also form the basis of the Trivers–Willard hypothesis, whereby females may alter the sex ratio of offspring to produce the more expensive offspring type when approaching the ends of their lives (Trivers, Willard 1973). For those interested in mechanisms, mammals may regulate in utero glucose or make use of processes associated with the maternal recognition of pregnancy to achieve these patterns (Cameron 2004).

Understanding the physiology of reproduction at the ecological level is an exploration of averages rather than individual differences. For instance, how litter size varies across species is a question about average litter size, not necessarily about the minimum or maximum litter size of an individual female. In this case, scientists are interested in species-specific averages. But we could also ask how does litter size vary with latitude? With that question, we would look at the average litter size of populations of a single species across the latitudinal range of that species. At a still smaller scale, we could ask about the effects of parity on litter size. For answers to this question, we would examine the average litter size for females in the same population at their first litter, second, third, and so forth. The ecological physiologist is asking about a reproductive trait measured across a group of individual females (population level), rather than what happens to and within an individual female. In contrast, molecular and cellular physiologists focus on the processes happening within females that result in the reproductive trait of interest.

Physiology: A Diverse Field

Physiology occurs at many levels, from molecular and cellular interactions to evolutionary consequences and constraints. At the molecular level, we reviewed hormones and their actions, and we used prolactin to exemplify hormonal diversity in structure, regulation, and function. At the systems level, we detailed the various ways in which a female recognizes that she is pregnant, making the point that the solution to a specific physiological problem may have many diverse strategies. Finally, we reviewed three topics (allometry, energetics, and trade-offs), important for understanding why species differ in their physiologies. A fourth aspect of understanding species differences is the concept of phylogenetic inertia. Simply put, the past constrains the solutions available at present and in the future. In particular, any species is largely restricted in what it can do by the genetic structure inherited from its ancestors. We discuss ramifications of this concept in later sections (box 9.1, see page 169).

The diversity of physiologies related to female reproduction is difficult to summarize, particularly given the limited information for most species. Not surprisingly, studies have focused on domesticated, farmed, and companion animals, or, for humans, abnormalities rather than "normal" females cycling repeatedly through gestation and lactation. In addition, a female's physiology operates under frequent, short-term hormonal and physiological changes rather than a long-term testosterone high as is more the case for males. Normally, a female's reproductive physiology is geared toward the continual metabolic flow of resources from food or fat into babies or milk. In this context, energetics, metabolism, body size, and storage are all key factors that shape the diverse physiologies of reproducing females.

In the next section, we move from what happens within individual females (her genetics, anatomy, and physiology) to aspects of reproduction in which females directly interact with mates or their own offspring. These interactions, like the physiology we've discussed above, are part of a series of reproductive events often termed the *reproductive cycle* (as opposed to the estrous cycle). The reproductive cycle has a chronological component. Ovulation, mating, conception, gestation, birth, and lactation usually occur in a cyclical sequence. Females interact with different partners during this cycle. Aspects of physiology will figure prominently in the different components of the reproductive cycle. Thus, reproductive physiology will be a thread throughout the next section of this book.

PART TWO

CYCLES

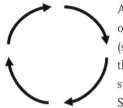A single female followed across her life undergoes a series of stages and events that, for the most part, compose a cycle (see figure). This cycle is the focus of the second portion of this book and is depicted as a series of arrows between stages. Of course, the cycle is not without interruptions. Starts and stops are frequent, and females may not transit linearly through the stages. Nonetheless, the concept of a cycle is a useful perspective.

In the five chapters of this section, we discuss the steps in this cycle as well as various modifications. An individual female enters the cycle as an oocyte, and oogenesis describes the maturational process of that gamete. Oogenesis is the beginning of a new female and occurs within her mother's ovarian follicle. Folliculogenesis describes the changes in the ovarian tissues that support the developing gamete (chapter 6, "Oogenesis to Conception"). Once a female mates and uses a sperm for conception, the daughter and mother together experience gestation (chapter 7). Within the protective environment of the uterus a daughter may compete with her mother or siblings for space or nutrients, but generally her needs are assured. She is warm, protected, and need do little but grow and develop. At birth, this all changes (chapter 8, "Birth and the Neonate"). With her first breath, she must cope with a cold and hostile environment. However, as a mammal she receives more aid from her mother in the form of a complex product, milk (chapter 9, "Lactation"). Making milk is demanding on her mother and eventually all good things come to an end. She is weaned, and her subsequent maturation leads to puberty (chapter 10, "Weaning and Beyond"). Although we just described each of these stages from the perspective of a developing daughter, on reaching sexual maturity a female repeats many of these stages, but from the other side—this time as a mother. When she revisits the cycle as a new participant, the costs and benefits of each stage change. Part 2 explores the stages of a female's reproductive life from both perspectives: mother and daughter.

CHAPTER SIX

Oogenesis to Conception

On the savannah, our female hyena is now pregnant. She is a low-ranking female, and, thus, the fraternal twins she carries are both daughters. Within her uterus, these developing daughters are undergoing an optimistic process: oogenesis. Thus, the pregnant female is currently provisioning not just her twin daughters, with which her belly is distended, but also her incipient granddaughters, as oocytes within her daughters' newly forming ovaries. Birth may be difficult, but, for now, multiple generations all safely inhabit one space.

Of course, mating precedes conception. Finding a mate is easy, and females are polyandrous, mating with multiple males, but mating itself is not simple. In fact, a female's elaborate, external genital morphology makes achieving the effective, precise alignment difficult. Partners require practice to coordinate their postures and movements. (East et al. 2003)

Embryogenesis begins with oogenesis.—Albertini 2015:61

This chapter explores beginnings, from the origins of female gametes to the formation of a new female. Oddly, these very different processes happen very close in time. Soon after conception, the first cells to differentiate in the new conceptus are those that become gametes for the next generation (Edson et al. 2009). Besides DNA, considerable additional material is deposited in these oocytes. This maternal input provides the building blocks and initial instructions not only for the next conception but also for the earliest development immediately after conception. Thus, oogenesis marks the initiation of embryogenesis. In effect, three generations coexist during pregnancy: the mother, her in utero daughter, and her daughter's incipient gametes.

In chapter 4, we described the static anatomy of the oocyte and follicle. In this chapter, we describe the dynamic aspects of their formation. Oogenesis is the growth and maturation of ova (female gametes), whereas folliculogenesis is the growth and maturation of the ovarian follicle in which oogenesis occurs.

Very few oocytes become part of the next generation. Most degenerate sometime between conception and puberty. After puberty, additional oocytes degenerate over the course of a female's reproductive life. However, a few ova are released (ovulated) at appropriate times. These ova are transported into the oviducts; the usual location for conception. Of course, prior to conception a female must mate successfully, select appropriate sperm, and render those sperm fit for conception (capacitation). All these processes, from oogenesis and folliculogenesis through to ovulation and conception, in addition to female choice of both mate and sperm, are reviewed in this chapter.

Oogenesis and Folliculogenesis

Creation of female gametes in mammals is the result of two coordinated processes: oogenesis and folliculogenesis. Oogenesis is the formation of female gametes from oogonia. Oogenesis creates cells that contain half the maternal nuclear genome, as well as all of the maternal mitochondrial genome, a plethora of cytoplasmic organelles, and accessory materials of maternal origin. Individual female gametes develop within a well-defined assemblage of ovarian cells, a follicle. Therefore, each oocyte is a single cell, whereas follicles are aggregations of maternal cells. The follicle starts as a one-cell-thick layer surrounding the oocyte, but a mature follicle has at least five named layers (e.g., granulosa, theca), each with a distinctive composition (figure 6.1). The follicle alters in size and morphology coincident with changes in the female gamete. Thus, folliculogenesis and oogenesis are tightly linked. Although the molecular and physiological details are understood for only a few mammals, the general timing of the various stages is known for a greater diversity of species.

Follicles are composed of maternal cells that regulate materials entering and leaving the maturing oocytes. The oocyte is no passive bystander. Molecules

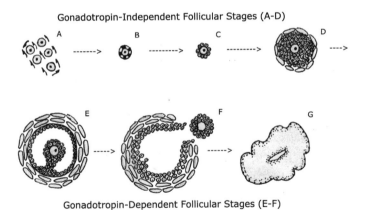

Gonadotropin-Independent Follicular Stages (A-D)

Gonadotropin-Dependent Follicular Stages (E-F)

Figure 6.1. Stages of folliculogenesis. A, primordial germ cells; B, primordial follicle; C, primary follicle; D, secondary follicle; E, antral follicle (also known as a Graafian follicle or preovulatory follicle); F, ovulation; G, corpus luteum. Not shown is an atretic follicle (one that involutes prior to ovulation). Stages A–D (*top*) are not under control of pituitary gonadotropin hormones (FSH or LH), whereas stages E–G (*bottom*) depend on gonadotropins. Modified by Abigail Michelson from Edson et al. 2009; Young, McNeilly 2010.

secreted by the oocyte influence the development of surrounding maternal cells. Consequently, oocytes can promote the growth and proliferation of the granulosa cells. For example, in lab mice, oocytes orchestrate follicular development from the primary follicular stage to the large antral stage (Eppig et al. 2002).

Oogenesis begins with the differentiation of primordial germ cells (PGC) just after conception and well before an embryonic ovary exists (McLaren 2003; sidebar on page 98). As the incipient ovary develops, the PGCs migrate to it in response to both attractive and repulsive cues along their path (Richardson, Lehmann 2010). The PGC nuclei divide repeatedly (by mitosis) but often without complete cytokinesis (cell division), and clusters of multinucleated cells (syncytia, germ-cell nests) result. As the embryonic ovary develops, these clusters break up and ovarian cells surround individual PGCs, creating primordial follicles. This is the beginning of folliculogenesis.

The initial pool of follicles is dynamic. Many follicles, along with their potential ova, degenerate early on, others stay dormant, and some continue folliculogenesis (Tingen et al. 2009). A major open question is exactly what regulates which primordial follicles will degenerate, which will stay dormant, and which will develop (Pangas, Rajkovic 2015).

As for the gametes, the development of mature ova has two components: genetic and cytoplasmic. The genetic maturation is a specialized process of cell division, meiosis. Meiosis is similar for females and males and across all sexually reproducing organisms, including plants, fungi, and animals. The process is complicated and a thorough description is outside the scope of this book; however, the figure in box 4.1 illustrates the major steps. The end product is a gamete that contains half the parent's genetic material. When female and male gametes fuse, the resulting offspring will possess a full complement of genetic material.

Meiosis has many stages, the timing of which differs between females and males. In male mammals, meiosis begins at puberty. In females, meiosis starts in utero and then stops for weeks or years. It stops quite soon after PGCs become isolated and surrounded by ovarian cells to form primordial follicles and only re-starts when a female is ready to reproduce, that is, at puberty. Consequently, the interval between the start of meiosis and its completion can be a matter of weeks for mice or years for humans, elephants, or whales. For long-lived females, decades can elapse between the formation of the first oogonia and the release of the last ovum in a final ovulation.

Although meiosis stops, the internal components of the oocyte continue to change and the developing oocyte is metabolically active. Furthermore, although the process (meiosis) has stopped, the genetic material remains active (Pan et al. 2005). In other words, oocyte genes are being expressed and directing cellular activity even though meiosis is said to be arrested. Changes in oocyte size are well documented, but changes in the biological constituents of the oocytes are poorly characterized. Most oocyte enlargement occurs during the early stages of

From Mice in Test Tubes to Changing People's Lives

Dame Anne McLaren (1927–2007)

Anne McLaren was a notable developmental biologist from the United Kingdom whose major contributions include the treatment of infertility and stem-cell research. Of her many honors the most impressive may have been as the first women to hold offices in the Royal Society (as vice president and foreign secretary), a society that had not had a women in office for 300 years.

Dame McLaren did a master's degree under J.B.S. Haldane. She studied rabbit genetics and murine neurotropic viruses as a PhD student with P. Medawar. She moved to the Royal Veterinary College, London, where she became interested in maternal effects on vertebral elements. As a result of this work, and the related nature-versus-nurture debate, she began to

perform embryo transfers and improve methods for implantation of embryos. By mastering these methods she cultured test-tube mice and later implanted them into uteri of surrogates.

She next moved to Edinburgh where she continued her work on mammalian reproduction at the Institute of Animal Genetics. Eventually, she became director for the MRC Mammalian Development Unit in London and continued her interest in understanding how primordial germ cells develop. During this time she authored two, now classic, books in her field: *Germ Cells and Soma: A New Look at an Old Problem* and a second book entitled, *Mammalian Chimaeras*.

Given the controversial topics of her research, stem cells and in vitro con-

ception, her work often intersected with ethical concerns of the time. This included reformation via the Family Law Reform Act and Human Fertilisation and Embryology Act. She did not shy away from discussions on embryonic stem cells and therapeutic cloning.

She was keen to communicate about science both to academics and to the public. She was president of Women in Science and Engineering (AWiSE), a fellow of the Royal Society, and the Royal College of Obstetricians and Gynecologists.

(Photo courtesy of March of Dimes.)

folliculogenesis, from the primordial to the primary and secondary follicular phases. These phases are independent of the usual pituitary hormones (follicle-stimulating hormone, or FSH, and luteinizing hormone, or LH). During these phases, both the oocyte and follicle grow in size. Of course, oocyte growth occurs by enlarging the single cell, whereas follicular growth is accomplished by increasing both the number of cells and the complexity of the constituent layers, including the granulosa, basal lamina, and thecal layers. Some follicular development is by mitosis (cell multiplication) of established follicular cells, but other development is by recruitment and subsequent differentiation of neighboring ovarian cells (Young, McNeilly 2010). The start of follicular growth is variable relative to birth. It is initiated well before birth in cows (*Bos*), ewes (*Ovis*), and women but is postponed until just after birth in mice, hamsters (*Mesocricetus*), and rabbits (*Oryctolagus*; Baker 1982; Eppig et al. 2002).

The later follicular stages (antral and preovulatory; figure 6.1) are hormone dependent. FSH initiates the formation of antral follicles, whereas LH initiates the resumption of meiosis in the oocyte. During these later stages, only the follicle enlarges, again by cell division. The oocyte does not change in size. Initiation of the later stages can occur well before birth, as in cows, ewes, and women, that is months or years before puberty. At the opposite extreme, antral follicles first appear at puberty in hamsters. More commonly, antral follicles appear days (mice), weeks (rats, rabbits), or months (sows, *Sus*) before the female first ovulates (Baker 1982; Eppig et al. 2002; Greenwald, Peppler 1968; Kanitz et al. 2001).

What is the timing of meiosis with respect to follicular stages? For example, is a primary oocyte tied to primary follicle? Unfortunately, no. The names given to the genetic changes in oogenesis (see figure in box 4.2) and those given to the stages of folliculogenesis are similar but do not correspond. Thus, a primary follicle does not contain a primary oocyte, nor does a secondary follicle contain a secondary oocyte. In short, the major named stages of oogenesis (oogonia, primary oocyte, secondary oocyte, ovum) are tied to the genetic process (meiosis) and not to the major stages of folliculogenesis (primordial follicle, primary follicle, secondary follicle, etc.).

Over the course of folliculogenesis, an increasing number of cellular and acellular layers separate the oocyte from other ovarian tissues. Just before ovulation, the oocyte is isolated by three acellular and four cellular layers. In order from innermost to outermost, they are the acellular zona pellucida, the cellular cumulus oophorus, the acellular follicular fluid, the granulosa cells, the acellular basal lamina, and finally the cellular theca interna and externa (see figure 4.3). The coordination of oogenesis and follicular maturation is regulated by materials that cross these layers to varying degrees, from both the oocyte to the outside and from ovarian tissue to the oocyte.

Why are there so many layers? One reason is that the oocyte and the mother are subject to different, evolutionary pressures. To anthropomorphize, each oocyte "wants" to be ovulated so as to contribute its genetic material to the next generation. Each oocyte competes with all the other oocytes in the mother's ovary for that privilege. In contrast, a mother is best served by ovulating only those oocytes that will produce the most viable offspring. A mother is equally related to all of her oocytes and does not benefit from oocytes competing with each other nor with her interests. Thus, the various layers between oocyte and the maternal circulation may negotiate these different selection pressures on the maternal versus the oocyte genomes. For instance, the zona pellucida may curb maternal influences on oocyte development, or the basal lamina may limit the impact an oocyte can have on the ovary. This complex layering is disrupted at ovulation.

After ovulation, the granulosa and theca interna grow together to seal the ovulatory opening. At this point, these cells also begin producing progesterone (Young, McNeilly 2010). The resultant post-ovulatory follicle is called a corpus luteum (CL). The longevity of the CL depends on neural or endocrine signals from multiple sources (see maternal recognition of pregnancy in chapter 5). When ovarian progesterone is no longer needed, the corpora lutea regress and are known as corpora albicans. If ovulation does not occur, the follicle involutes somewhat and is termed *atretic*. Becoming atretic finishes the life of a follicle and its oocyte. Very few oogonia are ever ovulated. For instance in women, 30–40 years of ovulations would generate 360–480 ovulations from an initial pool of approximately 7 million oocytes (Baker 1982).

Hormonally, ovarian activity is often separated into two phases: before ovulation (follicular) and after ovulation (luteal). The textbook characterization of the follicular phase is one of high estrogen and growing follicles. A spike in luteinizing hormone from the pituitary induces ovulation and estrous behavior. The subsequent luteal phase is one of high progesterone and uterine endometrial development. In women, some other primates, tree shrews (Scandentia), a few bats, and a rodent (*Acomys*; Bellofiore et al. 2017) the end of the luteal phase results in the sloughing of the newly created uterine tissue, a process called menstruation. As discussed in chapter 1 for rural Polish women, hormonal cycles as described in textbooks may not be the norm either for women or for other species (sidebar on page 101). For instance in giraffe, the so-called follicular phase (high estrogen) and the luteal phase (high progesterone) occur simultaneously (figure 6.2).

Estrous Cycles and Neo-oogenesis

The interval between periods of estrus (aka heat) is called the estrous cycle, and the interval between menses is the menstrual cycle. These cycles are easy to measure in captive animals, and, thus, we have much information on cycle lengths. For captive breeding programs, knowledge of estrous cycle length is critical for appropriate timing of many procedures, such as artificial insemination and embryo transfer. Such cycling is predominantly a by-product of captivity and few wild females undergo repeated estrous cycles. Instead, females are anestrous, pregnant, or lactating. Females may cycle briefly between these stages, but repeated cycling means the female has foregone a chance to bear young and thus may be at a selective disadvantage. "In natural populations the nonpregnant cycle is a rarity, and it is essentially a pathological luxury which cannot be tolerated" (Conaway 1971:239). Estrous cycle length is a measurable aspect of mammalian reproduction, but natural selection will operate to keep such cycles short and infrequent.

For most mammals, what we call the estrous cycle occurs in a hormonal context that is vastly different from that studied by reproductive physiologists. Such cycles occur either during lactation or during a transitional, hormonal milieu,

Figuring Out Ovarian Follicular Waves
Angela Baerwald (1975–)

Canadian physiologist Angela Baerwald has been a major force behind developments in our understanding of female reproductive biology, in particular ovarian dynamics. A native of Saskatchewan, Baerwald has been an assistant professor of reproductive endocrinology and infertility in the Department of Obstetrics, Gynecology and Reproductive Sciences at the University of Saskatchewan, Canada, since 2005.

She started her education at the University of Saskatchewan in 1993 and graduated with honors in 1997. As an undergraduate, Baerwald studied changes in ovarian function during the first trimester of pregnancy (Hess et al. 2000). In 1998, she began doctoral work in the Department of Obstetrics, Gynecology and Reproductive Sciences at the University of Saskatchewan, working to characterize ovarian function during natural menstrual cycles. Her groundbreaking finding that multiple waves of antral ovarian follicles develop throughout the menstrual cycle re-

ceived international recognition, as it challenged the traditional notion of a typical 28-day human cycle. These follicular waves resemble those previously documented for domestic farm animals and giraffes (chapter 6). Knowledge of follicular waves in women enabled proper timing of ovarian stimulation protocols to treat infertility, such that stimulation can be initiated at more than one time of the menstrual cycle. Baerwald also characterized follicular development in women using hormonal contraceptives (Baerwald et al. 2005). She found that the majority of antral-follicle development during hormonal contraceptive use occurred during the 7-day hormone-free interval. These findings contributed to the development of new hormonal contraceptive regimens in which the hormone-free interval has been reduced or removed. Baerwald also helped evaluate the transdermal contraceptive system, resulting in its approval by Health Canada (Pierson et al. 2003).

After completing her doctoral degree, Baerwald held a postdoctoral research associate position for 1 year at the Fertility Center, Ottawa Civic Hospital, University of Ottawa, Canada, where she evaluated the effects of synchronizing ovarian stimulation with follicle wave emergence in couples undergoing assisted reproduction.

Currently, Baerwald's research focuses on evaluating ovarian function in women of reproductive age, as well as those transitioning to menopause. Her recent work suggests that aberrant antral follicular dynamics during the transition to menopause leads to acute and atypical elevated estradiol (Vanden Brink et al. 2013). The clinical significance of these changes is under investigation. She also continues interdisciplinary research to characterize and compare follicular dynamics in women and mares.

(Photo courtesy of Angela Baerwald.)

such as after reproductive quiescence or after birth. In addition, ovulation without estrus (silent heat) may occur, as in several cervids, e.g., *Alces* (moose in the United States, elk in the United Kingdom, Europe), *Cervus* (deer), *Dama* (fallow deer), *Odocoileus* (white-tailed deer), *Rangifer* (reindeer-caribou; Jabbour et al. 1997). Even domestic mammals do not follow the textbook pattern. For instance, mares can develop new ovulatory follicles (and ovulate) during the luteal phase

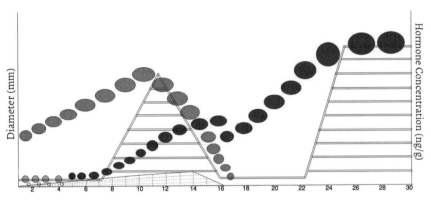

Figure 6.2. Follicular dynamics (follicular diameter, left *y*-axis) and steroid-hormone levels (right *y*-axis) in the giraffe for which the follicular and luteal phases occur simultaneously. Progesterone levels (*horizontal parallel lines*) are given for both the luteal phase and early pregnancy. Estrogen levels (*dotted lines*) are given only for the follicular phase, but estrogen is present during pregnancy. This figure illustrates the complex relationships between hormones and follicles during a luteal phase (a) without conception (days 0–17; *large gray ellipses*), (b) the overlapping follicular phase (days 0–17; small ellipses and dark ellipses) preceding conception, and (c) the early phase of gestation (days 18–30; *dark ellipses*) following conception. Note the corpus luteum of the preceding cycle and dominant follicle that will ovulate before conception form simultaneously. Ovulations occurred on day 1 (first cycle) and day 16 (second cycle), and conception occurred soon after the second ovulation. After ovulation, subsequent follicular waves are initiated (illustrated by small circles during days 2–4) and these waves occur throughout pregnancy (not illustrated). During gestation, individual follicles do not become dominant, but they do enlarge, and one will become dominant and ovulate if the embryo is rejected. Data from Lueders et al. 2009; figure by Abigail Michelson.

(Aurich 2011). As a result, the repeated estrous cycles observed in captive animals (and their complementary hormonal profiles) document a process that rarely happens in nature. This observation has ramifications for conservation and assisted reproduction, because hormonal data from repeated cycles of captive animals may be an artifact of captivity and not be valid for non-captive populations.

Another pervasive idea regarding oogenesis is that the number of oocytes a female has early in development is all that she will ever have (Zuckerman 1951). While this idea is entrenched in the literature, some data suggest that neo-oogenesis, the formation of new oocytes, may occur in mammals, even women (De Felici, Barrios 2013; Tilly et al. 2009; White et al. 2012). Oogenesis occurs in adult galagos (*Galago*), as well as in aye-ayes, *Daubentonia* (Gérard 1932; Petter-Rousseaux, Bourlière 1965). Certainly, cases like that of the female plains viscacha (*Lagostomus maximus*), which releases 400–800 mature oocytes at each ovulation, suggest neo-oogenesis may be possible. The traditional trajectory of oogenesis, with substantive follicular atresia (and lack of neo-oogenesis), may vary across mammals (Espinosa et al. 2011). Clearly, more data are needed.

Overall, adult ovaries are dynamic, with simultaneous waves of follicles at different stages of development, as well as numerous follicles in various stages of atresia. The development of female gametes is not a continuous process but, rather, a series of starts and stops. Gamete development may be put on hold during gestation. Similarly, lactation and oogenesis may cease for part of the year, e.g., during hibernation. Thus, the duration and frequency of oogenesis varies across mammals.

Diversity of Oogenesis and Folliculogenesis

The general process just described is a synthesis of studies on women, mice, and domestic animals, such as cows. What about other females? We will use the gray tree squirrel (*Sciurus carolinensis*), as a specific example of the process in a non-domestic species (Hayssen 2016), and then we will provide a few examples from other species to illustrate some of the diversity.

During the early stages of folliculogenesis in gray tree squirrels, the ovum enlarges to nearly full size (about 95 µm), while the follicular epithelium remains only one-cell thick and the follicle overall is about 150 µm. With subsequent differentiation of the cumulus oophorus, granulosa, antrum, theca interna, and theca externa, the follicle enlarges to a resting size of about 600 µm, while the ovum does not change in size. Thus, oocyte and follicular growth are out of synch.

Rather than the succession of a set number of follicles progressing towards eventual ovulation, waves of regular follicular growth occur during both gestation and lactation. Most of these follicles become atretic well before the antral stage. Therefore, as with most mammals, a regular cycle of successive ovulations does not occur. Atretic follicles are 400–600 µm; ovulatory follicles are 1,000–1,100 µm. After ovulation, luteal follicles early in gestation are similar in size to ovulatory follicles (1,000–1,300 µm), but they regress to about 750 µm before birth, with further reduction during lactation (Deanesly, Parkes 1933). In the fall, oogenesis and folliculogenesis slow down or stop completely. Ovarian activity resumes after the winter solstice in January when conditions suitable for rearing a litter (weather and food availability) are more likely.

Not all eutherians follow the pattern exemplified by gray tree squirrels. For instance, instead of ovulatory oocytes floating in fluid-filled antral follicles, ova of shrews and tenrecs are surrounded by a solid ball of granulosa cells (shrew: *Suncus*; tenrec: *Echinops, Hemicentetes, Micropotamogale, Setifer, Tenrec*; Enders et al. 2005; Kaneko et al. 2003). African elephants (*Loxodonta*) like squirrels, have waves of follicular growth, but, in elephants, some follicles bypass the ovulatory phase and go straight to the luteal phase (Hildebrandt et al. 2011), whereas in squirrels such follicles become atretic. Thus, pregnant African elephants have many more (2–8) CLs than embryos (1–2). Also, elephant luteal follicles are larger (3–6 cm), than ovulatory follicles (0.5–1.0 cm; Allen 2006). In manatees (*Trichechus*), a minimum of 21 CLs accompany each gestation (Marmontel 1988) and in dugongs up to 90 (Nishiwaki, Marsh 1985). Accessory CLs also occur in hystri-comorph rodents (Weir 1971b). Hyraxes are related to elephants and undergo continuous follicular development but only up to the antral stage (*Dendrohyrax*; O'Donoghue 1963). Similarly, giraffe follicular development (from primordial to pre-antral follicles) continues throughout the luteal phase and during gestation (figure 6.2; Lueders et al. 2009). Also in giraffes, complete follicular development, from primordial follicle to CL and thence to a corpus albicans, can occur in fetal and immature females (Kellas et al. 1958; Kayanja, Blankenship 1973; Wilsher et al. 2013). In addition to this diversity of follicular development, not all follicles have just one ovum. Polyovular follicles are common in some armadillos and tenrecs (*Cabassous, Chaetophractus, Chlamyphorus, Tolypeutes, Zaedyus*; Cetica et al. 2005; *Tenrec*, Nicoll, Racey 1985). These are just a few of the eutherian vari-ants on folliculogenesis.

Marsupials and monotremes also have variants on the process. In tammar wallabies (*Macropus eugenii*), migration of PGCs to the incipient gonad is com-pleted by the day the offspring crawls out of the vaginal canal and up to the teat. By the end of pouch life, 25% of PGCs are atretic (Alcorn, Robinson 1983). Unlike squirrels, oocyte and follicular growth in marsupials are synchronous up until the time of ovulation (Cesario, Matheus 2008; Frankenberg, Selwood 2001; Kress et al. 2001).

Monotremes and marsupials have larger oocytes than eutherians. The cyto-plasmic content of these oocytes includes material sometimes referred to as yolk, but this material is not as extensive as, nor compositionally similar to, the yolk of bird eggs. The material has a specific distribution within the oocyte and persists in the conceptus until the early cleavage stages (Menkhorst et al. 2009). The regu-lation and composition of the material as well as the details of its functions in conception and embryogenesis are poorly understood.

Given the variation in the development of follicles, variation in the release of ova from those follicles (ovulation) is also diverse.

Ovulation

Ovulation is the release of female gametes from the ovary. Each released gamete is surrounded by the same maternal cells with which it has been associated since early in oogenesis. Separating each ovum from the associated cumulus cells is the zona pellucida (occasionally referred to as an egg coat). As detailed in chapter 4, the zona is a matrix of acellular (glycoprotein) material produced both by the oocyte itself and by ovarian tissue. Thus, ovulation includes release of not only female gametes but also a complex array of cellular (the cumulus) and acellular (the zona) structures, which accompany the gamete as it journeys from the ovary to the site of conception.

Ovulation is a dynamic but not explosive process. Video footage of human ovulation (New Scientist.com, June 17, 2008) shows an indistinct, gelatinous plug on the surface of the ovary where it is brushed by the finger-like projections of the oviduct (fimbria). The fimbria eventually move the plug of cumulus cells and oocyte into the oviduct. The release is gradual, not, as usually depicted, an eruption (Gilbert 2014).

For ovulation to occur, the pre-ovulatory follicle must be near the surface of the ovary. The cell layers between the ovarian surface and the center of the follicle must lose their connectivity, and the acellular basal lamina must disperse or disintegrate. The complex biochemistry of these processes is not completely understood but involves large-scale feedback loops among the hypothalamus, pituitary, and ovary, as well as localized regulatory processes within the ovary (Richards et al. 2015). The details are beyond the scope of this chapter, but we will explore a few major taxonomic differences across mammals in the nature, control, and timing of ovulation.

How Many Ova Are Ovulated?

Species that give birth to singleton young are usually monovular; that is, they ovulate only a single ovum. Similarly, most females that routinely give birth to litters are polyovular and release multiple ova from multiple follicles. (Note that polyovular has two distinct meanings, either release of multiple ova at one ovulation or presence of multiple ova in a single follicle as with tenrecs.) Armadillos (*Dasypus*) provide an astonishing variant on these generalities as they ovulate a single ovum but produce litters of 4 to 8 perhaps even 12 identical offspring (Hayssen et al. 1993). In *Dasypus*, the early blastocyst divides to produce multiple offspring that use a single placenta attached at a single implantation site (Craig et al. 1997).

At the opposite extreme are females that ovulate many more ova than their average litter size. This superovulation occurs in diverse taxa, suggesting independent origins (Birney, Baird 1985; Wimsatt 1975). For instance, the plains viscacha, a rodent, usually gives birth to two young but ovulates 200 to 700 or more ova. Most of these ova degenerate but seven to eight achieve conception. All the conceptions implant, but only two go to term. The other embryos are resorbed (Espinosa et al. 2011; Weir 1971a, 1971b). Less extreme is *Elephantulus myurus*, an elephant shrew. Females ovulate 60–120 ova but only birth two young. After conception, some embryos have cleavage abnormalities and many degenerate as the uterus has too few implantation sites for all (Tripp 1971). *Hemicentetes semispinosus*, a tenrec, ovulates up to 40 ova, all of which may implant after conception, but as the maximum litter size is 10, most are resorbed during gestation. Some bats, such as *Eptesicus fuscus* (big brown bat) and *Pipistrellus subflavus* (eastern pipistrelle), shed three to seven ova, but only two young survive, one in each uterine horn (Wimsatt 1975). Similarly, many marsupials, such as the New World *Didelphis* and the Australian *Dasyurus viverrinus*, ovulate more ova than they have teats. In addition, some marsupials experience neonate losses after birth, when neonates are unable to find a teat. However, even before birth, failed conceptions or early embryonic cleavage decreases litter size. A final example is the pronghorn (*Antilocapra*), an ungulate that ovulates four to five ova but gives birth to twins due to in utero siblicide; we provide the details in chapter 7 (O'Gara 1969).

Several advantages might accrue to females producing excess ova. First, the resultant abundance of corpora lutea may aid in the feedback physiology of gestation. Second, competition among ova might present a selection mechanism for maintaining only the most viable offspring. Third, extra ova might be produced to hedge against future losses in subsequent risky stages. For marsupials, the ability of neonates to find a teat is risky, and the cost of gestating a few extra neonates may ensure a full litter, even if some are lost in the transition to the teat. Similarly, for eutherians, if conception failure and cleavage abnormalities are probable, then ovulating extra ova could compensate for that loss. Given that the production of extra ova is rare and occurs in distantly related taxa, its selective advantage may be unique to each instance.

What Triggers Ovulation?

Ovulation is regulated by both internal (hormonal or neural) and external (abiotic or conspecific) influences. The trigger is usually hormonal. Hormone levels may be fine-tuned to endogenous conditions, such as the state of the uterus, whether lactation is ongoing, what fat stores are like, and levels of stress hormones. Hormonal cycles will also be influenced by exogenous factors, such as predation, prey abundance, parasite abundance, and the presence of males or other females, as well as abiotic conditions, such as photoperiod, temperature, or drought.

Internally, the endocrine control of ovulation includes several key hormones. The time course and concentrations of these hormones vary by species. For some females, e.g., cows, women, and mice, the most obvious driver of ovulation is a surge in LH from the pituitary. However, the LH surge happens only if other hormones, such as estrogen, progesterone, FSH, prostaglandin, and prolactin occur in appropriate concentrations and at appropriate intervals. This endocrine ebb and flow differs across species. For instance, a progesterone peak precedes the LH surge in lab mice but occurs afterward in female dogs, women, and cows.

Ovulation also has a neural component. It must have a neural component, because the release of LH from the pituitary occurs only after the rest of the brain integrates various sensory and hormonal signals. When such integration occurs, neurons in the brain trigger the hypothalamus to secrete gonadotropin-releasing hormone (GnRH), which travels to the pituitary and causes LH release. Thus, ovulation is under combined neural and endocrine control. The neural component is altered by many signals, such as photoperiod, pheromones, mating, as well as general physiological condition and nutritional state (de la Iglesia, Schwartz 2006; Scaramuzzi et al. 2011). Thus, the brain synthesizes this complex external and internal input to control ovulation. Scientists can induce ovulation by injecting a female with the hypothalamic hormone, GnRH, or its placental variant, chorionic gonadotropin (CG).

Does Mating Induce Ovulation?

A commonly used but problematic dichotomy is that of spontaneous (or reflex) versus induced ovulation (Bakker, Baum 2000). Before scientists understood the complex neural and endocrine precursors to ovulation, mammalogists noticed that, for some mammals, ovulation reliably occurred a predictable time after mating, whereas for other mammals, including women, ovulation was not correlated with mating. The impression was that penile stimulation induced a "ripening" follicle to release its ovum and that mating was the direct cause of ovulation in these species. Thus, mating was said to induce ovulation in some species, e.g., cats, but occur spontaneously in others, such as sows and cows. The list of species with so-called induced ovulation is diverse and includes shrews, squirrels (Sciuridae), voles, lemmings, hares (*Lepus*), cottontails (*Sylvilagus*), rabbits, mink (*Neovison*), raccoons (*Procyon*), black bear (*Ursus americanus*), and camels (Camelidae; Bakker, Baum 2000; Boone et al. 2004). This diversity led some to conclude "induced" ovulation is an ancestral condition (Conaway 1971) and others to conclude it is derived (Bakker, Baum 2000).

To add another twist, the same hormonal surge that triggers ovulation can also precipitate mating behavior. For instance, the LH surge occurs four hours before mating in the diurnal Nile grass rat (*Arvicanthis niloticus*), as well as nocturnal laboratory rats and mice (Smale et al. 2005). In these cases, ovulation induces mating. In addition, female behavior, before and during coitus, can influence the pattern of mating and the outcome of copulation (Buzzio, Castro-Vázquez 2002; Erskine 1989).

The induced versus spontaneous dichotomy is also tricky because "induced" ovulators can ovulate spontaneously, e.g., cats, mink. Conversely, mating behavior can facilitate ovulation in mammals classified as spontaneous (e.g., *Mus, Rattus*; Bakker, Baum 2000). In addition, although gonadotropins, such as GnRH, regulate ovulation in all mammals, some authors use ovulation after an injection of GnRH or CG as evidence of mating-induced ovulation (Bedford et al. 2004).

In so-called spontaneous ovulation, ovarian steroids, not somatosensory stimulation, induce the LH surge that subsequently triggers ovulation (Bakker, Baum 2000). However, in some respects, no ovulation is truly spontaneous. All exogenous cues, including those from mating, are mediated by hypothalamic and pituitary hormones to influence (induce) ovulation. Orgasmic contractions, themselves, may facilitate ovulation, as in rabbits (Fox, Fox 1971).

Ovulation after mating is not immediate even in so-called induced ovulators. For example, female stoats (*Mustela erminea*) mate with multiple males, store sperm for up to 120 hours, and ovulate 72–96 hours after coitus. In addition, folliculogenesis continues for weeks after mating (Amstislavsky, Ternovskaya 2000).

This time frame gives females ample opportunity for mate choice, sperm selection, or perhaps choosing not to ovulate at all. Induced and spontaneous ovulation appear to be uncommon extremes on an extensive continuum (Conaway 1971; Jöchle 1975).

South American camelids are interesting in this context. Alpacas (*Vicugna pacos*) and llamas (*Lama glama*) are the domesticated progeny of vicuñas (*Vicugna vicugna*) and guanacos (*Lama guanicoe*), respectively. They are South American relatives of camels, and all camelids are said to be induced ovulators. Ovulation is spontaneous in about 5% of females but is usually associated with mating and occurs on average 2 days after mating. Thus, the reaction to mating takes a long time. In addition, ovulation may not be associated with physical stimulation of the genital tract (Ratto et al. 2005) but by a nerve growth factor present in seminal plasma (Adams, Ratto 2013; Silva et al. 2014). Also, 20% of females do not conceive after mating (Brown 2000; Vaughan 2011). Thus, ovulation may require more than just mating in alpacas and llamas.

Of course, for ovulation to occur, an ovulatory follicle must be available. In camelids, as with most mammals, follicular development proceeds in overlapping waves. In alpacas, individual follicles take 11–12 days to reach a size capable of ovulating. The largest, dominant follicle remains at that size for 3–5 days and subsequently regresses. Meanwhile, other follicles have developed, and, within 2–3 days, one of them becomes dominant. Ovulation will not occur until a female has a large follicle, regardless of when she mates, because the hormonal production by a large follicle is required for the hypothalamic-pituitary axis to be able to release LH. Although ovulation only occurs if females have follicles of sufficient size, behavioral estrus is independent of follicle size. When a female mates, ovulation occurs about 2 days later, so the time to create a dominant follicle and the time between mating and ovulation are similar. Consequently, mating might stimulate follicular growth rather than ovulation. Or perhaps a female has two options. (1) If she finds a mate when she does not have a large follicle she can use mating to stimulate follicular growth and ovulation. (2) If she already has a large follicle, she can use mating to stimulate ovulation. As mentioned earlier, mating could precipitate follicular growth rather than ovulation per se.

In any event, females that use the cues of mating to time ovulation do not have to wait for dominant follicles to develop, whereas females that ignore the neural input of mating must wait until ovulation occurs spontaneously. For women, the interval between ovulations is about 4 weeks, but, for laboratory mice, it is only 4 days. If timing is critical, a long delay may reduce reproductive success. Thus, females under time constraints could reproduce more efficiently, using copulatory stimuli to influence ovulation. Similar advantages would accrue to females when males are few and far between. Using the stimuli of mating allows more rapid reproduction. For instance, in alpacas and llamas, follicular development continues throughout gestation and lactation, so if a female loses her offspring, she can conceive again within a couple of days (Adams et al. 1990). Given that alpacas reject up to 80% of their embryos, the ability to conceive quickly may be advantageous (Silva et al. 2014). All told, timing ovulation with mating has potential advantages.

In light of this discussion we will retain the term *spontaneous* ovulation but substitute the term *facultative* in lieu of 'induced' ovulation. Abiotic and internal triggers are key to all ovulations, but for facultative ovulation, conspecific cues and the somatosensory consequences of mating (such as orgasm) also strongly influence the timing of ovulation.

Mating is one conspecific cue used for ovulation, but other conspecific stimuli may also influence ovulation. Females can use pheromones from conspecific females or males as exogenous cues to regulate mating. For example, dominant female marmosets (*Callithrix*) block ovulation in subordinate females in this way (Barrett et al. 1990). In addition, hormonal cycles in captive rodents can be suppressed or prolonged when females are housed in groups and isolated from males (Lee, Boot 1956). Thus, a high density of females in the wild could signal resources are scarce. That the effect operates via adrenal glands also suggests a stress-related phenomena (Ma et al. 1998). Although male pheromones stimulate estrus in captive female mice (Whitten et al. 1968), females also produce cues that result in specific and immediate responses in males (Rekwot et al. 2001). Thus, both females and males provide stimuli to each other to influence ovulation and mating.

Mating

A maximally swollen female chimp mates 1–4 times an hour with thirteen or more partners. Over her lifetime, she will engage in some six thousand or more copulations . . . to produce no more than five or so surviving offspring.—Hrdy 2000:80

Once her ova are ready for ovulation, a female needs to find a mate. To do so, females advertise, search, wait, or use some combination of these tactics. The mechanism a female uses will depend on features of the environment as well as factors that relate to predation, territoriality, and social interactions. Females advertise for partners across various sensory modalities, such as olfaction, vision, or sound. For instance, female Asian elephants (*Elephas maximus*) release a pheromone in their urine for several weeks before ovulation (Rasmussen, Schulte 1998). Because elephant females and males are widely separated, releasing an olfactory cue long before ovulation gives females time to attract a mate. In contrast, primate females often live in social groups that include males. Consequently, distinct visual cues are used by female chimpanzees (*Pan*) and baboons (*Papio*). Specifically, around the time of ovulation, the skin near the vaginal opening swells and may change color; these perineal, sexual swellings probably serve females in multiple ways from simple attraction to initiating male-male competition (Deschner et al. 2003). Visual cues may also be used by females that do not form social groups, such as European mink (*Mustela lutreola*), whose external genitalia enlarge considerably and become pinkish-lilac (Youngman 1990).

Other females use vocalizations. For instance, Asian chipmunk females (*Tamias sibiricus*) employ at least three distinct calls in interactions with males. One is an advertisement call while all three are used during courtship (Blake 1992). Other sciurids such as North American chipmunks (*T. merriami, T. obscurus*) and African bush squirrels (*Paraxerus cepapi*) vocalize to attract males (Callahan 1981). Female whales (*Eubalena*) may do the same (Schaeff 2007).

Some females use signals in combination. Mares attract stallions with visual signals, including a distinctive facial expression as well as turning her hind quarters toward him and moving her tail aside to display her perineal region. A rhythmic eversion of the labia (called *clitoral winking*) and the release of small quantities of fluid add to the olfactory and visual display. These actions prime the stallion for reproduction. Mares also display estrous behavior during gestation as well as during the seasonally quiescent phase of reproduction. Within social groups, these behaviors may facilitate long-term bonds between mares and stallions (Crowell-Davis 2007). Female camels also use multiple signals (Joshi et al. 1978). They vocalize and move their tail to expose their swollen vulval lips (labia).

Instead of advertising for a mate, a female may go to a specialized breeding area, such as a lek, where available males congregate. She then selects from that pool. Female deer, e.g., fallow deer (*Dama dama*) or red deer (*Cervus elaphus*, called elk in North America), use this strategy. Other females may do little advertising or seeking but instead let males come to them. Such is the case for female elephant seals (*Mirounga*; Leboeuf 1970). Yet another strategy is exemplified by a Malagasy carnivore, the fossa. Female fossa (*Cryptoprocta*) defend traditional mating sites high in trees. Individual females monopolize these mating locations to which males are attracted. Females have long copulations (up to three hours) and mate with multiple (4–5) males over a period of 1–6 days. The site is used by multiple females in succession and is used over multiple years (Hawkins, Racey 2009). Thus, females use diverse methods to find suitable mates. Once found, how does a female choose among them, and how many does she choose?

Polyandry

Polyandry is common in nature (Taylor et al. 2014). Even if time is short, females often mate with more than one male. For instance, California ground squirrel (*Otosperm ophilus beecheyi*) females solicit matings on average for only seven hours; yet, they mate with seven males during that time (Boellstorff et al. 1994). In mammals, polyandry occurs in taxa as diverse as marsupials (*Antechinus*; honey possum, *Tarsipes*), bats (big browns, *Eptesicus*), hedgehogs (*Erinaceus*), shrews

(*Sorex*), carnivores (Eurasian badgers, *Meles*; Ethiopian wolves, *Canis simensis*; dwarf mongooses, *Helogale parvula*; lions, *Panthera leo*), ungulates (deer, *Capreolus*, *Odocoileus*; pronghorn, *Antilocapra*), and rodents (ground and tree squirrels, *Cynomys*, *Sciurus*; beavers, *Castor*; guinea pigs, *Cavia*; deer mice, *Peromyscus*; mice, *Mus*; Asher et al. 2008; birdsall, Nash 1973; Carling et al. 2003; Crawford et al. 2008; DeYoung, et al. 2002; Engh et al. 2002; Haynie et al. 2003; Koprowski 1998; Stockley et al. 2002; Sale et al. 2013; Thonhauser et al. 2013; Vanpé et al. 2009; Vonhof et al. 2006; Wooler et al. 2000). These are just a few examples of the many species in which polyandry occurs. The extent of polyandry may be underestimated, because it is usually documented by paternity tests in species with litters, but even with singleton births, females may mate with multiple males.

Mating costs time and energy and increases risks of predation, injury, and sexually transmitted disease. Mating with multiple males is likely to exacerbate these effects (Thonhauser et al. 2013), but it also has potential benefits. Although the reasons behind multiple mating may be unique to each species, some benefits include increased genetic diversity of offspring, increased conceptions, insurances against a prior male being sterile, reduced male harassment, avoidance of infanticide, avoidance of inbreeding, and paternity confusion (Parker Birkhead 2013; Thonhauser et al. 2013). Three examples follow.

First, increased genetic variation in resulting offspring may drive multiple mating in Belding's ground squirrels (*Urocitellus beldingi*). These females live in unpredictable environments. With multiple mates, a female's offspring will have diverse phenotypes, some of which may survive better in different, environmental conditions (Stockley 2013).

Second, for elephant seal females, sperm availability rather than genetic quality may be more of a concern. Female northern elephant seals (*Mirounga angustirostris*) haul out onto beaches with many other females who may all mate with a single large male over a short period of time. These extensive matings may deplete the large male's stores of fully mature sperm. Thus, females may benefit by mating with smaller subordinate males who have not been depleting their sperm stores (Hoezel et al. 1999).

Finally, a small carnivorous marsupial in Australia (*Antechinus*) has an unusual mating system. All females solicit males during a short synchronous mating season after which all males die. As a result before females give birth, the only males in the population are in utero. Ejaculates contain low sperm counts and because females may not ovulate until after the males die, females have no chance to re-mate if conception does not occur. Females mate with multiple males as insurance against male infertility and to increase sperm competition. Females also store sperm in oviductal crypts for up to 2 weeks. They conceive and give birth to litters using sperm from on average three to four (up to seven) males (Kraaijeveld-Smith et al. 2002).

Coitus

The female is not a passive part of the mating bout.—Koprowski 1998:36

Copulation occurs wherever one finds mammals: underground, up in trees, deep in caves, in the water, even on the wing. The actual location will depend on the habitat females occupy; the risk of predation, parasites, and disease vectors; and the duration of copulation. Copulation ranges from a few seconds in dolphins (*Tursiops*; Puente, Dewsbury 1976) and peccaries (*Tayassu*; Tores 1993) to up to 18 hours in captive *Antechinus*, although antechinus matings in the wild average only four to six hours (Kraaijeveld-Smit et al. 2003). Antechinus are much smaller than dolphins and across 113 species of mostly rodents, carnivorans, and primates, copulations are shorter for larger mammals (Stallmann, Harcourt 2006). Of course, species with longer copulations will be more vulnerable to predation.

Mating is a risky business. Not only is predation an issue, but injury and death as a result of mating may occur, as in northern elephant seals (Le Boeuf, Mesnick 1991), monk seals, (*Monachus*; Atkinson et al. 1994), primates (Smuts, Smuts 1993), sea otters (*Enhydra*; Staedler, Riedman 1993), and feral sheep (*Ovis*; Rèale et al. 1996). Thus, many females only mate when the likelihood of producing offspring is high.

Estrus is a term given to periods when females choose to mate and refers more generally to female sexual behavior. As mentioned in chapter 1, textbook categorizations of female sexual behavior is often focused on the male point of view. Our example was "attractively" as "the stimulus value of a female to a male" and receptivity as "the stimulus value of a female for eliciting an intravaginal ejaculation" (Nelson 2011:289). We prefer *solicitation* and *facilitation*, respectively, because females are as driven to produce successful offspring as are males. During estrus, females actively search for and solicit mates. They may deposit pheromones, change their visual appearance, or vocalize to attract mates.

Often a set of interactions, broadly termed *courtship*, precedes coitus. During courtship and mating, key stereotypical behaviors may occur, and, in some taxa, these are driven by hormones. For example, many female rodents exhibit lordosis just prior to and during mating. Lordosis is a behavioral response and an associated posture that females use to facilitate copulation.

Like lordosis, many other female behaviors are often crucial for successful copulations. For instance, for bottlenose dolphins (*Tursiops*), female displays of her underbelly are critical to mating (Puente, Dewsbury 1976). Female American martens (*Martes americana*) control the timing and duration of coitus (Grant, Hawley 1991). In rabbits, coitus is usually achieved only when the doe actively aids the buck. Early superficial observations suggested the doe's part in coitus was entirely passive, but her movement is necessary to stimulate the male. If she is passive for more than a short period, the buck dismounts and resumes other activities (Rowley, Mollison 1955). In addition, a doe's orgasmic contractions may play a role in triggering ovulation (Fox, Fox 1971). Similarly, orgasmic contractions in women facilitate retention of sperm (Baker, Bellis 1993). In the African four-striped mouse (*Rhabdomys*), pleasure may drive female initiation of ventro-ventral coitus (Dufour et al. 2015). Overall, females are active participants in mating and probably have a larger input into the success of copulation than males. This key attribute is poorly represented in the literature whereby males are presented as the active sex and females the passive sex. We hope these assumptions and associated language come into disuse with time.

Vaginal Plugs

Various portions of the female tract accept sperm, such as the vagina, as in women, cows, ewes, and female rabbits; the cervix as in sows (*Sus*); or the uterus, as in many rodents and mares (Coy et al. 2012). Vaginal secretions can coagulate with semen to form a vaginal clot (vaginal plug or copulatory plug). These clots form in diverse mammals, including marsupials, bats, hedgehogs, moles (*Condylura*), carnivorans, primates, suids, and rodents (Delgado et al. 2007; Dixson, Anderson 2002; Eadie 1948; Gemmell et al. 2002; Hartman 1924; Hartung, Dewsbury 1978; Koprowski 1992; Poiani 2006). The composition, viability, and shape of the clot differs across species, as do the relative contributions of female and male secretions to the end product (Eadie 1948).

Theories regarding the function of the coagulum generally have had a male focus. The most common is that the clot is used by males to plug the vagina and prevent successful copulations by other males (Dixon, Anderson 2002). As the clots can be displaced by subsequent matings, such as in ring-tailed lemurs (*Lemur catta*; Parga 2003), and are often present when females mate with multiple males, their function as a "chastity belt" may be overrated. In addition, the coagulum is often a combination of female and male secretions, suggesting a female role. Females may benefit from a clot if it facilitates her choice of sperm or sperm competition among males. Females may even benefit directly by using nutrients from the clot (Poiani 2006). For instance, female tree squirrels often remove the clot and consume it (Koprowski 1992), as do female peccaries (*Tayassu*; Sowls 1966). The clots may also have antimicrobial properties (Poiani 2006). As with other reproductive features, vaginal plugs may benefit both sexes in different ways that may vary by species. We suggest that this area is one that could be carefully examined using proteomics and molecular methods to disentangle the functions of compounds in the plugs, their origins, and potentially their functions.

Gamete Transport

After mating, female physiology determines the timing and location of conception and even whether it occurs. That location is where every new life begins (Coy et al. 2012).

At copulation, a female receives not only sperm but also seminal fluids, an assortment of bacteria, and other potential disease agents. As long as one sperm cell reaches each ovum, a female benefits (increases her reproductive success) by destroying excess sperm, ejaculate, and associated debris quickly, thereby preventing transmission of disease vectors into her peritoneal cavity. During this process "most sperm are eliminated from the female tract" (Coy et al. 2012:1741). In women, 35% of sperm are ejected within 30 minutes of coitus (Baker, Bellis 1993). Complex morphological and physiological barriers exist to prevent the invasion and growth of bacteria and other microflora within the female reproductive tract (Tung et al. 2015). The vagina, uteri, and oviducts are separated by undulating mucus and/or muscle sphincters both at the cervix and at the utero-tubal junction. These barriers regulate the transfer of particulate material and fluids, including sperm and seminal plasma, between the segments (Koester 1970). Cellular (e.g., leukocytes and macrophages) and chemical (e.g., pH) defenses also exist to dispose of or destroy potentially harmful debris. Following coitus, leukocytes and neutrophils are released into the reproductive tract, and phagocytosis of sperm and debris follows (Aitken et al. 2015).

Once ovulation and mating occur, female and male gametes must meet and fuse for conception to happen. Although for tenrecs, conception may occur within an ovulated follicle (Nicoll, Racey 1985), its usual location is the oviduct. However, the oviduct is not just a passive site for conception. Oviductal motility, fluid composition, and tissue structure are dynamic. Furthermore, both oocytes and sperm induce changes in the composition of oviductal fluid.

Gamete transport can be multiphasic and complex. During coitus, females receive sperm in the vagina, cervix, or uterus and then deliver them to the site of conception (Coy et al. 2012). The traditional view is that conception is biased in favor of sperm that can swim the fastest. However, the hypothesis of a "sperm race" is no longer tenable (Holt, Fazeli 2016:105). Over 70 years ago, Hartman (1957) made the same observation when he said "it is highly unlikely that sperm motility has the slightest value for ascent through the oviduct." Sperm do not have the energetic resources or directional ability to travel under their own power to the site of conception, although some energy stores are provided in seminal fluid. Fortunately, the fluid dynamics and motility of the female tract obviate these deficiencies. For example, peristalsis of the highly motile female tract is sufficient to carry sperm to the uterus in opossums (Hartman 1924). In rabbits, damaged and immobile sperm occur near the ovary within one minute of mating, but functional sperm do not reach the upper oviduct until much later, well after the oviduct is cleared of early debris (Suarez 2015). In rabbits, strong orgasmic contractions may be responsible for rapid rise of sperm up the tract and for the resultant damage (Fox, Fox 1971). For mammals in general, contractions of the reproductive tract and biochemical recognition of sperm are needed for functional sperm to reach the oviduct (Suarez 2015). The female reproductive tract selectively regulates the passage of sperm via cervical and oviductal secretions as well as by the physical structure of the cervix and utero-tubal junction (Druart 2012; Suarez 2015; Tung et al. 2015). For example, the diameter of the utero-tubal junction in lab mice is only slightly wider than the sperm (Suarez 2015). In numerous ways, the female reproductive tract will "permit, assist, or block sperm passage toward the oocyte" (Holt, Fazeli 2016:108).

The oviduct is also a sperm reservoir and creates an environment that keeps sperm functional until oocytes are released (Gervasi et al. 2009). Given that ovulation and conception do not occur immediately after coitus, all females keep (store) sperm within their reproductive tracts for some time after mating (Orr, Brennan 2015). However, a few females store sperm for lengthy periods and extend the time between mating and conception, sometimes for months. Temperate zone bats, which mate in the fall but ovulate in the spring, are a prime example, but females from other species, such as marsupials, canids, and leporids, will store sperm for 2–4 weeks (Orr, Zuk 2014; Orr, Brennan 2015; also chapter 7, "Gestation"). In addition, the attachment and release of sperm from the oviduct provides a mechanism for sperm selection, because both attachment and release are regulated by the oviduct (Gervasi et al. 2009; Holt, Fazeli 2016). Even short-term sperm storage will allow sperm competition and female choice of which sperm to use for conception (Orr, Zuk 2014).

Just getting sperm to the oviduct is not sufficient to achieve conception. Mammalian sperm received upon mating are incapable of fulfilling their biological function (Nixon et al. 2011). They need help. Without the female, sperm are unable to fuse with ova. Sperm require a period of maturation and activation termed *capacitation*. Sperm maturation begins in males but after coitus; capacitation is under female control. The process involves chemical alterations of the sperm. Depending on the species, uterine or oviductal secretions may be key, as well as interactions of the zona pellucida or cumulus oophorus with sperm (Ickowicz et al. 2012; Kaneko et al. 2003). Unfortunately, "although recognized for more than 50 years, capacitation remains an ill-defined process, in which a series of morphological and biochemical changes render mammalian spermatozoa fully functional" (Ecroyd et al. 2009:998).

Besides capacitation, sperm need a second, chemical modification before conception is possible: removal of the acrosome, a cap-like structure that encases the sperm head. Cumulus cells and the zona pellucida interact with sperm to break down the acrosome. This disintegration allows the ovum and sperm to fuse. A subsequent cascade of molecular reactions changes the oolemma (the membrane surrounding the ovum) and prevents fusion of the ovum with other sperm (Coy et al. 2012; Kaneko et al. 2003).

Conception

Conception is the concatenation of maternal and paternal chromosomes within the oocyte for the formation of a conceptus (zygote). The process is regulated by the maternal genome. Although each oocyte contains only half the maternal genome, all the cytoplasmic contents, including transcriptional factors and RNA coding sequences, are maternal in origin. Occasional transmission of male-derived products, e.g., mitochondria is possible but rare. The ooplasm directs multiple components of conception, including breakdown of the sperm nuclear envelope, creation of the paternal pronucleus, and extensive restructuring of the chromatin (Mtango et al. 2008). For instance, maternal histones replace paternal protamines to influence the accessibility of the paternal genetic material (McLay, Clarke 2003). These extensive modifications to the sperm genetic material "affect later paternal genome function and embryo phenotype, and . . . this is under the control of genes expressed in the oocyte" (Mtango et al. 2008:257). In addition, although functional paternal mitochondria may enter the ooplasm at conception, they are usually selectively degraded and only mitochondria of maternal origin survive (Mtango et al. 2008; Luo et al. 2013). In humans, this process provides a mechanism for the concept of a mitochondrial Eve, a female whose mitochondrial DNA is currently present in all humans (Cann et al. 1987).

Potential interactions between males and their progeny can happen in utero after conception. The ejaculate may alter the biological characteristics of the female's reproductive tract with respect to pH, ionic concentrations, etc., such that the microenvironment is detrimental to the survival of the young who develop there. Mothers may counter these adverse effects. Females receiving sperm in the vagina might avoid extensive contamination of the uterus with cervical barriers, whereas females using the uterus for sperm deposition may allow the zygote to develop in the oviduct until the uterine environment can be adjusted and excess debris removed.

Offspring-male conflict may occur when a female has a postpartum estrus. In this case, the physical and hormonal remnants of the recent pregnancy may create a uterine or vaginal environment deleterious to sperm (Casida 1968). Mammals with two separate, reproductive tracts avoid these difficulties by transporting sperm up one tract after giving birth from the other (*Macropus*; Tyndale-Biscoe, Rodger 1978). For other mammals, the physical and chemical properties of seminal fluid may protect sperm viability. However, the lower fertility observed in relation to this estrus (Casida 1968) suggests these mechanisms may not be wholly effective.

As this chapter illustrates, the fusion of male and female gametes is an enormously complex event replete with potential conflict and cooperation among multiple players: females, males, offspring, and siblings. Natural selection will result in effective coordination among the participants. This is a highly interactive time frame of female reproduction and as such presents opportunities for extensive co-evolution among all parties.

In Sum: Female Choice

A female can potentially produce offspring with any ovum she ovulates. To do so she needs a single sperm, ideally one that is genetically compatible and that will best support her reproductive success. Copulation seldom leads directly or invariably to conception (Eberhard 1996), and female choice can operate before and/or after mating. First, a female can choose a male and her physiology can select among his sperm for an appropriate pairing. Alternatively, she can choose to mate with several males and select among the combined sperm. In addition, her reproductive tract can provide an environment to promote competition among all or a selected subset of sperm. The morphology and physiology of the vagina, uterus, and oviducts, as well as the composition and architecture of the cumulus cells and the zona pellucida provide potential vehicles for realizing a female's sperm choice, but we have only a superficial understanding of the mechanistic details of the process (Anderson et al. 2006; Holt, Fazeli 2016; Suarez 2015; Swanson et al. 2002).

Females modulate the behavior of sperm using biochemical and physical interactions with both the lining of the reproductive tract as well as the thick, viscous fluid contained inside. This fluid can alter the pattern and frequency of the sperm-tail (flagellar) movement as well as the direction of transport. In addition, the internal architecture of folds and grooves selects for the passage of some sperm while preventing microbes, ejaculate, and other material from gaining access (Suarez 2015; Tung et al. 2015). In fact, "only a privileged sperm population, selected on the basis of multiple criteria, is permitted to enter the oviduct, where they are subjected to even more selection processes" (Holt, Fazeli 2016:105). The reproductive tract of sows may even be able to differentiate between X-bearing and Y-bearing sperm (Holt, Fazeli 2016). Further understanding the dynamics of female choice, from biochemical mechanism to evolutionary consequences, will be critical to successful implementations of assisted reproduction as well as a clearer understanding of sexual selection.

After conception, the role of males in reproduction is severely limited, but that of females expands greatly. Gestation and the hallmark of mammalian reproduction, lactation, are both key to mammalian survival, and both are the purview of females. To these we now turn our attention.

CHAPTER SEVEN

Gestation

Conception to Birth or Hatching

After mating our female hyena undergoes a nearly 4-month (110-day) period of gestation. Her pregnancy is largely unnoticeable to the untrained human eye, but all the while much is occurring in utero. Her embryos implanted normally, and she is pregnant with twin daughters. A hemochorial placenta forms, unlike the endotheliochorial placenta of other Carnivora. The sex of her offspring may vary, depending on her social situation. Before group fission, her in utero offspring would more likely have been sons, but after fission, as now, daughters prevail. She is hungrier, and part of the food she eats supports her pregnancy while another part is deposited as fat to prepare for the even more energetically challenging stage of lactation. (Frank, Glickman 1994; Gombe 1985; Harrison Matthews 1939; Holekamp, Smale 1995)

One generation inside another.—Avise 2013:3

Gestation is a period of embryonic development and growth that occurs inside a mother's body. Gestation is not unique to mammals. Whenever conception is internal, some form of in utero development occurs. For instance, internal conception occurs in insects, some bony fish, sharks, snakes, lizards, birds, and presumably their kin, the dinosaurs. Many of these animals lay eggs, as do some mammals, e.g., the platypus and echidnas (monotremes). At least 150 vertebrate lineages give birth to their young (viviparity) as do the therian mammals: marsupials and eutherians (Blackburn 2015).

Gestation in therians has two distinct phases separated by implantation (the initial connection of the conceptus to the uterus). Most of gestation in marsupials occurs in the pre-implantation phase, whereas that for eutherians is predominantly after implantation. Besides phylogenetic differences, physiological and evolutionary processes also vary between the two phases. However, all therians have both phases, just with different emphasis. Across mammals, in utero development differs for mammals that lay eggs relative to those that give birth, and we will discuss the two modes separately starting with the egg-laying monotremes.

Conception to Hatching: The Monotreme Equivalent of Gestation

Conception to hatching in monotremes is considered the equivalent of therian gestation (Beard, Grigg 2000) and has two phases, but these phases are separated by egg-laying rather than by implantation. Before eggs are laid, they and the enclosed embryos develop inside the mother's uterus. After egg laying, incubation occurs outside her body, either within a maternal pouch, echidnas, or in an incubation nest, platypus. Although egg laying divides monotreme gestation into phases, egg laying is neither developmentally nor physiologically similar to implantation.

Developmentally, monotreme embryos mature well past the blastocyst stage before egg laying, whereas in eutherians extra-embryonic structures develop before implantation and the embryo proper develops after implantation. Physiologically, control of development in monotremes remains with the mother with little or no influence from the offspring both before and after egg laying. In contrast, for therians, implantation and placental development increase the fetal input into reproduction. Thus, in therians, greater fetal input changes the evolutionary dynamic pre- versus post-implantation.

Is egg laying or hatching comparable to birth? For mothers, egg laying, rather than hatching, is more similar to therian birth. After egg laying, a monotreme mother can abandon her eggs easily if conditions become unfavorable. But if conditions are favorable she keeps her eggs with her or is tied to her nest; if the latter her foraging and anti-predator options are limited. From the perspective of the offspring, egg laying also has similarities to therian birth. Before egg laying, embryos inside eggs are surrounded by a warm, moist environment protected from desiccation and microbial attack. After egg laying, the embryo is still nested within its shelled cocoon but the shell itself is vulnerable. However, the embryo is also trapped inside its shell and, in this regard, has not been born.

Overall, monotreme reproduction from conception to hatching bears only superficial similarities to gestation in therians. The egg-laying reproductive mode has its own suite of developmental, physiological, and evolutionary processes that differ from those of therians. Shelled eggs are the most distinctive feature of reproduction in platypus and echidnas. That shell filters interactions between mother and offspring both in utero and after egg laying. Unfortunately, the nature and extent of filtering is unclear because monotremes do not reproduce readily in captivity. However, preserved material, detailed field studies, and a few observations of captive individuals have illuminated features of monotreme reproduction.

After conception, shell deposition is a critical part of in utero development. As the zygote transits the oviduct, oviductal secretions wrap the developing conceptus in protective layers culminating in a solid shell. However, that shell must also allow nutrient transfer, because at conception, monotreme embryos do not have sufficient yolk to last through to hatching. Extra nutrients are transferred from uterine glands into the shelled egg (Hughes 1993). Thus, the egg shell must also be distensible. In fact, the egg expands about four-fold from conception to laying (4–5 mm to 15–20 mm; Jenkins 1990). During the 2–3 weeks the egg is within the mother (Grützner et al. 2008), the embryo develops a head with a brain and the beginnings of a segmented spinal cord, but it has little else: no eyes, limbs, internal organs (Hughes 1993), and this is the embryonic stage at which the egg is laid. Thus, much embryogenesis actually occurs after the egg is laid, during the 9–11 days of incubation and the 4–5 months of lactation (Ashwell 2013).

At the time of egg laying, a platypus embryo is the equivalent age of an 18- to 55-day-old human embryo. The great range in human age is because the structural features of a platypus embryo do not develop at the same rate or sequence as those of humans. For example the neural development of the platypus embryo at egg laying is similar to that of an 18-day-old human embryo, but the actual size of the platypus embryo (14 mm) is similar to that of a 55-day-old human embryo (Hughes, Hall 1998; Papaioannou et al. 2010).

What about placenta-like structures? Before egg laying, the only extra-embryonic membrane is the yolk sac. The amnion, chorion, and allantois are not identifiable. The yolk sac is the only functional organ in the entire in utero egg as none of the other embryonic systems are sufficiently formed. While the egg is within the uterus, the yolk sac rests next to the shell and transfers nutrients and water from the mother to the embryo. Thus, the yolk sac is the intermediary tissue between the mother and her offspring; if the shell were not present, embryologists would call the yolk sac a placenta (Hughes 1993).

Once laid, the egg is surrounded by air and not uterine fluid. Desiccation is a major threat. Some scenarios for the evolution of milk posit that early mammals sweated on their eggs, and this sweat eventually became milk. Moist eggs in a dark environment would also attract fungal infection. The anti-microbial properties of sweat underlie another hypothesis regarding the evolution of milk, that the proto-lacteal secretions started with an anti-microbial function and covered the eggs during incubation (Hayssen, Blackburn 1985).

The potential for material from the mother to cross the egg shell during incubation is clear. How much this happens in monotremes is not known, but the development of extra-embryonic membranes during incubation suggests some maternal transfer occurs. Certainly, the yolk sac is the major organ for gas exchange for the first half of incubation and becomes populated with blood vessels that transfer oxygen to the embryo and release carbon dioxide. Nutrients for development during incubation come from material within the yolk sac. This material was deposited by the mother during oogenesis and is augmented during in utero development. As incubation proceeds, the other extra-embryonic membranes, the amnion, chorion, and allantois, develop and take on functional roles in waste collection (allantois) and gas exchange (chorion). These membranes also rest against the egg shell where they receive materials from either the mother or the environment and filter them before they reach the embryo.

Platypus mothers lay their eggs in a burrow and thus the eggs rest against not only the mother but also the soil or whatever other materials the mother brings to the nest. In contrast, echidna mothers deposit their egg in a transient pouch formed on the mother's ventral surface. Thus, the eggs of echidnas are incubated while enfolded by maternal skin and may even hatch there (Beard, Grigg 2000).

For both echidnas and the platypus, young continue development inside the egg during incubation. Limbs develop. Eyes and ears form. Internal organs become prominent. But overall, the embryo is still just that, an embryo, with only enough functional tissues to break free of the egg and to lap up and then digest milk once out of the shell. As hatching nears, the embryo develops a bony protuberance on its upturned nose, which is capped with cornified tissues, something like a miniature claw or horn. This oscaruncle on the nose, in addition to a sharp, specialized tooth (an egg tooth) in the mouth, help the embryo to tear the membranous egg shell and allow the embryo to hatch. The proximate stimulus for hatching is not known but ultimately the yolk and maternal provisioning through the egg shell are insufficient for development: the embryo must hatch or die.

The hatched, 9–17 mm embryo is blind, deaf, and naked. Its forelimbs can grasp, but its hind limbs are mere paddles. The young will be completely dependent on its mother for about 4 months for platypus young (Holland, Jackson 2002) and from 4 to 7 months in echidnas (Rismiller, McKelvey 2009). Platypus mothers stay burrowed with their young for all of incubation and the first month or so of lactation (Holland, Jackson 2002). Thus, they fast for about 40 days. Echidnas, however, carry their pouched young with them and can forage as they incubate. Echidnas will also use nursery burrows after hatching, especially as the young develop their quills (*Tachyglossus*; Beard, Grigg 2000; Morrow et al. 2009; Rissmiller, McKelvey 2009).

Overall, embryogenesis in egg-laying mammals has three phases: in utero, incubation, and the first part of lactation. Neonates are exquisitely developed to hatch and can immediately imbibe and digest milk, but they are completely unable to do much of anything else. Their gonads have not developed, but they can breathe, suckle, and excrete wastes. They look as though survival outside the womb would be impossible, but they are precisely adapted to life in the pouch or nest. Egg-laying mammals are not "primitive," but extremely well adapted to their circumstances.

Whereas monotremes provision their embryos three ways; marsupials use four. For marsupials, as with monotremes, nutrients are deposited in the yolk sac during embryogenesis as well as through uterine secretions and during lactation. Unlike monotremes, marsupials also have direct transfer of nutrients from maternal blood to embryos via placental structures (Renfree 2010). Like marsupials,

	 🐢	 🦘	🐭
Yolk	✓	✓	
Secretions	✓✓	✓✓	✓
Placentation		✓✓	✓✓✓✓
Lactation	✓✓✓	✓✓✓✓	✓✓✓

Figure 7.1. Maternal nutritional provisioning of young by reproductive mode. Each of the three lineages of mammals has a distinctive combination of nutrient provisioning during reproduction. Check marks indicate that females use that type of provisioning. Number of checks is a qualitative assessment of the relative amount of provisioning. *Left to right*: Lineages—monotremes, marsupials, eutherians. *Top to bottom*: Provisioning—yolk, uterine secretions, placentation, and lactation (milk). Illustrations compiled by Teri Orr, images from Microsoft clip art.

eutherians use placental transfer to exchange gases and nutrients between mother and offspring. Unlike marsupials, eutherians do not use yolk to provide the significant starting materials for development before implantation (figure 7.1). Of course, another salient feature distinguishes marsupials and eutherians from monotremes: viviparity.

Conception to Birth: Gestation and Viviparity

In viviparous mammals, gestation has two major phases. The first phase is from conception to implantation and occurs without the aid of a placenta. The second phase (placentation) is from implantation to birth. During the first, pre-implantation phase, maternal physiology dominates both the sequence and the timing of development not only via oviductal and uterine secretions but also via the ovum. Well before conception, the major material needed to regulate and sustain early development was deposited into the ovum. Thus, the maternal genome regulates much of the first phase of gestation. After implantation, the embryonic genome and the maternal physiology interact to promote placentation and development. During placentation, fetal and maternal tissues intricately intertwine, but always remain separate to some degree. This intimacy gives the conceptus some influence over the course of gestation. In addition, pregnancy presents an opportunity for interactions among siblings in utero. These sibling interactions can have long-term effects, as we will discuss later. Overall, pregnancy involves complex interactions among multiple individuals.

For marsupials the majority of in utero development occurs before implantation. In contrast, for eutherians most development occurs after implantation. Nonetheless, both groups have both phases. Placental structures are present in marsupials and eutherians and thus both are placental mammals. Therefore, as we said earlier, identifying eutherian mammals as placental mammals is misleading at best.

What about the duration of each phase? First, the time between conception and implantation is much more similar across eutherian mammals than the time between implantation and birth. For instance, implantation occurs four days after conception in lab mice, at nine days in women, and between 40 and 50 days in African elephants (*Loxodonta*; Hildebrandt et al. 2007; Lee, DeMayo 2004). Thus, the longest pre-implantation period is about 10 times longer than the shortest. In contrast, the longest total gestation length is about 40–60 times longer than the shortest: the shortest is about 11–17 days in European hamsters (*Cricetulus, Cricetus*) and perhaps some shrews (*Blarina, Crocidura, Cryptotis*), and the longest is about 660 days in elephants (Hayssen et al. 1993). Looked at another way, across eutherians, pre-implantation occupies about the first 3% of gestation in women, the first 8% in elephants, the first 20% in lab mice, and the first 30% in dogs, but in metatherians. pre-implantation is the first 50–67% of gestation (Renfree 2010; Verstegen-Onclin, Verstegen 2008). These numbers demonstrate a second point about timing: that the time before implantation is not related to the time after implantation. This independence suggests that the two periods accomplish divergent goals for different mammals, as we will show. The question of timing has one complication: some females can delay implantation. This complication is explored toward the end of this chapter.

Implantation is not just a practical way to divide gestation. The timing of implantation has significant functional and evolutionary consequences for the physiology of gestation. This includes consequences for the timing of gestation as well as the types of interactions that can occur between mother and offspring. In all therian mammals, environmental stress may occur before and/or after implantation, the consequences of which may differ across mammals. Rodent mothers that experience stress before implantation tend to reject their embryos, whereas mothers who are stressed after implantation tend maintain litter size but have smaller neonates (Brunton 2013). Thus, implantation changes the dynamic of gestation in many ways.

Conception to Implantation

Pre-implantation is a period of embryonic cell division, much of which occurs while the zygote is still enclosed within a maternally derived, protective membrane, the zona pellucida. The zona is a non-cellular, shell-like coat that surrounds the developing zygote and keeps it from falling apart during the earliest cell divisions. We described the zona and its functions in chapter 6.

A case study will give a feel for the complex interactions between the conceptus and the mother before implantation. Let us examine *Equus*. Not only are horses well known as companion animals, draft animals, and racing animals, but because of the economics associated with breeding horses, their reproductive biology is well studied and synthetically reviewed (Davies-Morel 2008). We start just as the ovum is released from the ovary.

After ovulation, the ovum is transported by ciliary action and by current flow a short distance down the oviducts. Assuming mating has occurred, uterine contractions transport sperm from the cervix through the uterus and into the oviducts. Chemical attractants may be produced by the ovum to direct sperm motility. Uterine secretions also activate sperm and not only allow sperm to move through the zona pellucida but also facilitate sperm-ovum adhesion and conception. For the next 4 to 5 days, the new conceptus undergoes multiple cell divisions using material deposited by the mother. At this point, the conceptus still has its maternal coat, the zona pellucida and, thusly protected, remains within the oviduct while the uterus is cleared of excess sperm and ejaculate. The conceptus transits to the utero-tubal junction where it secretes a small bit of prostaglandin, which acts locally to relax the smooth muscle of the junction and allow the conceptus to pass into the uterine horn. At this stage, day 6, the conceptus begins to enlarge as opposed to just multiply in cell number. It also secretes a non-cellular capsule of its own for protection as the zona disintegrates, no longer able to contain the larger conceptus. The conceptus, floating in the uterus, can now absorb nutrients from uterine secretions tailored to fuel embryo development. But not until day 8 is the conceptus differentiated enough for scientists to detect which part of the ball of cells will become the embryo as opposed to trophoblast. Before day 8, each cell in the conceptus is capable of becoming a foal; after day 8 individual cells have specific functions and fates (Davies-Morel 2008).

The chemical nature of the non-cellular capsule secreted by the conceptus attracts proteins and other uterine secretions. This attraction moves nutrients closer to the developing tissues of the conceptus. For the next 7–10 days, the blastocyst remains unattached in the uterus and moves between uterine horns and around the central body of the uterus (Davies-Morel 2008). The conceptus may travel between uterine horns 12–15 times per day (Bazer et al. 2009). This mobility is necessary for the continuation of pregnancy, but exactly how conceptus mobility translates to continued gestation is unknown (Klein, Troedsson 2011). The mobile phase is regulated by both localized secretion of prostaglandin from the conceptus as well as by general secretions of progesterone from the maternal ovary. Thus, both conceptus and mother regulate the activity of the uterus. As the conceptus enlarges, its reduced mobility is exacerbated by internal changes in uterine water balance causing swelling of the uterus and restricting the internal space available for movement. Finally, 18–20 days after conception, the now stationary conceptus tenuously attaches to the uterus, usually at the junction of one uterine horn and the body of the uterus. This initial attachment is fragile and not until about days 25–35 of pregnancy do placental growth, development, and integration of the conceptus with the uterus occur (Davies-Morel 2008).

So when exactly is implantation in *Equus*? Is it at day 18, with the initial, fragile attachment, or at day 35, when attachment is obviously substantive? Like much of reproductive biology, questions such as "When does implantation occur?"—seemingly clear and concise questions—become muddled with caveats and specific cases. Implantation marks a transition from the relative autonomy of offspring and mother to an intricate fetal and maternal interdependence. The exact timing of the transition is less important than its functional consequences. Before implantation, the maternal physiology alone, more or less, determines if pregnancy will continue, after implantation, the conceptus and its siblings can influence that process via placental attachment.

Not only horses but other females can restrict implantation by physically changing the uterus. For instance, in short-tailed fruit bats (*Carollia*), as the conceptus travels down the oviduct, females close the uterine cavity (by allowing the uterine tissues to swell) so that only the area closest to the oviduct is open. This forces implantation in that particular region. Once implantation occurs, the uterus re-opens allowing a larger area for fetal development (Oliveira et al. 2000). Thus, for both horses and *Carollia*, uterine swelling restricts where implantation will occur.

As described for horses, the very earliest stages of development in mammals are regulated by materials of maternal origin within the ovum. The contents of the ovum at conception create, activate, and control the genome of the conceptus (Mtango et al. 2008). As development proceeds, the conceptus genome regulates its own maturation. Therefore, initial development is under maternal control, but subsequently, the genome of the conceptus takes over. Before implantation, the conceptus is mostly trophoblast with only a minority of cells destined to become an embryo. Nonetheless, the pre-implantation blastocyst is not simply an inert ball of cells. For many species, the blastocyst must chemically signal its presence to the mother and in doing so maintain pregnancy. Tangentially some of these early chemical signals may be sexually dimorphic, perhaps allowing a mother to choose the sex of her offspring. The details of the maternal recognition of pregnancy were explored in chapter 5 on physiology.

Finally, although the conceptus is tiny and the mother may not even recognize that she is pregnant, conditions during this period can alter the physiology of the offspring. For instance, if female mice are put on restricted diets during implantation, their offspring develop high blood pressure (Lesse 2012). Embryo rejection before or during implantation is common and often a result of fetal defects or environmental stress (Hayssen 1984). It may also result from pheromonal cues (Bruce 1959; sidebar on page 126).

Conceptus Changes: Blastocyst Structure and Function

Before we start this section on changes in the conceptus, a note on terminology is warranted. Here we use the term *blastocyst* to broadly refer to the conceptus from conception until implantation. Embryologists have a more narrow definition of the

Bruce Was a Woman
Hilda Margaret Bruce (1903–1974)

Hilda Bruce devoted much of her life to biological research. Her educational background was indicative of her times. Her terminal degrees were a bachelor of science in Household and Social Sciences followed by a bachelor of science in Physiology from Kings College for Women (Parkes 1977). Her career started in teaching, but in 1930, she began work at the National Institute for Medical Research (NIMR) in Hampstead, England. Her initial work was on vitamin D (eight co-authored articles). She also was one of the first to do research with golden hamsters (*Mesocricetus*) newly obtained from the Mideast and destined to become a commonly used laboratory rodent. Between 1933 and 1944, she continued work on vitamin D in several laboratories before returning to the NIMR at the request of A.S. Parkes to coordinate work using diverse laboratory animals. Early publications with titles such as "Cereals and Rickets: The Role of Inositolhexaphosphoric Acid" would not lead the reader to suspect a substantive contribution to reproductive biology.

In coordinating breeding colonies of laboratory animals of various species and strains, Bruce knew that issues of sexual dominance, preferences, and differences in breeding systems altered the success of reproductive efforts. She also knew that having a cohort of females at the same stage of gestation would be useful, but to achieve that outcome, females would need to have synchronous periods of estrus. In the 1950s, she began to examine some of these issues. She conducted a series of mating trials with and without oral steroids and noticed that the presence of a new male a day after a female mated with a different male could synchronize estrus in female mice; no steroids needed. When presented with a new male, a female stops the first pregnancy and starts a new one with the new male. Even just the olfactory cues of a new male are sufficient for a female to block implantation. This olfactory block to implantation is now known as the Bruce effect. Her finding "was a major discovery and an outstanding example of disciplined research" (Parkes 1977:2) and brought her the Oliver Bird Medal.

In several sole-author articles, Bruce described the olfactory stimulation of a neuro-response in mice. Additional work by Bruce explored the suckling response of lactation in rats, the role of the thyroid in reproduction, and the effect of current lactation on development of embryos (again in mice). In 1963, Bruce retired from the NIMR, but her research continued as a part-time investigator at the Department of Investigative Medicine at Cambridge until 1973, where she once again focused on nutrition, development, and, of course, pheromones. Her long career occurred when being a woman in the sciences was uncommon, and through that career she substantively advanced the field of reproductive physiology.

(Photo courtesy of Journal of Reproduction and Fertility, used with permission.)

blastocyst and use additional terms for the developmental stages between conception and implantation. Whatever the name, the early structure eventually differentiates into two components: first, the embryo proper and, second, other tissues collectively called trophoblast. Trophoblast cells differentiate into extra-embryonic membranes and subsequently placental tissues.

During very early development, the blastocyst changes its biochemistry to facilitate the maternal recognition of pregnancy. In addition to chemical activity, the blastocyst physically changes shape. The earliest cell divisions occur within the confines of the zona pellucida and occur without any additional nutrients. Thus, the conceptus increases in cell number but not size. However, once the zona disintegrates, the conceptus can absorb uterine secretions and cells can increase in size as well as in number. As the conceptus enlarges it changes. For instance, the blastocyst of the short-tailed fruit-bat (*Carollia*) has abundant microvilli (small projections) covering its surface, and these microvilli change in number and character. Why? The obvious answer is that they aid in implantation, but the microvilli may simply increase surface area for absorption of nutrients or be a membrane reservoir either for further cell division or for elaboration of internal cellular structures (Oliveira et al. 2000). The take-home message in this case is that when thinking about the functional relevance of a specific structure, the obvious function may not be the only one.

Despite being commonly depicted as such, the blastocyst need not remain a sphere. For instance, in pigs the spherical conceptus elongates to become a mere thread at implantation (Bazer et al. 2009). The thin (1 mm in diameter) filament can reach well over a meter in length (figure 7.2). The thread-like conceptus twists and folds during elongation and extends along the elaborate, rippled internal architecture of the uterus; a uterus with so much internal structure that the meter-long conceptus occupies only 10–20 cm of uterine space (Englehardt et al. 2002; Perry, Rowlands 1962). Most of the thread is trophoblast, cells that will become placenta; the actual embryonic portion of this thread at implantation is only the tiniest bulge somewhere along its length.

Also amazing is that the elongation of the blastocyst, from a sphere to a 1-m-plus filament, occurs over only four days, which, if you do the math, converts to at least 10 mm per hour. Actual measurements of filaments elongating from 10 to 150 mm, suggest even faster (30–45 mm per hour) rates. This increase in length occurs without an increase in the number of cells (Geisert et al. 1982). Thus, early in development, the conceptus increases in cell number but not in overall size but later the conceptus increases in overall size but not in cell number. Eventually both cell number and overall size increase. With all the activity involved in elongation, the metabolism of the conceptus changes from nearly inactive at conception to extremely active at implantation (Lesse 2012). Much of this metabolic activity is fueled by mitochondrial action, mitochondria that are likely entirely of maternal origin but which, presumably, are now under embryonic control.

Filamentous blastocysts are not confined to pigs but also occur in other artiodactyls such as cattle (*Bos*), deer (*Cervus*), and pronghorn antelope (*Antilocapra*), although they may not reach the lengths of the suid threads (Clemente et al. 2009; Demmers et al. 2000; O'Gara 1969). In cattle, filament elongation

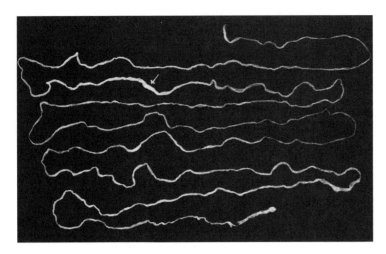

Figure 7.2. The elongated, thread-like blastocyst of a 13-day domestic pig (*Sus*) conceptus. The arrow marks the position of the embryo. The original image is stated to be actual size, and, to provide an idea of scale, the arrow was 4 mm in length. From Perry, Rowlands 1962; reproduced with permission.

may occur even if the embryo is missing. Thus, growth is either regulated by the trophoblast, by uterine physiology, or by some combination (Clemente et al. 2009).

Before we leave filamentous blastocysts, one unusual case deserves mention. The filaments in pronghorns have a murderous function. Pronghorns give birth to twins but ovulate three to seven ova, some from each ovary. Conception occurs for all these ova and the resultant blastocysts elongate into threads that move through the long uterine horns. Some become tangled and die as they elongate. Others implant. The first blastocysts to implant lengthen their filaments. These filaments pierce and effectively kill any and all other blastocysts in that uterine horn; however, the maternal uterine architecture prevents blastocysts from crossing into the opposite uterine horn. Thus, one embryo survives in each horn (O'Gara 1969). In utero competition may be enhanced because 44% of these twins have different fathers (Carling et al. 2003). In the end, pronghorns routinely give birth to twins by the unlikely mechanism of in utero siblicide.

Implantation

Implantation marks the transition between a somewhat autonomous, free-floating blastocyst and a blastocyst with a firm attachment to, and dependency on, the maternal uterus. We have discussed the timing of implantation but not the developmental condition of the conceptus. As with timing, developmental condition also varies. In elephant shrews (*Elephantulus*), implantation occurs when the conceptus is extremely small—when it has only four cells (van der Horst, Gillman 1942)—although more details on this interesting observation are needed. In contrast, the bandicoot (*Perameles*) embryo may have a crown-rump length of 1 cm at the time of implantation (Padykula, Taylor 1983), and the horse conceptus is already 3–8 cm in diameter 10–14 days before implantation even begins (Betteridge et al. 1982). Besides timing and development, implantation varies according to how the conceptus is oriented when implantation occurs as well as where in the uterus the conceptus implants.

The relationship between the conceptus and uterus has three different orientations: centric, eccentric, or interstitial. In centric (central) implantation, the blastocyst enlarges and makes contact with large areas of the uterine surface, as occurs in rabbits (*Oryctolagus*), dogs (*Canis*), cattle (*Bos*), and pigs (*Sus*). In eccentric (off-center, partial) implantation, only part of the uterus enfolds a portion of the blastocyst, as in mice (*Mus*), rats (*Rattus*), and hamsters (*Mesocricetus*). Finally, interstitial (enveloped) implantation involves a complete encapsulation of the blastocyst by the uterine wall, for example in guinea pigs (*Cavia*) and humans (Lee, Demayo 2004). Although differences in placement are species specific, their functional consequences are unclear. The placement may simply be a by-product of uterine morphology, growth of the placenta, or development of the embryo. Also, as the embryo grows the initial orientation is obscured. Probably, the cellular interactions between fetal and maternal tissues are of more functional consequence than the initial orientation at implantation although they may be interrelated.

In regards to the location of implantation, in many species the uterus is unable to support embryos along its entire length. Instead, the uterine lining (endometrium) may have a specific, and limited, number of implantation sites. For instance, in ewes and cows the endometrium is broken up by a number of raised caruncles that are the sites of implantation (Gray et al. 2001), whereas in rabbits, implantation can occur almost anywhere in the uterus (Bautista et al. 2015). Also as mentioned earlier, in mares implantation preferentially occurs at the junction of the uterine horns and the body of the uterus. Some of the differences across species may be due to species-specific growth rates or trajectories or to uterine circulatory patterns and the position of other organs in the peritoneal cavity. None of that is very well studied, but we do know that some differences are due to litter size.

Equids and bats are monotocous, that is, they generally conceive one foal or pup at a time. For bats, multiple births are rare and only occur in a few species. For twinning mares, embryo death is common. Sixty-one percent of twin pregnancies in mares result in the resorption of one twin, and in an additional 10% of twin pregnancies, both twins are resorbed (Chevalier-Clément 1989). However, females of many other species carry multiple offspring in a single pregnancy (box 4.2).

In females with litters (polytocous), implantations could be balanced across both uterine horns or asymmetrically positioned on one side or the other. Having similar numbers of embryos on each side may improve embryo survival by providing a less-crowded space for growth. It also might be a mechanism for females to control litter size. In addition, equal numbers of embryos may equalize weight distributions during pregnancy and make locomotion more efficient or predator evasion more possible, especially for females with saltatorial (jumping) or arboreal locomotion (Baird, Birney 1985). Balanced spacing might reduce competition among siblings for maternal resources. Species for which embryos are often more or less evenly spaced along the uterine horns include shrews (*Sorex*), dogs, rabbits, and guinea pigs (Baird, Birney 1985; Bruce, Wellstead 1992; Tsutsui et al. 2002).

Spacing can happen quite quickly. For example, in hares (*Lepus*), blastocysts enter the uterine horns five days after conception and peristalsis of the uterus moves them to and fro until they are equally spaced about two days later at which time the blastocysts implant (Drews et al. 2013). What exactly causes the equal spacing is unknown.

How can females ensure that approximately equal numbers of embryos implant on each side? What might influence the process? A major factor is the shape and floor plan of the uterus (see figure 4.6). An open bicornuate uterus, such as in shrews, carnivorans, and hyraxes, would be more conducive to transuterine migration. Migration of blastocysts across uterine horns results in a balanced distribution of embryos in shrews (*Sorex*), hyraxes (*Procavia*), and probably most other polytocous species with bicornuate uteri, e.g., raccoons (*Procyon*), mongooses (*Herpestes*), moles (*Scalopus*; Baird, Birney 1985). However, if the uterine horns are physically separated, as in the duplex uterus of rabbits, then migration of embryos from one side to the other would be impossible. Even partial separation might impede migration. For instance, some taxa, such as sheep, cattle, and pronghorn, have a septum between the uterine horns that makes migration difficult.

If migration of blastocysts is prevented by uterine architecture, then balancing embryos between left and right may rely on ovulation numbers in each ovary or on "programmed" embryonic mortality as described earlier for pronghorns. Embryonic mortality is routine in species that ovulate many more ova than the resultant litter size at birth. Besides pronghorn, these include elephant shrews (*Elephantulus*), tenrecs (*Hemicentetes*), striped hamsters (*Cricetulus*), and the plains viscacha (*Lagostromus maximus*; Wimsatt 1975; details in chapter 6). A similar phenomena occurs in those polytocous marsupials who give birth to more young than the number of teats available. In the marsupial case, neonatal mortality determines litter size. Embryo mortality can be, and often is, a natural component of reproduction.

Besides using uterine morphology or embryonic mortality to balance the left-right distribution of embryos, females could achieve equal numbers of embryos on each side by having equal numbers of ovulations from each ovary. For this to happen, the two ovaries must be in synch so that each ovulates the same number of follicles simultaneously. Therefore, either the ovaries must be in communication with each other or some other factor could be involved that would synchronize the ovulations. For instance, mating influences ovulation in rabbits and subsequent changes in the reproductive tract associated with mating could influence both ovaries equally. However, the black agouti (*Dasyprocta fuliginosa*) has a duplex uterus, a litter size of two, and only one ovulation per ovary, but mating does not influence ovulation (Mayor et al. 2011). Therefore, some other trigger must be synchronizing left and right ovulations in black agoutis.

For evolutionary biologists, these differences pose interesting questions regarding the relationship between physiology and morphology. In rabbits, did the presence of a duplex uterus, which prevents blastocyst migration, promote the development of mating-influenced ovulation? Or did a prior influence of mating on ovulation promote the evolution of a duplex uterus? Which came first? Why does the black agouti have a duplex uterus when ancestral rodents such as beavers have bicornuate uteri? Was selection operating in the agouti to reduce litter size and thus reduce the number of ovulations and having a duplex uterus reinforced that litter size reduction? Generally, evolutionary biologists assume morphology constrains physiology, in that physiological processes are more dynamic than morphological structures. However, the dynamic nature of physiology could lead to selective advantages for specific morphological variants. As neither physiology nor soft-tissue morphology fossilize well, the stages through which reproductive adaptations occur are difficult to discern, thus they are a fertile field for investigation.

Of course, monotocous species must also coordinate ovarian function, not for controlling uterine balance, but for ensuring that only one ovulation occurs at a given cycle. Thus, monotocous species must have ipsilateral (same side) as well as contralateral (opposite side) control of follicular development. In other words, one ovary must prevent follicular maturation in the ovary on the opposite side (contralateral inhibition) as well as preventing all but one of its own follicles from maturing (ipsilateral inhibition). In some monotocous species, females alternate different pregnancies first on one side and then on the other. Alternatively, some just use one side, e.g., cetaceans and some bats. Bats are exceptionally variable in this regard and include species with either partial or complete dominance of either the left or right ovaries or the uterine horns (tabled in Rasweiler, Badwaik 2000). For instance, molossid bats have a bicornuate uterus, but only the left horn is functional (Crichton, Krutzsch 1987). Similar to bats, in the toothed whales, odontocetes, implantation is nearly always in the left uterine horn and the left ovary ovulates more often than the right (Slijper 1966). Other species also have reproductive asymmetries (left or right dominance), including, but not limited to, the platypus; several African antelope, *Aepyceros* (impala), *Kobus* (kob), *Madoqua* (dik-dik), *Nesotragus* (*Neotragus*, suni), *Sylvicapra* (duiker), and the mountain viscacha, *Lagidium* (Ashwell 2013; Loskutoff et al. 1990; Mossman, Duke 1973). An interesting example of this variation comes from cattle. Although in beef cows implantation is more often in the left uterus than in the right, in dairy cows the distribution is equal (Gharagozlou et al. 2013). The physiological complexities of coordinating ovulation and implantation with litter size are one reason why litter size (polytocous vs. monotocous) often characterizes mammals at high taxonomic levels. For instance, most whales, bats, and primates are monotocous, but most carnivorans, lagomorphs, shrews, and rodents are polytocous.

In addition to regulating the number of offspring in each uterine horn, some female physiologies can influence the placement of embryos by sex. For example, in Mongolian gerbils (*Meriones unguiculatus*), more male fetuses are located in the right horn and more female fetuses in the left horn. Transplantation experiments, moving ovaries from one side to the other, indicate that the ovary is driving this bias, but the details of how this occurs are an enigma (Clark et al. 1994). Regardless of the proximate causes, a key consequence of this segregation is that males are surrounded by males and females by females while developing in utero. As a result, hormones are often shared by embryos of the same sex.

Hormonal, or even cellular exchanges, among siblings in utero will depend not only on uterine morphology and location of embryos but also on the degree of crowding and potential mixing of placental tissues. Such intermixing of placental tissues is well documented in cattle twins. When the twins are of different sexes, the female twin develops behaviors and growth patterns associated with steers. These females, called freemartins, are the normal result of mixed-sex twins in cattle, but this same type of female has been observed in other farmed ungulates, such as sheep, goats, deer, pigs, horses, and camels (Padula 2005).

Thus, even the seemingly simple step of implantation has lasting consequences. Uterine position and the sex of adjacent siblings may influence neonatal size, degree of sex dimorphism, postnatal growth, and behavior, as well as adult physiology (Bautista et al. 2015). Implantation is a critical event in the reproductive cycle, as it marks the end of simple offspring-mother interactions and sets the stage for the subsequent intricate intimacy of the placental phase of pregnancy.

Gestation

Gestation is usually defined as the interval between conception and birth. As only limited growth occurs before implantation in eutherians, the duration of gestation is mostly determined by the time between implantation and birth. How long that interval lasts primarily reflects the growth of the offspring and their ability to survive after birth.

Faster growth will lead to shorter gestations as well as an increased ability to survive outside the womb. Neonates differ greatly in their developmental condition at birth and their environmental protection after birth (chapter 8). For instance, the "embryonic" neonate of a kangaroo (*Macropus*) lies protected in a pouch after a 27-day gestation, while the neonatal wildebeest calf (*Connochaetes*) runs after her mother after a 270-day gestation. The ability of mothers to support young at different developmental states also varies. For instance, a wildebeest mother has no pouch to support an "embryonic" neonate; likewise, a kangaroo mother does not have the appropriate internal anatomy to keep a neonate within her womb to a mature state. Across a broad range of species, gestation length tends to be longer when well-developed neonates are the result (Case 1978). But neither body size nor the simplistic altricial-precocial dichotomy explain the large variation in gestation lengths.

Both fetal development and maternal physiology and anatomy have been co-evolving for a very long time. As such, the length of gestation has a large genetic component. An obvious example is the generally shorter gestations of marsupials compared with eutherians of the same size (Hayssen et al. 1985). Although, much of the genetic component is due to selection pressures on different taxa, some has to do with other constraints, such as body size. That being said, gestation length is not entirely explained by body size (Clauss et al. 2014). For instance, elephants (*Elaphas*, *Loxodonta*) have a longer gestation length (650 days) than the much larger blue whale, *Balaenoptera musculus* (330 days; Hayssen et al. 1993). Lest you think the marine environment is a factor, the blue whale's pregnancy is shorter than that of the smaller killer whale (*Orcinus*; 517 days; Robeck et al. 2004). At the opposite extreme, bats, despite their diminutive size, have long gestation lengths; for instance, the 10-g *Myotis moluccarum* has a gestation length of about 80 days without a delay, whereas gestation in the 10-g shrew (*Sorex araneus*) is under 21 days (Barclay et al. 2000; Hayssen et al. 1993; Lloyd et al. 1999). A final example that illustrates the independence of size and gestation length is a comparison of leporids and canids. In the 1–5 kg range, cottontails (*Sylvilagus*), rabbits, and hares (*Lepus*) overlap in body size with some of the smaller canids (e.g., foxes, *Vulpes*), but gestation in leporids is 38 days, whereas in canids, it is 55 days (Pielmeier et al., submitted). In fact, hares produce well-developed precocial young faster than canids generate their altricial young. The lagomorph-canid example suggests that diet may be a factor but leporids and canids also differ in number of litters per year, litter size, and neonatal mass; all factors that may alter gestation length.

Not surprising, then, parsing out how gestation lengths relate to environment, physiology or ancestry is difficult. To determine these relationships, one would need to study species within a single taxonomic group that differ in ecology, physiology, or both. One such group, the pikas (*Ochotona*) is a diverse genus with species living in talus slopes as well as species that live in more generalized alpine areas. The talus-dwelling species have smaller litter sizes, fewer litters per year, lighter litter mass, and longer gestation and lactation lengths, suggesting that habitat can also alter gestation length (Hayssen, in prep; Smith 1988). Thus, gestation length can be influenced by environmental circumstances across species.

Regardless of differences across species in the length of gestation, variation among individuals within a species still occurs. For instance, lactation can extend gestation in females that simultaneously nurse one litter while being pregnant with the next, as discussed in the following section on delays. In addition, females within a social group may alter gestation length such that births are more synchronous. For instance, red deer hinds (*Cervus elaphus*) that conceived early in the mating period had longer gestations than those who conceived later in the mating period (Asher 2011).

The environment may also alter the gestation length of individuals. Females may respond to adverse weather or climate with small changes in gestation length. For instance, in a 35-year study of red deer, gestation was three-quarters of a day shorter for every degree that March temperatures were warmer (Clements et al. 2011) and pronghorn females had longer gestations and slower prenatal growth rates after dry summers (Byers, Hogg 1995). A different study of pronghorn, suggests that nutrient availability may also alter in utero growth. In this case, spring-summer nutrition resulted in bigger offspring as well as reduced neonatal mortality, but only for some females (Barnowe-Meyer et al. 2011). Fur seals (*Arctocephalus*) also have longer gestations when less food is available (Boyd 1996).

Does gestation length differ between fetal sexes? For wild populations, data are difficult to obtain. A few studies have obtained valuable, but limited, data all of which suggest that the answer is no. First, neither bison (*Bison*; Berger 1992), reindeer (*Rangifer*; Loison, Strand 2005), nor fur seals (*Arctocephalus*; Boyd 1996) exhibit sex differences in gestation length. Second, although early studies indicated that red deer gestate male offspring for longer than female offspring, no effect of progeny sex was found in a later, more extensive study (Clements et al. 2011). The question of sex differences in gestation length is part of the larger question of sex differences in maternal care, an intensely studied topic outside the scope of this book.

Pre- and Post-Implantation Delays and Diapause

From a human perspective, pregnancy proceeds without interruption from mating to birth. You have sex, conceive, and 9 months later you give birth. This relatively invariant link between conception and birth exists for most other mammals, but for some, the timing between mating, conception, and birth is unhinged. In these species, females can insert delays into the sequence. The delays can be either facultative, that is, variable in presence and duration, or obligate, that is, an established component of every pregnancy. Delays come in three forms: (1) females delay conception by storing sperm after mating, (2) they delay implantation by holding blastocysts in abeyance, and (3) they prolong gestation by retarding development of the conceptus. We will briefly discuss each method.

Delays between mating and conception involve the storage of sperm. Of course, all animals with internal conception store sperm however briefly. Thus, even if only for a few hours, all female mammals store sperm (Orr, Brennan 2015). The more interesting issue is longer-term storage on the order of days to months. For instance, female eastern quolls (*Dasyurus viverrinus*) and brown antechinus (*Antechinus stuartii*) store sperm in oviductal crypts for 14–16 days, whereas female little brown bats (*Myotis lucifugus*) store sperm in their uterotubal junctions for up to 138 days (Orr, Brennan 2015). Most mammals lack obvious sperm storage structures; however, some bats with long periods of sperm storage use pre-existing structures without substantial modifications. The lack of obvious sperm-storage structures suggests that more females may store sperm than we have documented and that females may store sperm on a short-term basis, probably to serve a variety of functions. For instance, females could use sperm storage to gather sperm from different males and enhance the genetic diversity of a litter.

Once female sperm storage ends and conception occurs, additional delays are possible. Delays between conception and implantation are a prominent feature of gestation in at least 130 species: 30–40 marsupials and 95 eutherians (Renfree 2006; Fenelon et al. 2014). Although often termed *embryonic* diapause, the barely differentiated blastocyst (i.e., pre-embryo) is the usual stage at which development is arrested. Delays in implantation can be long and may last up to 11 months as in the tammar wallaby (*Macropus eugenii*) or the European badger (*Meles meles*; Renfree 2010; Yamaguch et al. 2006). In an ovariectomized tammar, diapausing embryos may be viable for 2 years (Renfree, Shaw 2000). Delayed implantation can also be facultative.

Facultative delays are often due to nursing older progeny. In kangaroos, blastocyst cell division and growth may be arrested either by the suckling of an older sibling or by maternal exposure to harsh environmental conditions (Renfree 2010; Sharman, Berger 1969). The delay can be terminated by photoperiod or by removal of an older offspring's suckling stimulus, and progesterone is the hormonal key (Renfree 2006). In eutherians, developmental arrest due to the neurohormonal correlates of suckling is also common in laboratory rodents, *Mus* and *Rattus* (Gidley-Baird 1981)

After implantation yet another (but less common) delay may occur. A few bats, e.g., *Artibeus, Cynopterus, Haplonycteris, Macrotus*, as well as some of the Carnivora, e.g., *Phoca*, slow the development of their offspring, a process termed *delayed development*, or *post-implantation diapause* (Heideman 1989; Wilson et al. 1991; Banerjee et al. 2009).

Pregnancies with and without diapause may alternate. For instance, in two bats, the Jamaican fruit bat (*Artibeus jamaicensis*) and the short-nosed fruit bat (*Cynopterus sphinx*), diapause occurs in just one of the two annual pregnancies (Banerjee et al. 2009; Wilson et al. 1991). Thus, females alternate pregnancies with normal fetal growth rates with pregnancies in which early embryonic growth is slowed or stopped for 2–3 months.

The length of diapause can change over time as the environment changes. For instance, over a 35-year period, harbor seals (*Phoca vitulina*) from the North Sea have shortened their period of diapause and now give birth about 3.5 weeks earlier, probably in response to an increase in their food base (Reijnders et al. 2010).

Obligatory delays uncouple the timing of mating and birth in all females of that species rather than in a few individuals. In eutherians, they are a feature of the reproductive patterns of the roe deer (*Capreolus*), armadillos (*Dasypus*), some bats (*Eidolon, Miniopterus*), the Siberian mole (*Talpa altaica*), the giant anteater (*Myrmecophaga*), and many carnivorans, including skunks (*Conepatus, Mephitis, Spilogale*), mustelids (*Arctonyx, Enhydra, Gulo, Lontra, Martes, Meles, Mustela, Neovison, Taxidea, Vormela*), otariid seals (*Arctocephalus, Callorhinus, Eumetopias, Otaria, Zalophus*), phocid seals (*Cystophora, Erignathus, Halichoerus, Leptonychotes, Lobodon, Mirounga, Ommatophoca, Pagophilus, Phoca, Pusa*), walrus (*Odobenus*), and bears (*Ailuropoda, Tremarctos, Ursus*; Fenelon et al. 2014). These delays probably evolved for different reasons in different groups, and some delays may no longer fulfill their original function.

The usual explanation for the evolution of delays is that they coordinate the timing of reproduction when the environment provides only a narrow, predictable window of resource abundance (Hayssen 1984). Theoretically, the timing conflict occurs because females, offspring, and males have their peak energetic demand at different points in the reproductive cycle. For females, peak demand is during lactation; for offspring, peak demand is after lactation; and for males, peak demand is at mating. In seasonal environments preparations for winter survival add a competing demand. Uncoupling the timing of mating and birth is one way to alleviate the situation. For instance, bears mate in the spring then put gestation on hold for the summer while they fatten up. They are able to forage over wide areas without the encumbrance of tiny neonates to slow them down. In fall, when food is scarce, they den up, renew their pregnancies, give birth to tiny cubs, nurse them throughout the winter, and come out during the spring with hearty, large mobile cubs (box 9.1).

Delays may have other benefits. For instance, delays allow a female to have multiple mates over a period of time and so increase the genetic diversity of her progeny. Delays also give females a wider time window over which to find mates when a low density of food forces large home ranges and large distances among individuals. Thus, delays have the potential to increase polyandry and female choice. For instance, female bears undergo estrus repeatedly before implantation; they mate with multiple males and have litters with multiple paternity (Spady et al. 2007). For grizzly (*Ursus arctos*) and polar bears (*U. maritimus*), the delay allows females make the transition from gestation to lactation at will. Heavier females give birth earlier and lactate longer than lighter mothers (Robbins et al. 2012). Finally, in European badgers (*Meles meles*) with a delay of up to 11 months, repeated mating during the delay may reduce the risk of infanticide by confusing paternity (Yamaguchi et al. 2006).

Of course, adding delays prolongs gestation and thereby reduces the number of pregnancies a female can have over her life span. This reduction in reproductive output provides a counter-selective pressure to delays (Lindenfors et al. 2003). Thus, reproductive patterns that include multiple litters per year are unlikely to incorporate long delays. In a similar vein, long delays would be selected against in species in which females have short life spans.

Mother and Young during Gestation: Placentation

The most obvious attribute of gestation is the intertwined fetal-maternal physiologies. After implantation, maternal and fetal interactions regulate gestation sometimes in concert, sometimes with a push-pull dynamic. The placental phase of gestation begins with the incorporation of the fetal trophoblast into the endometrial folds of the maternal uterus and the elaboration of taxon-specific placental structures. From this point, the eutherian trophoblast functions as an extensive endocrine system that contributes to the hormonal regulation of maternal metabolism. As most neonates are lean with minimal energy reserves, mothers may be minimizing the mass or size of the fetus until after birth (Pond 1977). If so, mothers are retaining significant control over gestation.

The uterus provides the fetus with warmth, food, shelter, and protection. One popular metaphor is that the female plays host to her parasitic offspring throughout the rest of gestation, but in reality, the proximate control of pregnancy is the joint neuroendocrine system, which includes the maternal brain, the fetal brain, the placenta, and endocrine organs, such as the thyroid, adrenals, ovary, and uterus, of both mother and offspring (Voltolini, Petraglia 2014). Biochemical feedback and interactions are occurring among all these components and the degree of interaction changes over time. Even without multiple offspring the system is quite complicated. Unfortunately, most studies examine each component in isolation or only at one time period, for example exploring the regulation of maternal blood pressure by the conceptus during early pregnancy (Bany, Torry 2012). But maternal-offspring interactions are intricate and understanding proximate mechanisms requires a holistic synthesis. Such studies are sorely lacking.

Understanding the evolution of maternal-offspring interactions is also complicated. Functional interactions between mother and offspring are complex with multiple levels of cost and benefit to mother and young. Natural selection operates to aid the survival of offspring as well as to allow females to hold resources in abeyance to support a future reproduction. Although the placenta is often considered an organ of cooperation or conflict between the offspring and the mother (Haig 1996), both "cooperation" and "conflict" are metaphors and neither extreme will fully characterize the evolution of gestation. If we focus just on uterine-placental interactions, here are some of the many components of that interplay: the placenta protects the mother from her offspring, protects the offspring from the mother, provides offspring with material for growth and development (immune transfer), protects offspring from the environment, provides temperature control for offspring, gives offspring a mechanism to influence maternal physiology, gives mothers an avenue to control offspring growth and development, allows offspring to compete with each other in close quarters, permits mothers greater spatial freedom but in later stages reduces maternal mobility, provides mothers with a greater inertial mass to reduce thermoregulatory costs, provides heat to mothers as a result of offspring growth, and allows mothers to detoxify their bodies by moving toxins into the placenta or offspring. Clearly, the consequences of placentation are complex and have led to diversity in placental and uterine morphologies (chapter 4) as well as the associated physiologies (chapter 5).

As yet, we lack a theoretical or physiological synthesis that incorporates ancestral constraints on gestation with environmental effects. Furthermore, current theory does not integrate the influences of litter size, growth rates, and developmental condition of neonates into an understanding of gestation. The body of theory that most closely examines this synthesis, life-history theory, generally looks at reproductive effort without teasing apart the separate influences of gestation and lactation, nor the interplay of the length of gestation with litter size and pre- versus post-natal mortality. We suggest a focus on the latter may be key for a better understanding of the evolution of gestation. Much is yet to be done.

Gestation Condensed

Gestation is the in utero phase of mammalian reproduction. For monotremes, this phase is quite short and culminates in egg laying. Monotreme development continues outside the mother's body, within eggs, until the embryos hatch. Although traditionally the period of monotreme reproduction from conception to hatching has been considered equivalent to therian gestation, the production of shelled eggs rather than thinly protected embryos, as well as the lack of placental implantation, makes the physiological and evolutionary dynamic between mother and offspring in monotremes much different from that in therian females. Thus, the reproductive modes are not equivalent.

For therians, the additional stage of implantation has significant functional and evolutionary consequences for the physiology of gestation, for differences across mammals in the timing of gestation, and for the types of interactions that can occur between mother and offspring. Marsupial-eutherian differences in gestation have been examined but such comparisons may have limited value as within eutherians the diversity is huge. In eutherians, gestation varies in length, presence and types of delays, in utero growth rates, embryo retention-rejection percentages, litter size, neonatal development (both overall and with respect to specific organ systems), in utero weight distribution, length of time between pregnancies, and presence of concurrent lactation. This diversity reflects natural selection operating over millions of years. Although selection pressures on males and offspring may influence gestational diversity, from the female perspective, potential factors that may have played a role in gestation include phylogeny; uterine morphology; placental morphology (both gross and fine structure); ovarian-uterine coordination; female body size, diet, mineral and water needs or restrictions; habitat (aquatic, terrestrial, arboreal); climate (predictability, seasonality, temperature, altitude, pressure); foraging and anti-predator behaviors; and disease and predation pressures on mothers or neonates. Even this short list hints at the staggering matrix of dependent and independent variables that generated the suite of co-evolved characters we call gestation. As with understanding the proximate mechanisms that affect any given gestation, further understanding of the ultimate factors that resulted in the diversity of gestations across species will require a holistic synthesis.

Two key elements contribute to gestation, litter size and neonatal development, both of which are often assessed at birth. Thus, our next chapter deals with the transition between gestation and lactation and features of mammalian reproduction, such as placentophagy, that accompany this transition.

CHAPTER EIGHT

Birth and the Neonate

Hyena birth is a dangerous affair. Both mother and pups may die. Difficult births (dystocia) and stillbirths are common, especially for a female's first pregnancy. Sixty percent of pups born to first-time mothers are stillborn. Why? First, the pups are sizable (1.1–1.6 kg), but more importantly, these pups must pass through a very narrow clitoral opening. Finally, the path a pup must take to exit the uterus includes a sharp angle. This convoluted path prolongs the time needed to exit the birth canal and increases the risk of suffocation. Females can have up to four pups in a litter, and the suffocation risk increases for the later-born pups. During the first pregnancy, the birth canal stretches and even tears. Once healed, these changes are permanent. Therefore, births from successive pregnancies are easier and have more successful outcomes. Not everything is stacked against a successful birth, as the reproductive tract is modified to help with this difficult passage even for the first birth. For instance, a mucus-secreting gland provides lubrication during birth and makes the passage slippery. Furthermore, extensive smooth muscle aids in expelling pups. Yet, birth is a period of high mortality. Our female is fortunate; this is not her first birth, and both she and her daughters have survived. She may even have obtained extra nutrients if she ate the afterbirths.

Her new daughters are dark brown with open eyes, erect ears, and prominent incisors and canines. They respond to sound and are vocal! With well-developed claws, they are quite active. They crawl within an hour of birth, even if, for a couple of days, they cannot quite stand up. They are hidden in dens away from predators but not away from danger, because sibling interactions can be aggressive, even leading to siblicide in twins of the same sex. Once above ground, these aggressive interactions are reduced, but life as a young hyena is difficult both during birth and beyond. (East et al. 1989; Frank, Glickman 1994; Frank et al. 1991; Pournelle 1965)

The Internet abounds with videos of mammalian births. In an oceanic kelp bed, a sea otter mom (*Enhydra lutris*) twists and rotates as contractions occur. A head appears and mom pulls the baby out and puts it onto her belly as she floats. Then she proceeds to do a thorough cleaning of her infant. A dolphin mom swims, with the tail-half of her baby exposed. The tail half is sucked in and out of the birth canal for a few minutes, and then, in a rush, the front half of a new calf is expelled in a cloud of blood. The baby immediately swims to the surface for a breath and swims off with mom as she expels the afterbirth. Also in the water, a hippo neonate is released headfirst and, like the dolphin calf, quickly swims to the surface for a breath of air. Up in the trees, a three-toed sloth mom (*Bradypus*) hangs upside-down. Her baby is extruded onto her belly and, after some cleaning of herself and her infant, the mom slowly takes a leafy snack off a nearby branch. Also upside-down, a bat mom gives birth. The mother flying fox (*Pteropus*) uses her wing to keep her pup from falling and cleans her baby's head even before the birth is complete. A pipistrelle mom (*Pipistrellus*) lies against a rock face so her infant is born between her and the rock wall—between a rock and a soft place. Meanwhile, on land, a giraffe calf enters the world front feet first and hangs half in half out, almost reaching the ground. After a time, its weight helps pull the calf all the way out, and it flops onto the ground. The calf rests a few moments still enfolded in its previously protective fetal membranes before wobbling upright to suckle.

Each of these descriptions is of singleton births. Births of litters are usually in less exposed circumstances. Mothers bearing several young often choose underground dens, rock crevices, tree cavities, or arboreal nests, which offer some concealment from the hungry senses of predators or the invasive lenses of amateur natural history videographers.

Singleton or not, from just these few vignettes, patterns emerge about mammalian birth. First, the process is usually fast. Second, mothers and pups interact in complex ways both during and after the event. Third, births of singletons differ from births of litters. Finally, birth is a messy affair. Clearly, birth is a complex, carefully regulated interaction between a mother and her young. The interaction is controlled by endocrinology, behavior, and morphology, as well as by the evolutionary selection pressures on mothers and offspring (Naaktgeboren 1979).

The three mammalian reproductive modes (chapter 2) have different strategies for birth (aka parturition or whelping) or the equivalent. We frame our review of the process in monotremes and marsupials with the question, Given that the process is at least a two-party event, who controls the timing of birth? For eutherians, the answer to that question is well beyond the scope of this book, especially considering that a third neuroendocrine organ, the placenta, is extensively involved. For instance, human placental tissue has at least 34 neuroactive agents that could, directly or indirectly, alone or in combination, influence birth (Voltolini, Petraglia 2014). Accordingly, we frame our review of the diversity of eutherian birth around a series of other questions, such as how long does birth take, when and where does it occur, and what key events happen afterward. We end the chapter by describing the outcome of parturition: the neonate.

Monotreme Egg Laying and Hatching

For monotremes, the timing of hatching would appear to be determined by the fetus, who must break out of its shell using a trait it shares with reptiles, a deciduous egg tooth. Meanwhile, the act of egg laying might naturally seem to be under maternal control, as signaling through the egg shell would be difficult. If birth is equivalent to egg laying, maternal control is paramount, but if, instead, birth is equivalent to hatching, then fetal control is paramount (Hayssen 1984). Unfortunately, the data required to disentangle the details of who controls birth, the egg or the mother are lacking. For the platypus, egg laying is poorly described, but during egg laying in echidnas, the mother curls up with her cloaca facing her pouch and the eggs are thereby deposited. These eggs are further incubated, and once hatched, the puggles must make their way to the milk patch on their own.

Marsupial Birth

Birth is necessarily a cryptic process given the vulnerability of mother and pups. Consequently, it has been observed in very few species. For example, in Australian marsupials, birth has been observed in only 11 of about 250 species. However, even from these few observations, some patterns have emerged. For marsupials, newborns move on their own from the urogenital sinus to the mammary area. These actions take one of three different directions, depending on the presence and morphology of the mother's pouch. In marsupials with a forward-opening pouch (e.g., kangaroos, *Macropus*), the mother positions her urogenital sinus below the pouch. The newborn climbs up and into the pouch using a swimming motion with alternate strokes of the forearms and the head moving from side to side. In marsupials with a backward-facing pouch (e.g., bandicoots, *Isoodon*), the mother positions her urogenital sinus above the pouch and the young slither the short distance down into the pouch using a snake-like wriggle. In marsupials without a pouch (e.g., quolls, *Dasyurus*), the mother stands on four legs with her hips raised so that the urogenital sinus is above the mammary area and the newborn young crawl down from the sinus directly to the exposed teats. In all species, birth takes under 10 minutes. Newborns use olfaction and their sense of gravity to orient to the pouch or mammary area (Gemmell et al. 2002). Clearly, for marsupials, we have some information about how young get from the urogenital sinus to the mammary area, but they start in the uterus. What about the first part of their journey? Unfortunately, we have no information on how embryos initially get from the uterus to the urogenital sinus, but one can easily envision that similarly diverse motions may be used as might movements of the uterine wall.

After birth, marsupial mothers, even herbivorous ones, lick up the amniotic fluids. The placenta may be retained and reabsorbed (*Perameles*) or, if expelled, may be removed, as in the northern quoll (*Dasyurus halluctatus*), or eaten, as in the common brushtail possum (*Trichosurus vulpecula*; Gemmell et al. 2002).

Just as with the "who controls birth, the egg or the mother" quandary in monotremes, in marsupials, theories abound about who controls the timing of birth. Unfortunately, the hypotheses are not in synch with each other. Lillegraven (1975) suggested that the maternal immune system forces the fetus to crawl out of the uterus, but data on females exposed to paternal antigens prior to pregnancy refute this hypothesis, indicating that the immune system is not the trigger (Walker, Tyndale-Biscoe 1978). Also, removal of the maternal pituitary gland prevents parturition in some kangaroos (Hearn 1974), suggesting that birth in marsupials may be driven by the mother. In contrast, prostaglandins from the placenta trigger birth in tammar wallabies (*Macropus eugenii*) and indicate that the fetus controls birth (Renfree 2010). Supporting the embryo-control hypothesis, a review of progesterone in marsupial reproduction concluded that the embryo initiates not only its own birth but also determines the length of gestation. In some species, the embryo shortens the life of the corpus luteum (Bradshaw, Bradshaw 2011). Thus, recent reviews take the side of fetal control.

Eutherian Birth

From a eutherian perspective, a marsupial neonate's ability to initiate birth at such an early developmental stage is remarkable. Eutherian neonates at an equivalent stage are incapable of such coordination (nor would they benefit from it). Even immediately before birth, when they are much older than any marsupial neonate, many eutherian offspring are incapable of initiating parturition. For instance, in rabbits and laboratory rodents, fetal decapitation does not alter the timing of birth, implying that parturition is controlled by maternal, not fetal, factors (Nathanielsz 1978). Of course, one could envision problems from drawing too many conclusions from studies of such incapacitated offspring. In other species, late-term young are well developed with a mature neuro-endocrine system. For these young, initiating their own birth is possible. For instance, sophisticated analyses on parturition in sheep indicate that maturation of the fetal adrenal-pituitary axis is necessary for the induction of labor (Liggins et al. 1973; Thorburn, Challis 1979). Work in primates is similar to that on domestic ruminants and indicates that the fetus controls parturition (MacDonald et al. 1978).

Taken together, these studies suggest that birth is fetally controlled in mammals with precocial young and maternally controlled in mammals with altricial young. Ignoring the inherent problems with the terms *precocial* and *altricial* (which we discuss later), this hypothesis has another flaw. The data behind the generality are confounded by both litter size and maternal size. Precocial species that have the most data on birth are sheep with small litters (1 or 2), and the altricial species with the most data are mice, rats, or rabbits with large litters. Thus, developmental state may not be the key factor. Rather, neonates from singleton litters might control their own birth, but the process may be maternally controlled when a litter is involved. This hypothesis has the benefit of an obvious evolutionary advantage, that is, to reduce sibling competition. Alternatively, maternal size may be an influence such that mothers control birth for small species and offspring control birth for large species. The evolutionary drivers behind this hypothetical pattern are less clear. In any event, examining the control of parturition in the leporids, a family that contains both precocial (hares, *Lepus*) and altricial (e.g., cottontails, *Sylvilagus*; rabbits) species, may help clarify questions of fetal development and control of parturition because the confounding effects of body size, litter size, and evolutionary history could be circumvented.

Regardless of who triggers the process, eutherian birth involves components of the maternal and fetal immune systems, hypothalamic-pituitary-adrenal axes, the neuro-muscular systems, and the placenta. Birth is accompanied by an immune response (Gomez-Lopez et al. 2013), because the process of birth provides a major avenue for pathogens to enter the mother's body. For example, tissues may rip and be exposed to pathogens entering from the vaginal canal. Birth also alters the microbiome of the vagina, cervix, and uterus. The placenta has its own microbiome, and during birth all these microbial communities, maternal and placental, come into contact and provide the offspring with a starting microbiome. This microbial community will be augmented throughout life (Mysorekar, Cao 2014; Prince et al. 2014; chapter 13).

Overall, fetal and maternal systems interact in different combinations and with variable levels of control to regulate birth in different species (Petraglia et al. 2010). No single proximate mechanism explains this complex process for all mammals. For example, in brown bears (*Ursus arctos*), the timing of birth is related to maternal fat stores; fatter females generally give birth earlier than females in poorer condition (Friebe et al. 2014). For brown bears, some part of the physiological system that regulates fat storage also interacts with the timing of parturition. Most eutherian species show a decrease in progesterone with birth, but humans maintain a high level of progesterone throughout the process (Wagner et al. 2012). These few examples illustrate the diversity of proximate mechanisms that control birth.

Returning to the big picture, birth is a complex series of steps requiring the interaction of a female and her offspring. Each species must balance various physiological processes, such as uterine contractility versus quiescence or cervical opening versus closure, as well as the potentially different demands of individual young in a litter. Many things must happen for both the offspring and the mother to survive. In roughly chronological order, the process includes the following steps: the cervix must dilate; placental structures must separate from the uterus; offspring must be oriented appropriately; if several pups are to be born, a birth order needs to be established; offspring need to be expelled/released one at a time; placental structures must be expelled/released; uterine contractions need to occur without crushing any offspring still in utero; the afterbirth and other debris must be removed; the newborns must be cleaned, and the first suckling must begin. How all these steps are physiologically coordinated in diverse mammals is not known. Thus, rather than navigating each of these complex steps that vary by species, a broader approach presents a better place to start. We next ask when and where does birth occur, how long does it take, what happens to the placenta, and how do these aspects of birth connect to the female's biotic and abiotic environment.

When Does Birth Occur?

In nocturnal mammals, birth generally occurs during the day, whereas in diurnal mammals, birth often occurs at night (Gemmell et al. 2002; Olcese 2012). Thus, birth usually occurs when a female would normally be resting in a safe place rather than during a time when she might be physiologically motivated to do other tasks, such as forage.

How is diel (24-hour) timing achieved? In addition to the brain, peripheral tissues, e.g., the uterus, have circadian clocks (molecules with consistent temporal oscillations). These peripheral clocks allow physiological processes to be coordinated such that antagonistic processes are out of phase (e.g., sleeping and foraging) and synergistic processes are in phase (e.g., sleeping and digestion). External cues, e.g., sunrise, entrain the clocks via internal hormonal signals, such as melatonin or glucocorticoids (Olcese 2012). Genes related to known clock function are expressed in the uterus, suggesting that the uterus has a molecular clock.

Births can be synchronized in populations of mammals, e.g., Arctic (*Pusa hispida*; Kelly et al. 2010) and Antarctic seals (*Leptonychotes weddellii*; Hastings, Testa 1998). Such synchrony may occur for different reasons. For instance, temporal cycles of food availability may select for birth at a particular time of year. A second selection pressure may be to minimize the probability of predation by swamping predators with an abundance of morsels. Timing might also be related to those abiotic features of the environment that may be more conducive to offspring survival, such as warm versus cold waters.

Of course, the definition of synchrony varies. For instance, in quolls (*Dasyurus hallucatus*) all births occur within 7 days (Gemmell et al. 2002), whereas in roe deer (*Capreolus capreolus*) all births are within 45 days (Plaard et al. 2013). Interestingly, in roe deer, individual females consistently give birth earlier or later in the birthing season. Furthermore, as roe deer age they give birth earlier in the season (Plard et al. 2013). Additional examples of the seasonal timing of reproduction are given in chapter 11.

How Long Does Birth Take?

The duration of birth depends largely on how one defines birth. If defined as the time when the first body part of the baby appears to the time the entire infant has left the mother, then birth usually lasts but a matter of minutes. However, if birth includes labor (the contractions leading to birth), then the process is much longer. The initiation of contractions in females is subtle, making accurate as-sessment difficult in wild animals. Of course, moms with litters have a series of births, one after the other with some period of recovery between each birth. So is birth the time a single offspring is released or the time from first to last?

Across mammals, birth is a vulnerable period for both mother and offspring. External events, such as the appearance of a predator, may lead females to suppress labor, at least temporarily, and prolong birth while the mother tries to escape predation (Naaktgeboren 1979). In general, selection will operate to make parturition as short as possible, especially if females and young are in exposed locations.

Where Does Birth Occur?

The physical location of birthing mothers is as varied as the postures mothers and neonates use during the process. Bats and sloths give birth while hanging from trees; sea otters while in kelp beds; whales, manatees, and hippos in water; seals (Cystophora) on ice floes; polar bears in snow caves; moles in subterranean nests; foxes, wild dogs (Lycaon), hyenas (Crocuta), and meerkats (Suricata) in dens.

The aquatic lifestyle of cetaceans and sirenians requires birth at sea, which presents numerous complications, such as thermoregulation in water, giving birth while floating, the risk of predation, the risk of neonatal drowning, and complications due to nursing while at sea. Tail-first births are one accommodation to birth at sea. By keeping the baby's head within the mother's body, the risk of drowning during birth is reduced. Another accommodation to life at sea in bottlenose dolphins (Tursiops), killer whales (Orcinus), and perhaps other odontocetes, is a postnatal period of about a month in which mothers and calves go without sleep (Lyamin et al. 2005). The extended motor activity helps maintain body temperature of the neonate and the 24-hour responsiveness reduces predation (Lyamin et al. 2005).

Looking at other marine mammals, sea otters (Enhydra) have surmounted these challenges but not seals. Seals give birth on land, often on isolated islands or stretches of beach, which have become traditional birthing grounds. Terrestrial birthing may have been an ancestral condition. More likely, the tie to land is in part because seal pups present nicely wrapped, calorie-rich nuggets for killer whales, sharks, and other aquatic predators. The tie to specific birthing areas at specific times of year has allowed the evolution of highly polygynous social organizations and extreme sexual dimorphism, such as in elephant seals (Mirounga). It may also explain the presence of the extended gestation (via delayed implantation) of seals if seasonal timing is sufficiently important (Bartholomew 1970).

Many mammals give birth in special natal dens or nests. Nest building is a common sign of pending birth (Naaktgeboren 1979). Hormonal control of this behavior may be even more sensitive to the onset of birth than control of the uterine myometrium. In any case, natal nests are often located away from the usual burrow or den. Indeed, many females give birth alone, and, if possible, away from her social group if she has one. If physical separation is not feasible, a female may give birth at a time when members of the group are likely to be asleep, for example, during the day in nocturnal rodents or at night in diurnal ungulates. Giving birth at some distance from the group may allow a female to be less conspicuous if predators are looking for groups of animals to hunt, but giving birth when conspecifics are asleep seems less helpful in the event of a predator's attack. However, this, too, could be adaptive, if group members are also predators of her young. For example, marmosets usually give birth away from their social group. However, in captivity with confined spaces and altered light cycles, females are forced to give birth near group members. These conspecifics are extremely excited by the arrival of the baby and especially the delicious afterbirth, excitement that has dire consequences. Group members happily ingest the afterbirth and, when overzealous, also eat the baby (Shaw, Darling 1985).

Aftermath and Placentophagy

Birth is accompanied by debris. Many mothers clean the vagina or urogenital sinus before, during, and/or after birth. Fluid and tissues are released before birth (the water breaking), as well as during and after the young emerge from their internal haven. All these fluids are likely to have distinctive odors that may facilitate mother-offspring bonding (Uriarte et al. 2012). Such mother-offspring recognition and bonding is important: for instance, Mexican free-tailed bats (*Tadarida brasiliensis*) must identify one pup out of thousands in their crèches (Loughry, McCracken 1991). However, the aftermath of birth may also attract predators.

Consumption of the placenta and/or afterbirth (placentophagy) is common in mammals. Even some fathers will eat the placenta (Gregg, Wynne-Edwards 2005). The list of mammals that eat the placenta is both diverse and long and includes herbivores as well as carnivores, for example, cattle, goats, rabbits, dwarf hamsters (*Phodopus*), and canids. A few exceptions include horses, elephants (*Elephas, Loxodonta*), giraffes, and many marine mammals (Gregg, Wynne-Edwards 2005; Melo, Ganzález-Mariscal 2003; Rameriz et al. 1995; Virga, Houpt 2001; von Keyserlingk, Weary 2007).

Why might females ingest these materials? Several hypotheses have been proposed. For example, by eating a placenta, the ever frugal mother may save the energy, water, and nutrients in the placenta, material that would otherwise be lost. Alternatively, the materials themselves, including associated hormones, may play a role in recovery. For instance, prostaglandins, oxytocin, or other hormones when ingested may (a) facilitate the return of the uterus to normal size or (b) prepare the mammary glands for nursing (Naaktgeboren 1979). As endogenous opioids are also present in the placenta, placentophagy may make the delivery of subsequent offspring in a litter less painful (Corona, Levy 2014). Obviously, removing the placenta also cleans the birthing area reducing the probability of attracting pathogens, scavengers, or predators. Indeed, predators are likely to have evolved to detect the odors left behind after birth has occurred. A female may attempt to eliminate these smells through disposing of the afterbirth by whatever means possible.

Placentophagy may have arisen or be maintained by any one or a mixture of these possible functions. The basal nature of this trait suggests either a highly advantageous strategy or a fixed action pattern that can no longer be lost, even in herbivorous mammals (Naaktgeboren 1979). Of course, the key output of birth is not the placenta but the neonate.

Neonates

A fetus makes a huge transition when it becomes a neonate. Before birth, it was warm, wet, protected, and did nothing but float in its encapsulated, seemingly homeostatic, darkness. That changed with birth. Now the neonate is exposed to the environment outside its mother, to air and light, to cold temperatures, and to hard and slippery surfaces. Suddenly, it has to protect itself from desiccation, disease, predators, parasites, and potentially siblings. Neonates vary in how they meet the new challenges outside the womb. Naturally, they also vary a great deal in shape, size, and developmental condition.

Some marsupial neonates are exposed to the environment only for the short time they crawl up or down their mother's abdomen and into the pouch, but, of course, some marsupial mothers do not have a pouch, and these less fortunate neonates are dragged while nursing under the mother as she forages. The constant supply of milk keeps these neonates warm and sustained, but they also must have tough skin to protect themselves from abrasions while on mom's teat.

Many eutherian neonates are born to a softer life in protected natal dens furnished with a cozy nest prepared by the mother. For instance, nests of pygmy rabbits (*Brachylagus*) lie about 12 cm below ground and are constructed of fine grasses, shredded bark, and maternal hair. Although fleas and mites occupy the nest as well as neonates (Rachlow et al. 2005), the location provides protection from weather and allows the neonates to huddle, reducing the costs of thermoregulation.

Many other eutherian neonates are immediately exposed to the cold and can suffer a rapid decrease in body temperature as a result. For instance, the core temperature of neonatal lambs falls by up to 4.5°C immediately after birth (Faurie et al. 2004). Mammal neonates have brown adipose tissue (BAT), a special tissue that helps with this transition. Heat-generating BAT is unique to mammals and helps neonates survive the cold stress of birth. In adults, BAT functions to keep mammals warm in cold weather or during periods of torpor (Cannon, Nedergaard 2004). In contrast to neonatal lambs, the body temperature of walrus calves (*Odobenus*) is 1–2°C higher than that of their mother (Fay 1985). The reasons for this added thermoregulatory cost are unknown.

Behaviorally, life as a newborn is highly varied. It may follow its mother within hours of birth, be hidden in a burrow, be attached to a teat in a pouch, or carried on a maternal back or belly. Consequently, newborns, themselves, are highly varied in size, behavior, color, abilities, and perceived maturity.

Neonatal Size

Newborn mammals vary in size from well under the weight of a dime for kangaroo neonates to as heavy as a car for neonatal baleen whales (Hayssen et al. 1993; Laws 1959). Naturally, much of this variation relates to maternal size. Across all mammals, bigger females have larger neonates (486 species, $R^2 = 94\%$; Hayssen 1985). However, such large-scale generalizations, even if quantitatively accurate, are not always terribly informative.

Many mammalian mothers produce several young at a time rather than only one. Across 486 species (including bats and whales), litter size was known for 482 (168 singleton litters; 314 litters of more than 1). For mothers of similar size, those producing singletons have neonates about three times larger than the neonates of females with litters. Therefore, across mammals, neonatal mass is also dependent on litter size. This is consistent with predictions from life-history theory.

For mothers, number of young as well as neonatal size is important. For instance, the amount of energy designated to reproduction can be split multiple ways: one pup of 10 grams or two of 5 grams or four of 2.5 grams, and so on. Thus, litter mass, rather than neonatal mass, may be important. Using 482 species (Hayssen 1985), this conclusion is confirmed statistically, as litter mass for species with singletons does not differ from litter mass of species with larger litter sizes.

Other factors are also important in determining neonatal size. To illustrate this, we can look at species with singleton litters. For example, zebra (*Equus*) and white rhino (*Ceratotherium*) neonates are both 30–40 kilograms at birth, but a rhino mother is more than 10 times larger than a zebra. In addition, a rhino's gestation is 50% longer (16 vs. 11 months; Hayssen et al. 1993). Thus, the much larger rhino takes much longer to produce the same-sized neonate. Clearly, growth rates also influence neonatal size (Huggett, Widdas 1951; Frazer, Huggett 1974).

Another example of the relationship between maternal and neonate size comes from a large group of mammals that nearly all have singletons: bats. Bats are the only flying (volant) mammals. The heavier you are, the harder you must work to fly. Wing loading (the amount of weight for a given wing area) is an estimate of that relationship. For female bats, the additional mass she carries during gestation increases the energy she needs to sustain flight. The added encumbrance could hamper her pursuit of fast-moving insects or her ability to hover while gathering nectar, but the extra burden would be less intrusive for females whose diet is fruit. For bats, neonatal mass is also related to foraging and flight (Hayssen, Kunz 1996).

Overall, neonatal size reflects not just maternal size, litter size, and growth rates but also diet, phylogeny, and behavior. Mothers are faced with a slew of variables within which to construct a neonate of the "ideal size," and the ideal size may be a moving target as she ages. Other aspects of the neonate are equally complex.

Neonatal Behavior

Many ungulates, such as deer, antelope, hippos, giraffes, horses, and rhinos, give birth to singleton young with well-developed sensory, thermoregulatory, and locomotory abilities. In some respects, these young "hit the ground running" soon after birth. These neonates have been put into two rough categories, hiders or followers, according to their behavior in the few days after birth. Neonates in the hider category remain immobile, often concealed in vegetation, while their mothers forage. This strategy may reduce predation in closed habitats. However, in open areas, or for species with migratory habits, a hider strategy would be untenable. Thus, neonates in the follower category stay close to their mothers after birth (Ralls et al. 1986). Young in the hider category might be selected for cryptic behavior and morphology, whereas young in the follower category might be selected for especially long legs and rapid locomotion after birth.

Is the hider-follower dichotomy valid? A quantitative assessment using the behavior of mothers and young for 22 species of ungulates provides some support for the classification. However, depending on which exact criteria were used to define hiders versus followers, the same species (e.g., giraffe) might be put into either category. In addition, maternal behavior, rather than neonate behavior, might determine the categorization. For instance, mother hippos remain close to their young during the day (as happens with followers), but then leave them concealed when they forage at night (as for hiders; Ralls et al. 1986).

The hider-follower dichotomy is useful because it simplifies complex observations. However, natural selection did not create hider and follower strategies; humans created these categories. Natural selection operated on neonates and juveniles such that those young animals that avoided predation, either by hiding or by following, had more offspring as adults than those that died. The hider-follower dichotomy is only a description of the suites of behaviors and appearances that are obvious to us and, in this case, described by us as anti-predator strategies. Thus, the dichotomy, though useful, has a human, not biological, origin and may well be a continuum.

Other aspects of neonatal behavior are related to their maturity at birth, but before we examine neonatal development, we want to review one aspect of the morphology of neonates and juveniles that is related to predation and other aspects of a neonate's social environment. This feature is key to mammals, namely, their fur.

Neonatal Pelage

Some mammals exhibit a distinctive neonatal or juvenile pelage. Neonates and juveniles are more vulnerable to predation than adults (chapter 12), but adults are vulnerable to predators as well. Thus, a cryptic pelage could benefit both adults and young. Some neonates are carried by or on their mothers, such as colugos (*Cynopterus*), giant anteaters (*Myrmecophaga*), and sloths (*Bradypus, Choloepus*). The fur of these neonates matches that of their mothers and presumably allows the neonate to blend into its mother's belly or back (figure 12.1).

Some primate neonates are also carried by their mothers, but their pelage may be different from that of their mothers, sometimes to a flamboyant degree, for example, the bright orange neonatal fur of Francois' leaf monkey (*Trachypithecus francoisi*). The pelage of these baby primates gradually changes to the adult glossy black coat with white markings over the course of 6 months (Burton et al. 1995). While this is an unusually striking case, in fact, neonatal-adult pelage differences are common in primates. In one study of more than 135 species, neonatal pelages differed from those of their mothers in over half of the species. Usually, the differences were subtle or inconspicuous, rather than flamboyant, but some were eye-catching. Conspicuous colors could be risky if they attract predators; however, they may also act as social signals to promote infant care by social group members or to increase the defense of infants against infanticidal males. A third function of an unusual natal coat in social primates may be to obscure clues to paternity (Treves 1997). The Treves study suggested the primary function was to reduce the risk of infanticide.

Besides primates, other neonates have distinctive juvenile pelages. These neonates are distributed among many taxa, including bats (e.g., Phyllostomidae: *Carollia*; Vespertilionidae: *Lasiurus, Miniopterus, Myotis*; Molossidae: *Nyctinomops*), rodents (e.g., *Microtus, Myodes, Ochrotomys, Peromyscus*), carnivorans (e.g., *Chrysocyon, Cryptoprocta*), pinnipeds (both otariids and phocids), and ungulates (e.g., cervids, tapirids; Caro et al. 2012; Christianson et al. 1978; Cloutier, Thomas 1992; Ecke, Kinney 1956; Köhncke, Leonhardt 1986; Linzey, Linzey 1967; Milner et al. 1990; Padilla et al. 2010; Timm 1989). Even for neonates born without fur, the first pelage may differ from that of the adult. For instance, in white-footed and deer mice (*Peromyscus*), the juvenile pelage is a uniform gray, whereas that of the adult is a cinnamon rufous (Gottschang 1956).

Besides color differences, neonatal or juvenile fur may be darker, duller, longer, and/or less dense than the adult pelage (Kunz et al. 1996). Walruses have two early pelages, a fetal pelage and a natal pelage. The fetal fur is a fine, white, wooly pelage that is shed and ingested by the fetus some 2–3 months before birth (Fay

1985). That pelage is replaced by a natal pelage that lasts through the juvenile's first summer (Fay 1985). Fetal harbor seals (*Phoca vitulina*) also have a white coat that is usually shed in utero, but the white coat is maintained in subspecies that give birth on ice (Boulva 1971). Wolverine (*Gulo*), kits also have white coats but are not exposed on ice but rather reside in snow dens (Mehrer 1976).

When neonates have a different pelage, adults and young may also differ in behavior or habitat. This is true for otariid and phocid seals. While adults are primarily aquatic, pups are terrestrial for their first weeks of life. Pups that are born in caves or on predator-free islands are dark in color or look like adults, whereas pups that are born in Arctic regions, where polar bears are common predators, have white neonatal pelage (Caro et al. 2012). For ungulates, the hider-follower categorization of neonates suggests that young in the hider category might be selected to have a neonatal pelage that allows for greater concealment and that follower young might look like adults. This hypothesis works for spotted white-tailed deer fawns (*Odocoileus*), who remain concealed for several days after birth but not for the striped calves of tapirs (*Tapirus*), who follow their mothers after birth but have a pelage as cryptic as that of deer fawns.

Of course, many neonates have an adult pelage from birth or immediately develop one, such as ground squirrels (*Ammospermophilus*, *Urocitellus*; Maxwell, Morton 1975). In phocid seals, the pelage contributes strongly to insulation in neonates, more so than in adults (Kvadsheim, Aarseth 2002). Unfortunately, the selective advantages behind newborn and juvenile pelages in mammals have only been substantively explored for primates and pinnipeds. Clearly, we have a lot more to understand about the adaptive features of the newborn.

Neonatal Development: The Traditional Altricial-Precocial Categorization

Neonates also vary in how well they can function immediately after birth. Neonates differ in their sensory abilities (hearing, seeing, smelling) as well as how well they can locomote. Some offspring can regulate their body temperatures immediately after birth; however, many cannot. Newborn harp seals (*Pagophilus*) have a full coat of dense fur and stay warm even on the ice floes on which they were born, whereas most neonatal bats are not well furred at birth and cannot maintain a stable body temperature (Kurta, Kunz 1987). This simple physiological parameter, thermoregulation, is likely to be important to mothers who wish to leave offspring in dens or nests while foraging. It may also relate to how much milk is needed by her young on her return.

Many biologists use the terms *altricial* and *precocial* to describe the developmental condition of neonates. The terms denote extremes on what is a developmental continuum. Broadly, altricial neonates are less developed than the well-developed precocial neonates. The relative nature of the terms means that the same word is used for widely different developmental conditions. For instance, neonatal mice,

Figure 8.1. Neonate-juvenile collage. Young mammals vary in shape, developmental stage, and size. *Left to right*: *Row 1*, White-faced capuchin (*Cebus capucinus*), Roatan, Honduras; kiang (*Equus kiang*). *Row 2*, Hippo (*Hippopotamus amphibius*), Masai Mara, Kenya; brown-throated three-toed sloth (*Bradypus variegatus*), Costa Rica. *Row 3*, Wild boar (*Sus scrofa*), Burgundy, France; Galápagos sea lion (*Zalophus wollebaeki*), Gardner Beach, Espanola Island, Ecuador. *Row 4*, Thomson's gazelle (*Eudorcas thomsonii*), Serengeti National Park, Tanzania; monjon (*Petrogale burbidgei*), Mitchell Palteau, Kimberley, Western Australia; lab mice (*Mus*). Images by Jurgen and Christine Sohns, FLPA; Roland Seitre; Anup Shah; Suzi Eszterhas; Pierre Vernay, Biosphoto; Tui De Roy; Winfried Wisniewski, FLPA; Martin Willis; Michel Gunther, Biosphoto; used with permission from Minden Pictures.

hyenas, and humans are all termed *altricial*; yet, their state of maturity is very different (figure 8.1). Wolverine kits are referred to as altricial (Banci, Harested 1988) yet are fully furred at birth (Mehrer 1976). Similarly, precocial neonates are those that exhibit advanced sensory and locomotor abilities, but marsupial neonates have advanced olfactory abilities, and, with respect to locomotion, they crawl, unaided, up to the teat at birth. However, marsupials, along with monotremes, are generally considered the most altricial mammals. Clearly, use of the terms depends on context.

To add rigor to the dichotomy and to make it more of a continuum, scientists have used a laundry list of yes-or-no questions about the neonate where "yes" answers suggest a precocial neonate and "no" answers an altricial one. Such questions include Are the ears and eyes open? Do they have fur or hair? Can they turn over, crawl, walk, or swim at birth? These questions can be used collectively to establish a larger number of developmental categories rather than a simple dichotomy (Derrickson 1992).

These developmental questions have both biological relevance and taxonomic bias. For instance, although the olfactory system of all newborns is functional it is more pronounced in many marsupials that must find their way to the pouch. Meanwhile, the condition of eyes and ears reflects a degree of sensory ability and associated predator avoidance. For instance, age at eye opening is not especially relevant to fossorial mammals, nor to mammals, such as bats, that rely on sound rather than sight. In addition, olfactory development might be exceptionally important in mammals, but as humans, we do not typically assess that sensory mode. Similarly, the presence of abundant fur indicates an ability to thermoregulate, but the lack of fur might not mean a neonate cannot maintain its body temperature, especially if that neonate uses blubber or fat as insulation rather than fur, e.g., whales, hippos, or elephants. A final example is the golden mole, *Amblysomus*, that lacks external eyes and ears, thus has no age at eye or ear opening (Kuyper 1985), making these categories irrelevant.

Another issue is that use of these criteria is inconsistent. For instance, human neonates cannot turn over but do have open eyes and ears as well as hair. However, human infants are often considered altricial, which they are when compared to a newborn dolphin (*Tursiops*), hooded seal (*Cystophora*), or Arctic hare (*Lepus*).

Undeniably, the condition of the newborn influences the amount and kind of care a mother must provide to promote the survival of her offspring in the face of abiotic (e.g., cold) or biotic (e.g., predators) challenges. The difficulty is in finding criteria that assess development across a wide range of species. Mammals have different lifestyles, such as living in the ocean or flying, that may result in unique conditions and influence how we categorize their development. Even marsupial neonates show a range of neonatal developmental stages. Phalanger and kangaroo neonates are relatively well developed "with prominent external ear and eye primordia, retinal pigmentation, significant differentiation (e.g., digits present) in the hind limb" (Smith 2001:122). In contrast, the dasyurid neonate is ultra-altricial; the "ear and eye primordia are barely visible and the head is virtually all nose and mouth . . . the forelimbs are robust, the hind limbs are barely beyond bud stage" (Smith 2001:122). Thus, developmental state is context dependent. Because researchers are trying to summarize complex factors and turn a continuous variable into a categorical one, the terms *altricial* and *precocial* are fluid and extremely context dependent.

In an effort to simplify the developmental continuum, at least for eutherians, Clauss et al. 2014 (and others, e.g., Martin, McLaron 1985) have used litter size as a proxy for precocial or altricial. Species with singletons are classified as precocial, whereas those with more than one offspring per litter are classified as altri-cial. Unfortunately, this proxy means that species that frequently give birth to twins, such as sheep (*Ovis*), moose (*Alces*), and pronghorn antelope (*Antilocapra*), are considered altricial, as are the decidedly well-developed neonates of hares. Bats are the second most diverse group of mammals. Using this litter size proxy, most bats would be considered precocial; yet, bats are functionally immature at birth. Neonatal bats are not well furred, cannot maintain a constant body tem-perature, cannot fly, are helpless, and are completely dependent on their mothers (Kurta, Kunz 1987). Litter size is also confounded with body size as many large-bodied species have singletons (pigs are an exception) and many small-bodied species have litters (bats are again an exception). Clearly, the use of litter size as a proxy for neonatal development is problematic.

Regardless of the challenges, contradictions, and difficulty, the altricial-precocial developmental categorization has extensive use. Developmental condition of the neonate influences subsequent maternal care. Unfortunately, we do not currently have a consistent and reliable means of assessing that variable across mammals. Clearly, a general overhaul may be long overdue. Alternatively, comparisons across developmental categories should be done within a limited taxonomic range. As far as we can tell, leporids (rabbits, cottontails, and hares) are the only mammalian family with a large range of neonatal development. Thus, leporids may be ideal for examining the influence of neonatal development in mammalian reproduction.

Birth in Brief

In various ways, birth is complex and precarious. Mothers and their offspring are vulnerable to predation during the process, and the remnants of birth can also attract predators or parasites. Therefore, birth must be fast, but it also must be carefully regulated between the mother and her litter.

The birth process differs in monotremes, marsupials, and eutherians. For monotremes, one question is, Which process is equivalent to birth? Egg laying or hatching? Egg laying is under maternal control, but hatching is determined by the young. For marsupials, the birth we observe takes about 10 minutes and differs across species in maternal posture and neonatal locomotion, but what happens in utero before we see the neonate exit the birth canal is not known. Also, although recent reviews suggest a fetal signal, what cues the initiation of birth in marsupi-als is not clear.

Birth in eutherians is even more complicated and diverse because of extensive placentation and the larger number of species, respectively. Many different physi-ological systems in both the mother and the offspring are involved, but which systems and to what extent varies. Not only does the control of birth differ but so does the timing, the duration, the location, and the aftermath.

Of course, the major result of birth is a neonate or neonates. Not surprisingly, neonatal mammals are as varied as their mothers. They vary in size, behavior, color, and development. Some traditional dichotomies used to describe neonates, such as hider versus follower and altricial versus precocial, are in constant use, but may not account for the enormous diversity of neonates.

Overall, birth marks not only the end of gestation but also the transition to lactation. Birth and the characteristics of the neonate are the starting place for the hallmark of mammalian reproduction: lactation.

CHAPTER NINE

Lactation

Birth to Weaning

She has two mouths in addition to her own to feed, and this will continue for well over a year (18 months on average). At least after 6 months, the pups can feed on items besides her milk. Regardless, lactation finds our hyena mother testing her limits. Her energetic demands are much higher than during periods when she is not reproductive and are well over the demands of males of her species. Admittedly, her body prepared a full month before parturition when her mammary glands began to swell and she started storing fat. However, much of the demands of lactation are met by her daily foraging. Her pups meanwhile are housed in a den waiting for their weary mother's return. Nursing first occurs in a natal den but after 2–3 weeks, she will move her cubs to the communal den. To identify her pups, all she needs to do is call and listen for their vocal cues, which she quickly recognizes. These pups and no others will she nurse. As her pups age, she will bring food back to the den (~6 months). Aggression between siblings is intense, especially during the first week in their natal den and may result in siblicide. Once at the communal den, play wins out over aggression, probably much to the mother's relief. Even more relief occurs at around 14 to 18 months when those hungry mouths are officially on their own and her pups are finally weaned. In those prior demanding months, hormones were active: oxytocin for milk letdown and prolactin for general regulation. Lactation is no easy affair, even for robust hyena females. (Drea et al. 1996; East et al. 1989; Golla et al. 1999; Hill 1980; Hofer, East 1993, 1995; Holekamp, Smale 1990; Holekamp et al. 1999b; van Jaarsveld et al. 1982)

> The greatest metabolic challenge that mammalian females must face: lactation.
> —Woodside et al. 2012:301

Lactation and its product, milk, are the quintessence of mammals. Gestation is not unique to mammals, but only mammals provide milk to their young via mammary glands. Mammals are defined by this multifaceted component of their reproduction, which has influenced their biochemistry, physiology, anatomy, behavior, sociality, ecology, and, in short, every aspect of their evolution. When females first started providing their newborns with milk over 200 million years ago, they delayed the need for their young to catch their own food (Pond 1977;

Lefèvre et al. 2010; also, chapters 2 and 7). Thus, developing young could use the energy of milk to grow rather than to find, catch, chew, and sometimes detoxify an adult diet (Vernon, Pond 1997; Pond 2012).

The implications of lactation and nursing are broad and sometimes surprising. Neonatal lips and cheeks are modified for nursing in monotremes and marsupials (Pond 1977). Neonatal tongues may also have special modifications. For instance, those of whales (e.g., *Mesoplodon*) are muscular with scalloped or villous edges so that the teat can be firmly pressed against the roof of the mouth and direct milk down the throat without seeping around the sides (Cross 1977; color photo in Shindo et al. 2008). Although nursing does not require teeth, neonates need powerful facial and respiratory muscles for retrieving milk. These are the same muscles we, as humans, use for conveying emotions.

What changed for females with the advent of lactation? From a female's perspective, lactation exchanges the physical burden of carrying embryos with the metabolic burden of feeding and caring for young outside of the womb. After birth, a female's mobility is no longer impaired, but she must produce enough milk to feed hungry mouths until weaning. Neonates may be getting a free lunch, but mothers are paying the price.

Cost of Lactation

How much and in what way does lactation present a cost to females? To create milk, mothers must find food; catch, or otherwise process it; digest it; and transport the components to the mammary glands where they must synthesize the appropriate proteins, fats, and carbohydrates; add water; and accumulate and store the end product until needed. All of this takes energy. So from the get-go the cost of milk includes its production. Biologists often estimate the cost of lactation by measuring the percentage of extra food or energy a mother takes in per day during lactation over and above what she uses when she is not producing milk. These studies have presented a two-fold take-home message. First, lactation does cost something. Second, the actual cost varies widely. One can appreciate this variation by considering some of the published data on energy increases for milk-producing females (table 9.1). These values range from 44% to 300% greater than the amount of energy or food needed by non-reproductive females. To accommodate the increased intake, the digestive tract and associated organs may increase in length or in absorptive capacity (Pond 1977).

One difficulty with food-intake measures is that they assume females are relying completely on energy from food they eat while lactating. Were this true, it would lead to the interesting contradiction that lactation costs nothing for females such as bears, seals, and whales that fast during lactation. The zero cost is assumed because, as females are not eating, their percent daily food increase is zero. But clearly, these numbers do not tell the whole story. They underestimate the cost of lactation when females use reserves stored during gestation or even

Table 9.1. Increases in energy expenditure, food consumption, or foraging in lactating versus non-reproductive females

Increase	Species	Citation
26% energy	Women, *Homo sapiens*	Dufour, Sauther 2002
44–80% energy	Vole, *Microtus pinetorum*	Lochmiller et al. 1982
45% energy	Spiny mouse, *Acomys cahirinus*	Degen et al. 2002
60% food	Sifaka, *Propithecus verreauxi*	Saito 1998
50–325% food	Kangaroo, *Macropus giganteus*	Gélin et al. 2013
63% food	Coypu, *Myocastor coypus*	Gosling et al. 1984
66–236% energy	Squirrel, *Sciurus niger*	Havera 1979
77–226% food		
74% energy	White-footed mouse, *Peromyscus leucopus*	Millar 1978
78% energy	Bat, *Myotis lucifugus*	Kurta et al. 1989
92–126% energy	Guinea pig, *Cavia porcellus*	Künkele 2000;
		Künkele, Trillmich 1997
112, 120, 165% energy	Shrew, *Crocidura olivieri, C. viaria, C. russula*, respectively	Genoud, Vogel 1990
135–170% food	Deer, *Odocoileus hemionus*	Sadleir 1982
155–293% food	Grasshopper mice, *Onychomys leucogaster*	Sikes 1995
200% foraging	Red panda, *Ailurus fulgens*	Gittleman 1988
240–260% food	Deer, *Cervus elaphus*	Arman et al. 1974
285, 300% energy	Shrew, *Sorex coronatus, S. minutus*, respectively	Genoud, Vogel 1990
300% food	Hedgehog tenrec, *Echniops telfairi*	Poppit et al. 1994
323% energy	Vole, *Microtus brandti*	Liu et al. 2003

Note: Ranges reflect different stages of lactation. Single values are usually maxima.

earlier. Thus, the question, How much does lactation cost females merges into the next question: How do females pay for lactation?

Accommodating the Burden, or Paying the Price

Although birth technically marks the beginning of lactation, many females start preparing for lactation during gestation, not only by increasing their stores of fat and protein but also by physiologically priming the mammary glands for milk production. As we mentioned in the preceding section, mammals that routinely fast during lactation must deposit all the energy needed for milk synthesis prior to lactation. The energy needed may be substantive, for instance, during lactation, fasting elephant seal (*Mirounga*) mothers lose 42% (~100 kg) of their body mass. All of that mass had to be deposited earlier, during gestation or even before (Costa et al. 1986). In contrast, females of many species use both stored reserves and daily food intake to support milk synthesis. Examples are spiny mice (*Acomys*), hamsters (*Phodopus*), and cotton rats (*Sigmodon*; Degen et al. 2002), as well as humans (*Homo*; Dufour, Sauther 2002). Yet a third strategy is to rely solely on food available during lactation to sustain milk production. For example, many marmots (*Marmota*) and ground squirrels (*Urocitellus*) hibernate over the winter.

To survive winter, females must use the fat they stored during the summer and, therefore, have very little left to devote to reproduction (Broussard et al. 2005). These mothers spend the first half of the summer in gestation and lactation. They convert all the food they eat into their progeny. Then, after they wean their young, they spend the rest of the summer accumulating fat reserves to survive the next winter (Hayssen 2008c).

Some mammals use a completely different approach. These mothers store fat during lactation to fuel the next pregnancy, for example, reindeer (*Rangifer*). These denizens of the Arctic give birth in the spring, just before the short abundance of summer vegetation. Throughout the summer, they provide milk for their calf while also building up adipose tissue for the barren winter to come. Females wean their calves in the fall and will only conceive if they have accumulated sufficient fat (Pachkowski et al. 2013). Thus, fat stores laid down while lactating fuel gestation over the winter. Overall, mammalian females use various combinations of stored reserves and current resources to fuel gestation and lactation. Food availability and climate may dictate the exact strategy a female uses and even this need not be constant over the life of a female.

Energy is certainly an important component of milk production but so too is water. Water from milk may be the only source of water for developing neonates. As a consequence, water needed for lactation may be significant. The milk of naked mole-rats (*Heterocephalus glaber*) is the most dilute among rodents. Naked mole-rats are unusual mammals. They have a hive-like social system similar to that of bees. Only one female in a colony is reproductive, the queen, and her sole occupation is gestation and lactation. All other individuals provide the queen with food and they also tend to the young. With an average of 11 young per litter, the demands of lactation are tremendous. To meet that demand, a queen mother must daily produce an amount of milk equivalent to half her body mass. Given that her life span can exceed 30 years with several litters per year, she could produce 900 offspring. Thus, her water needs are persistent and substantial, and so, the food provided to her by the colony must have high water content (Hood et al. 2014).

Most mammalian mothers do not have colony members to secure food and water for them. Consequently, females may need to remain close to sources of water or find other ways to obtain more water. For instance, some Australian desert mammals (two rodent genera, *Notomys*, *Pseudomys*; the dingo, *Canis*; and kangaroos, *Macropus*), as well as many other mammals, consume the urine and feces of their young, thereby recycling roughly one-third of the water they put into milk (Baverstock, Green 1975).

As illustrated by the special case of the colonial naked mole-rats, one way to ease the costs of lactation is to share the burden. Group members in other social mammals, e.g., lions (*Panthera*), wild dogs (*Lycaon*), and meerkats (*Suricata*) also provision reproductive females or their young (see chapter 13 on conspecific interactions for more details). What options are available for other species? Can males help? With one or two possible, but hotly debated, exceptions (the Dayak fruit bat, *Dyacopterus spadiceus*, and the masked flying fox, *Pteropus capistrastus*), male mammals do not produce milk (Francis et al. 1994; Hosken, Kunz 2009; Kunz, Hosken 2009; Racey et al. 2009). But males can provide food to mothers during gestation or lactation and to offspring when they begin eating solid food (Rasmussen, Tilson 1984). Such food provisioning by males is common in monogamous canids (Moehlman, Hofer 1997). For most species, offspring, rather than males, supplement milk by finding food on their own. We will have more to say about the timing of first solid food in a bit.

The high cost of lactation has provided researchers with an opportunity to test for the existence of metabolic ceilings. These ceilings are relevant for any metabolic aspect of an animal's biology, but lactation is a particularly demanding process and has served as a focal point of physiology. The topic of metabolic ceilings is focused on a long-standing issue in physiology: What limits maximal performance? Numerous researchers have investigated this question, but the work on lactation has provided some of the best insights. The result has been the formation of two main hypotheses regarding the limits of milk production—the heat-dissipation hypothesis (Speakman, Krol 2010, 2011) and the central-limitation hypothesis (Hammond, Diamond 1992).

According to the heat-dissipation hypothesis, the necessity of losing excess heat produced by creating milk may set limits on the duration of lactation or the quantity of milk (Simons et al. 2011; Speakman, Krol 2010, 2011). In laboratory mice, when temperatures are warm, lactating mice won't eat more or nurse more pups, even if food is provided ad libitum. However, in cold conditions, mice will consume more food, especially lactating females, and these cold mothers produce more pups. The heat-dissipation hypothesis suggests that, at low temperatures, getting rid of excess heat is easy, and more energy can be devoted to milk production. The energetics have been modeled, measured (field metabolic rate), compared across a range of taxa, and even experimentally selected for, but the conclusions and detailed mechanisms of how heat loss is important remain unclear. The competing hypothesis is similarly under heavy investigation.

The central-limitation hypothesis posits that mammalian intestines have a finite capacity to absorb nutrients and that this capacity sets maximal absorption rates and therefore the amount of resources available for lactation (Hammond, Diamond 1992). Once this maximum is reached, females are unable to up their milk output. Certainly, female mice are unable to rear an infinite number of pups at once (26 at most according to Hammond and Diamond's experiments), but does heat dissipation or digestive absorption set the limit? The jury is still out in part because separate research groups provide support for each argument. Nevertheless, these two hypotheses about why metabolic ceilings exist came from observations of lactating females and have provided an exciting framework for future studies.

Ultimately, whatever is going on, the dairy industry may be interested in the result. Keeping cows in cold barns to increase milk production is feasible, but giving them larger or longer intestines is probably not. One problem is that extrapolating what happens in small mice to large cows may not be realistic given different surface area to volume ratios and gut architectures. But using both cows and mice may provide examples along a size continuum within which to investigate the role of digestive tract morphology for limits on energetically demanding processes, such as lactation.

Other Burdens of Lactation

We have discussed the cost of lactation solely in terms of milk production, but lactation has other consequences. Besides energy, milk supplies vitamins and minerals to developing young. Providing these nutrients can deplete maternal stores and weaken bones (Kwiecinski et al. 1987; Wysolmerski 2002). Calcium is especially important as it is a major component of bones and teeth, as well as a major player in muscle contraction, coagulation of blood, and transmission of nerve impulses. However, calcium is deficient in diets that consist solely of insects or fruit, such as those of anteaters and many bats, respectively (Barclay 1994). Also, because of acid rain, calcium may be limited even in deciduous forests (Battles et al. 2014). Another cost experienced by lactating females is that milk synthesis may result in oxidative damage (Fletcher et al. 2012).

Nursing, the time spent actually suckling young, can increase vulnerability to predators and take time away from foraging. If young are kept in a nest or a den, the need to return to the young restricts the area a female can cover while foraging. The cost of moving young from one den to another is an additional burden and may incur additional risks of predation. Alternatively, if young are constantly carried, a mother's movements and travel may be limited. Some of these burdens can be taken up by males, or relatives, as happens in small primates, such as titis (*Callicebus*), marmosets (*Callithrix*), and tamarins (*Saguinus*; Tardif 1994). In these Neotropical primates, fathers, or other troop members may carry young and return them to the mother for nursing.

Although nursing is a key part of lactation, the total time spent nursing can be quite short. For instance, tree shrews (*Tupaia*) nurse their young once every 2 days for 2–10 minutes per session (Emmons, Biun 1991; Martin 1968). Given that lactation only lasts about 30 days, total nursing time is under 2 hours. Similarly, rabbits (*Oryctolagus*) and hares (*Lepus*) nurse their young once each day, but, like the tree shrews, only for short periods (2–6 minutes), and do so over perhaps 25 days (Broekhuizen, Maaskamp 1980; Broekhuizen et al. 1986). Again, this adds to about two hours total nursing for the entire lactation period. At the opposite extreme, the young of tammar wallabies (*Macropus eugenii*) are attached to their mother's teats without interruption for 100–125 days. After the teat attachment phase, they detach, but suckle frequently for 75–100 days. Finally, they alternate grazing and suckling until the end of the 350-day lactation (Trott et al. 2003). As an intermediate example, sika deer (*Cervus nippon*) fawns nurse for 20–30 minutes per day for the first 14 days of lactation (Fouda et al. 1990). As evidenced by this diversity, the actual time spent nursing is far less than the length of lactation, at least in eutherians. This reduction in nursing time lowers the time mother and young are vulnerable. It may also increase the absolute length of lactation.

How Long Is Lactation and What Determines This Duration?

When defined as the time between birth and weaning, lactation ranges over almost three orders of magnitude, from 4 to 5 days in hooded seals (*Cystophora cristata*), casiragua (*Proechimys guairae*), and elephant shrews (*Macroscelides proboscideus*), to more than 900 days in chimpanzees (*Pan troglodytes*), and up to 6.5 years in orangutans (*Pongo pygmaeus*; Hayssen 1993). Although extremely short lactation lengths (<10 days) are rare, long lactation lengths (>500 days) are more common, especially for large-bodied species with singleton offspring, such as kangaroos, great apes, walruses, sirenians (dugongs and manatees), elephants, and rhinos. However, half of known lactation lengths range from 29 to 125 days (Hayssen 1993).

Briefly, let's examine lactation in monotremes and marsupials. In general, the mean lactation length is longer for the egg-laying monotremes (150 days, 3 species) and marsupials (120 days, 75 species) than for eutherians (50 days, 675 species). Why is lactation longer?

Lactation in monotremes and marsupials, but not eutherians, encompasses the exponential period of offspring growth. In marsupials, lactation has a phase of continuous teat attachment and a phase of intermittent nursing. Occasionally, the teat attachment phase has been equated with the post-implantation phase of gestation in eutherians. Superficial similarities exist, but this synonymy ignores the vastly different proximate mechanisms (metabolic and physiological control) and ultimate causes (selective pressures and evolutionary constraints) that distinguish gestation and lactation from each other. Thus, the longer length of lactation is because marsupials and monotremes use lactation instead of gestation to support most offspring development and growth.

Body size influences the length of lactation across mammals. For example, rearing a large white-rhino calf (*Ceratotherium*) takes longer than supporting a small ferret pup (*Mustela*). Females of larger species have longer lactations, with some interesting exceptions (Hayssen 1993; van Noordwijk et al. 2013). On the shorter-than-expected side are earless seals (phocids) and baleen whales but not other marine mammals. The extremely short lactations of phocid seals may be an adaptation to the unpredictable locations where they nurse their young

(Bonner 1984) combined with the ability of these seals to store fat (Schulz, Bowen 2005). For instance, hooded seals give birth and nurse on melting ice floes. Over 4 days, they pump 30 kg of super-fatty milk into their neonates and then leave their bloated pups on the melting ice to fend for themselves (Bowen et al. 1987). Baleen whales are the largest mammals; yet, lactation lengths for these giants are not commensurately long, only 5–7 months for *Balaenoptera* (Brodie 1969). The relatively short lactations of baleen whales reflect the timing of their annual migrations. These females move away from their primary food source at high latitudes to more equitable low latitudes to calve and nurse while simultaneously fasting. After a few months, mothers travel back to colder, food-rich waters. A lactation length proportional to their large mass would not allow reproduction to be synchronized with the seasonal nature of their food supplies and migration.

Other mammals, such as bats, primates, and marsupials, nurse for longer than expected. Marsupials are born at a very early stage of fetal development, and mothers use lactation rather than gestation to support the physical growth and development of their young. Primates, on the other hand, use lactation to support psychosocial growth. Similarly, for bats, the extensive neuromuscular coordination and skeletal development required for food acquisition in flight may demand a longer post-natal development. Perhaps newborn bats and primates are as "fetal" in their neurological development relative to their adult needs as neonatal marsupials are physically.

Although body size is important across mammals as a whole, it may not be relevant within smaller taxonomic groups. Only for primates, artiodactyls, and rodents is lactation related to female mass (Purvis, Harvey 1995). Even within these orders, the relationship between lactation and female mass is weak. Body size is important in big-picture comparisons, for example, elephants provide milk for their calves for much longer than mice provide milk for their pups. But for comparisons at smaller taxonomic levels, such as within families or genera, body size is much less important. What other factors influence the length of lactation?

To get to a full answer, we need to understand the functions of lactation. If the sole function of lactation was to provide sustenance, then mothers might just stuff their young as full of milk as possible, as fast as possible, and then leave, as hooded seal mothers do. But lactation often plays a larger role in mammalian reproduction than simple nutrition.

Lactation may help with keeping offspring warm. Milk can have either direct or indirect influences on thermoregulation. Indirectly, the high fat composition of ursid, cetacean, and phocid milk serves a thermoregulatory function when the fat is converted to insulating blubber. More direct thermoregulatory advantages may also accrue. Many altricial neonates are unable to maintain high body temperatures, and milk with a high specific heat coming from endothermic mothers may serve to warm the young from the inside out, especially in newborns for which ingested milk can be a relatively large proportion of neonatal mass. For instance, when mother tree shrews provide milk every other day to their young they also provide a bolus of heat even if the warmth may not last long (Fuchs, Corbach-Söhle 2010).

Lactation also lends itself to intimate social interactions with multiple facets. Some interactions between mother and offspring are pragmatic, such as the transfer of immunity, calories, water, or minerals. Others have to do with secondary qualities of milk, such as thermoregulation. However, in some cases, a mother's help is more prosaic, as with the offspring of small rodents, who depend on maternal stimulation to induce urination and defecation (Numan, Insel 2003). Other functions of lactation also reflect the proximity of mother and young. As a consequence of this proximity, mothers offer their young socialization and protection from predators, as well as opportunities for learning about food, migration routes, and safe havens. These consequences may be at least as important as the nutritive role of lactation. For example, Indian Ocean dolphins (*Tursiops aduncus*) have non-nutritive (milk-less) suckling of their calves for up to 2 years after weaning (Boness et al. 1996). *Tursiops aduncus* is an extreme example of suckling past the age of nutritional independence. At the opposite extreme, the hooded seal stops nursing her young 4 weeks before that pup first eats solid food. Lactation may have different functions, evolutionary constraints, and physiological control, depending on whether young first eat solid food before weaning, near weaning, or well after weaning. Just as implantation separates two phases of gestation, first solid food is a pivotal point in the function, physiology, and evolution of lactation (Langer 2008).

First solid food eaten near weaning occurs in polytocous (having a litter rather than one young at a time) species with altricial young; in this case, lactation has a clear energetic role. For example, first solid food is nearly simultaneous with weaning in many sciurid (e.g., squirrels) and muroid (e.g., rats or gerbils) rodents. In contrast, first solid food well before weaning is common for mammals with single, precocial offspring. The first solid food a neonate eats may be on the day of birth or shortly thereafter. Examples of this early milk supplementation include hystricomorph rodents such as porcupines (*Erethizon*), and coypus (*Myocastor*), as well as other mammals, such as sloths (*Bradypus, Choloepus*), zebras (*Equus*), and various bovids. For these species, the energetic and nutritional constraints on lactation may be less important than the benefits of maintaining contact between mother and young, benefits such as reduced juvenile mortality and increased opportunities for learning social or foraging patterns, as with *Tursiops aduncus*. First solid food can also occur well after weaning, as in phocid seals. For instance, elephant seals nurse their young for 23 days and then leave them. Those young fast for 6 weeks before they first hunt for food (Carlini et al. 2000).

In general, the longer the period of lactation, the sooner first solid food is ingested by young relative to weaning. For instance, for most (90%) eutherians with lactations of at least 1 year, first solid food occurs within the first third of lactation. In contrast, when lactation is under 50 days, only a few (11%) eutherian young eat solid food in the first third of lactation. One exception is the short-tailed shrew (*Blarina brevicauda*). These mothers regurgitate food onto their nipples, which is lapped up by their young before they open their eyes (Miller-Ben-Shaul 1963). However, in general, infants depend completely on milk for a larger proportion of lactation when lactation is short, whereas offspring may rely less on milk when lactation is long (Hayssen 1993). With early first solid food, lactation has benefits besides nutrition.

Many well-developed, singleton young eat solid food early in lactation. At birth, they are furred, mobile, with open eyes and ears, but they remain with their mothers for long periods. For these species, concurrent gestation and lactation seldom occurs. Overall, the relationship among these reproductive characteristics—early first solid food, precocial singleton young, and small relative mass at birth—is distinctive and may be a highly derived condition. But not all species with precocial young fit this pattern, for instance, guinea pigs (Cavia), hares (*Lepus*) and hyraxes (*Dendrohyrax, Heterohyrax,* and *Procavia*) have precocial young and have them in litters.

At the opposite extreme are species with litters of altricial young. Although reproductive patterns are highly variable, in these cases, the mass of the litter is often a large proportion of the mother's mass, and the age at first solid food is often late in lactation, usually closer to weaning. In addition, gestation and lactation are commonly concurrent. The production of multiple offspring with a large litter mass at birth, altricial development, and late first solid food may be the ancestral eutherian condition. Of course, this pattern, too, has exceptions. Bears are certainly one. They have two to four altricial young that are a very small percentage of their mothers mass.

Each unique combination of reproductive characteristics may result in different consequences for females. At weaning, litter mass of mice and voles (altricial with late first solid food) is often larger than that of the mother, whereas a calf or human infant at weaning (single precocial offspring with early first solid food) is still only a small fraction of maternal mass. In addition, the mouse and vole weanlings have relied almost entirely on their mothers for nutrients and energy during lactation, whereas the human infant and the calf supplemented their diets with solid food for a large portion of lactation. Not only are the demands of lactation enormous when multiple young are dependent for all their nutrition and energetic needs on milk, as with voles and mice, but, with concurrent gestation, the female may also be meeting the nutritional needs of a second, in utero, litter. Such a mother's energetic budget and potential for future reproductive success might be tightly constrained, even though these species tend to be small, with high metabolic rates and, consequently, able to process energy more quickly than larger mammals.

For mothers with single precocial offspring, milk production may not be the core of lactation; instead, nursing may be central to maintaining the mother-offspring bond. Non-nutritive nursing is common for large mammals, and the benefits are varied. Predators may be less apt to attack a small juvenile if a nearby mother is ready to intervene. Young have time to learn the location and quality of food, nest sites, or potential threats. Finally, when young are born into a larger social arena of colony, troop, tribe, or herd, their survival and reproductive success may depend on their integration into appropriate social roles. Long lactations that continue after nutritional or energetic need may facilitate this social integration. Parenthetically, the hormone oxytocin is involved in facilitating both milk letdown, social bonding, and parental care (Finkenwirth et al. 2016).

The timing of weaning may differ with nutritive versus non-nutritive lactation. The cost to a mother of continuing to nurse a single offspring a fraction of her size, which already obtains a proportion of its nutritional requirements on its own, will be much less than the cost associated with supporting multiple young that collectively weigh as much or more than the mother does. Overall, the cost-to-benefit ratios associated with weaning are different, and mothers of singletons may allow repeated nursing longer than mothers of litters. Mothers of litters are also more likely to start a second litter immediately after the birth of the first. These females are pregnant and lactating simultaneously. When this happens, the birth of the second litter forces the weaning of the first.

Overall, the duration of lactation is influenced by its various functions, by the size of the mother and the size of the neonate at birth, by litter size, by developmental state of the neonate, and by the fact that the first solid food an offspring eats can be either before, at, or after weaning. For individual females, food availability, health, season, social interactions, age, and many other aspects of her body condition may influence lactation. The diversity of functions for lactation and the central place of lactation in mammalian reproduction have developed over 200 million years, but it all started with milk. Thus, a chapter on lactation would be incomplete without a discussion of that material unique to mammals.

Milk: Static and Dynamic

Milk composition varies among mammals and has a copious literature aimed at understanding this variation (Kuruppath et al. 2012; Oftedal 2013; Skibiel et al. 2013). The major constituents of milk are fats, carbohydrates, proteins, minerals, vitamins, and water, but the relative amounts vary a great deal. Take fat, as an extreme example. Black rhino (*Diceros*) milk has only 0.2% fat, whereas the milk we are most used to, that of dairy cows (*Bos*) has 20 times more fat, or 4%. The milk of harp seals (*Pagophilus*) is 57% fat; thus, harp seal milk has 285 times more fat than black rhino milk (Oftedal et al. 1987; Skibiel et al. 2013), and milk of gray (*Halichoerus*) or hooded seals may reach 61% fat (Boness, Bowen 1996). Other components have less extreme ranges. For instance, sugars (carbohydrates) range from near zero in some seals, such as the Weddell seal (*Leptonychotes*) or the South American fur seal (*Arctocephalus*; Oftedal et al. 1987) to a high of 10% to 12% in the tammar wallaby (Trott et al. 2003). Our own, human, milk is 3% to 5% fat, 6% to 7% sugar, and just about 1% protein (Jenness 1979).

What accounts for the diversity? Ancestry (evolutionary history) accounts for a great deal of the variation (Skibiel et al. 2013). In other words, bear milk differs from cow milk because bears tend to be bear-like just as cows and their relatives tend to be cow-like. But that answer is somewhat frustrating; we want to know why bear milk differs from cow milk, not just that the ancestral milks also differed. For instance, what exact feature of bears results in the consistent production of low-sugar and high-fat milk? Although the answer "the ancestors differed" is not very satisfying, it does tell us that the differences in milk are ancient, not recent (box 9.1). Thus, the reasons for the differences have to be ancient as well—the differences are not recent adaptations to current conditions. To find answers, we have to look at other aspects of the biology of particular groups.

Diet is one such defining feature of milk composition. Milk needs building blocks that come from the diet; therefore, differences in diet can lead to varied milk composition. Herbivores not only eat different materials than anteaters or carnivores but also process their food differently. For example, herbivores rely on their intestinal microbiome to digest materials, such as cellulose, that mammalian enzymes are unable to break apart. Herbivores then absorb the products of microbial digestion, as well as the microbes themselves. Carnivores skip the microbial fermentation step. Thus, the material available to the mammary gland for synthesis differs according to diet. Not surprisingly, the end product—milk—differs.

Another large-scale influence on milk is the function of lactation in different species. For marsupials, lactation is the major component of their reproductive biology, and milk is central in the growth and development of young. For many primates, lactation has a significant social function, and milk is less important than the nursing and maternal proximity that accompany it. Our earlier discussion of when young first eat solid food explored this functional component of lactation, as well as how other key variables, such as litter size and neonatal size, alter the length of lactation. Overall, milk compositions are consistent with the different functions of lactation.

Dynamic Milk

Up to now, we have discussed milk composition as though it were a static product, but milk composition usually changes over the course of lactation. Milk composition in marsupials is especially dynamic, particularly for the tammar wallaby. Lactation for tammar young has three stages each about 100 days long (Trott et al. 2002). For the first 100 days, the neonate is permanently attached to a teat and receives, on demand, a dilute milk that is low in fat but high in sugars. As these young grow, the teat also enlarges to continually fill the neonate's oral cavity. For

BOX 9.1. Why Do Bears Have High-Fat, Low-Sugar Milk?
AN EXAMPLE OF THE LONGEVITY OF REPRODUCTIVE PATTERNS

The eight extant species of bears provide an excellent example of the persistence of reproductive traits in spite of great differences in diet and habitat. Today's bears have diets that may be highly specialized (sloth bears, *Melursus ursinus*, eat ants; giant pandas, *Ailuropoda melanoleuca*, eat bamboo) or strictly carnivorous (polar bears, *Ursus maritimus*, eat only other animals), although most bears are omnivorous and opportunistic dietary generalists (e.g., black and brown bears, *U. arctos, U. americanus*). Habitats for bears are also diverse, ranging from lowland jungles to the high altitudes of the Andes, and from the equator to the Arctic (see figure). In spite of these great differences, bears share a reproductive pattern that may be 20 million years old.

As far as we can tell, here is the ancestral reproductive pattern of bears from the female perspective. The mother of all current bears looked for a male, ovulated when she found one, conceived, and then put that conception on hold by delaying implantation while she looked for another male. She may also have used short-term sperm storage before she ovulated. She repeated this process until she had up to four embryos. Next, she looked for a secluded den in which to wait out a food-restricted season. She held her embryos in abeyance by continuing to delay their implantation until an environmental cue, probably photoperiod, signaled the time for implantation. How she obtained a photoperiodic cue while nestled in her dark den is a bit of a mystery, but embryo growth ensued nonetheless. About 60 days later she gave birth to one to four, blind, bald, and toothless cubs. These were exceptionally tiny as each was perhaps 0.3% of her weight (equivalent, in a human frame of reference, to a 100-pound mother giving birth to 5-ounce infants). The small size of the neonates reduced the demands of early lactation.

This reduction was very important, for although she was nursing, she neither ate nor drank throughout the weeks or months of her denning period. The only material she may have ingested was the excreta of her cubs, possibly recovering some water and nitrogen lost through nursing. Her milk was high in fat and low in sugar as adaptations to the physiological constraints of fasting as well as the constraints of not drinking while still producing milk. This ancestral-bear pattern is retained with minor modifications in all eight species of extant bears with their different diets and habitats (Farley, Robbins 1995; Garshelis 2004; Oftedal 2000; Ramsay, Dunbrack 1986; Spady et al. 2007). Thus, bears present an excellent example of evolutionary holdovers in reproductive patterns and why ancestry may influence reproduction more than current habitat.

Polar bear (*Ursus maritimus*) mother and cubs, Hudson Bay, Canada. Image by D. Meril and M. Manon, Biosphoto; used with permission from Minden Pictures.

the next 100 days, the young releases the teat but remains in the pouch and suckles frequently. For the last 100 days, the young leaves the pouch, begins to eat grass, and suckles vigorously but less frequently. For these older young, milk is low in sugar and high in protein and fat. In addition to changes in major nutrients (protein, fat, and sugar), marsupial milk composition has subtle changes tailored to the developmental needs of the young. For instance, hair and claws are made of proteins called keratins, and these proteins include a large number of sulfur-containing amino acids (e.g., cysteine). Marsupial milk is rich in these sulfur-containing amino acids exactly when marsupial young are developing hair and claws. Thus, specific materials are secreted at particular times and, thereby, precisely regulate key stages of neonatal development (Lefèvre et al. 2010; Renfree 2006, 2010; Trott et al. 2002). As Marilyn Renfree put it, marsupials "exchange the umbilical cord for the teat" (Renfree 2010:S26).

Marsupials have a second complex aspect to milk production. Not only does milk composition change over time, but milk of different compositions is provided to young of different ages simultaneously (Renfree 2006). Low-fat, high-protein milk is given to teat-attached pouch young, whereas from an adjacent nipple, high-fat, high-protein milk is provided to her joeys at heel (Renfree 2006). To prevent milk from overloading the small pouch young when the large joey suckles, the two mammary glands respond differently to concentrations of mesotocin (the marsupial variant of oxytocin) to stimulate of milk letdown. The gland supplying the pouch young is very sensitive to mesotocin, whereas the gland supplying the joey requires higher concentrations of the hormone for the same level of response (Renfree 2006).

One last aspect of lactation in tammars deserves mention. Cross-fostering experiments, exchanging an older pouch young for a younger one, demonstrate that mothers regulate the rates and development of their pouch young by both the composition and the amount of milk (Trott et al. 2002). Overall, lactation in marsupials is exceedingly complex, with exquisite timing to mesh the needs of young of different ages with the needs of the mother. What about other mammals?

Monotreme females also use lactation rather than a placenta as the conduit for nutrition and developmental signals. Monotremes do not reproduce well in captivity, but what little we know about milk composition in egg-laying mammals suggests that echidna (*Tachyglossus*) milk changes over the course of lactation, but platypus milk does not (Sharp et al. 2011). Reasons for this difference await further study.

Although marsupials have the most dynamic milk changes over lactation, eutherian milk is by no means monotonously constant. The most obvious example is the difference between early milk, colostrum, and so-called mature milk. Colostrum is well known for providing protective antibodies to newborns as well as an array of other compounds related to immune function. Comparison of colostrum in about 20 species suggests that differences in colostrum reflect the degree of prenatal transfer of immune components (Langer 2009). The anti-microbial functions of milk continue past the colostrum stage and are also an ancestral feature of milk. Anti-microbial components are present in monotreme milk (Eniapoori et al. 2014) and milk may have originated from anti-microbial secretions (Hayssen, Blackburn 1985; Vorbach et al. 2006).

Although the change from colostrum to milk is obvious, other subtle changes in eutherian milk composition occur. During the early part of lactation, short-term milk composition changes while the mammary glands are adjusting to suckling; for example, the amount of sugar in milk at first suckling is lower than that a few nursings later (Jenness 1984). Milk composition also changes over the course of lactation. For instance, the vitamin A content of elephant seal milk is six times higher in late lactation, whereas the amount of vitamin E drops. The elephant seal vitamin E changes are similar to those in terrestrial females, but the vitamin A changes are not (Debier et al. 2012). Similarly, the sugars in milks of grizzly bears (*Ursus arctos*) and black bears (*Ursus americanus*) are two to six times higher early in lactation (1–3%) when mothers are denned up with their cubs and fasting, but after hibernation, sugar drops to 0.5% when mothers are able to forage (Farley, Robbins 1995). These are just two examples, but more nuanced changes occur, not only over the course of lactation but even during a single nursing bout. These changes may regulate both neonatal development and provide feedback for maternal physiology.

Control of Lactation: Mother, Offspring, Siblings

At the physiological and molecular levels, the control of lactation is complex both during a single nursing bout as well as over the entire period of milk production. Behavioral and physiological changes in mothers are associated with both increased food intake and milk synthesis, and they involve multiple endocrine systems, including metabolic hormones, such as leptin, ghrelin, insulin, thyroid hormones; hormones involved in mineral regulation, such as calcitonin, vitamin D, and parathyroid hormone; and hormones involved in suckling, such as oxytocin and prolactin. All these (and more) messengers interact with various neuroendocrine pathways to permit or prevent milk synthesis, suckling, and other aspects of lactation. The scientific emphasis on humans, cows, and laboratory mice and rats has great practical value but is of little help in understanding the proximate control of lactation in other mammals.

The building blocks provided by milk and the act of being nursed clearly change neonates but not just because of the nutritional or immunological effects. Other components in milk include hormones and molecules that can alter the timing or scope of neonatal development. In this regard, the mammary gland can be viewed as an endocrine organ, and the term *lactocrine* is now used to describe some of these non-nutritive or immunological functions (Bartol et al. 2013). Lactation and milk may have an active, not passive, role in neonatal development.

Here are examples of ways mothers may influence their young through milk or nursing.

Through milk, mothers have the potential to alter the behavior of their young. For instance, in macaques (*Macaca*) milk cortisol is associated with the temperament of offspring; higher cortisol levels lead to higher weight gain but more nervous, less confident offspring (Hinde et al. 2014). In addition, recent speculation posits a complex interaction between mother's milk, intestinal microbes of the neonate, and infant behavior (Allen-Blevins et al. 2015).

Nursing is also a means to influence behavior. Rabbit mothers secrete a pheromone (an odor that influences the behavior of another individual) during nursing that triggers neonates to quickly locate nipples and begin suckling. This odor provides a conditioning stimulus (think Pavlovian response) and promotes learning of nearby environmental cues. Thus, maternal odors are organizing neonatal cognition (Coureaud et al. 2010). For rabbits, nursing and lactation are a mechanism to provide neonates with information about the environment while they are still in the nest.

Milk and nursing also set yet another stage for sibling competition. Because it occurs outside of the confines of the female's reproductive tract, conflict during lactation is easier to observe relative to earlier stages. The finite number of mammary glands may result in conflict if litter size exceeds the number of teats, as is the case in many dasyurid and didelphid marsupials. For these species, only the first young to find teats will have a chance to survive. Young that survive this initial reduction in litter size, as well as those that are born as singletons, attach to a teat constantly for the first phase of their long lactations.

For mammals without permanent teat attachment, sharing teats is possible. However, timing is important. Milk composition probably changes during the course of a single suckling period, and subsequent young may not receive milk of the same quality. Young that suckle later may not be able to suckle as long, especially if females control when nursing stops. However, stimulating the teat may enhance milk letdown in subsequent sucking. These challenges may produce sibling conflict for specific nipples. Aggressive interactions between kittens and piglets for teat ownership result in the establishment of a dominance hierarchy among siblings (Hudson, Distel 2013).

Sibling competition may be reduced if, as soon as the mother appears, siblings quickly attach to a nipple and not let go. This tenacious nipple attachment is known for more than 40 species of muroid rodents (mice, voles, deer mice, gerbils, hamsters, and others). The young of a few of these species even have specially developed incisors to help with the attachment (Gilbert 1995). Females of different species have different ways of dislodging their young at the end of a nursing session. Prairie vole (*Microtus ochrogaster*) mothers use their teeth to pull their tenacious young off the nipples, whereas pine vole (*Microtus pinetorum*) mothers spin in tight circles to release the young, presumably either taking advantage of centrifugal force or perhaps just getting them dizzy (McGuire, Sullivan 2001). For marsupials without pouches, such as short-tailed opossums (*Monodelphis*) females may be able to retract their nipples and leave their young in a nest (Fadem et al. 1982). Some mothers don't use any method and simply drag their determined young with them as they forage (Alligood et al. 2008). Ouch.

Although the negative effects of siblings on one another during lactation are well documented, having siblings may also be positive. Rhesus macaque (*Macaca mulatta*) mothers have greater milk yields after previous litters. Thus, by priming the mammary glands, older siblings indirectly help subsequent siblings (Hinde et al. 2008). In addition, siblings huddling in a nest can reduce the costs of thermoregulation so more energy can be used for growth (Nicolás et al. 2011). As part of a litter, siblings may benefit through the dilution effect, whereby the chance of being the individual eaten by a predator is decreased purely by being in a group.

Finally, What Do Pups Do with Milk Once They Have It?

What do pups do with milk? Play or pile on fat? Learn skills or get bigger? Once milk is delivered, mothers have little to no control over how their young use that milk. A naive assumption might be that milk is only used for growth. Babies suckle and get bigger over time. But this view neglects to account for the fact that even just living takes calories. The energy from milk has to be used for daily physiology (maintenance) as well as for growth. From the offspring's perspective, milk can be used for growth, for storage (e.g., put into fat, including blubber), or to support activities, such as play or learning how to forage (Arnould et al. 2003). Further complicating things, not all components need to be used the same way. The calories from milk could be used for activity, but the mineral components (e.g., calcium) might be stored or used for growth.

Pups from even closely related species living in the same habitat may opt for different strategies. For instance, Antarctic fur seal pups (*Arctocephalus gazella*) use milk for growth and neurological development, e.g., learning to swim and dive, whereas subantarctic fur seal pups (*Arctocephalus tropicalis*) direct resources to adipose tissue (Arnould et al. 2003). These differences occur in the context of different weaning ages. Antarctic fur seal pups must forage independently at 4 months of age, whereas subantarctic fur seal pups have access to milk for 10 months, but only for 3–4 days every 2–4 weeks (Georges et al. 2001). Therefore, pups of subantarctic fur seals must survive long periods of fasting (Arnould et al. 2003).

Before leaving the pup side of lactation, the proposed blowhole nursing of sperm whale calves (*Physeter macrocephalus*) deserves mention. The head shape of sperm whales, with its huge rostrum and reduced, narrow, and under-slung jaw, may make grasping a teat between tongue and palate a difficult maneuver. In a set of underwater observations in the Caribbean and Sargasso Seas, calves appear to push their blowholes up against the recessed slit in which the teat lies and allow mothers to release milk into what is essentially the pup's nasal passage. The logistics of how calves keep the nasal and digestive passages separate are both speculative and complicated (Gero, Whitehead 2007). However, underwater observations of two calves in the Mediterranean Sea suggest that the more usual oral nursing can occur (Johnson et al. 2010). Certainly suckling by mouth is the more likely mode of nursing in sperm whales, but if blowhole nursing does happen, it would be yet another example of the diverse ways in which lactation adapts to the myriad needs of females and their offspring.

Lactation: The Essence of Mammals

Lactation defines mammals (quite literally). Mammals were first named by Carl Linnaeus (1758) based on the Latin *mam* for milk-producing glands. But lactation is much more than anatomy. It includes the physiological components of nursing and the associated behavioral feedback between mother and offspring. The German name for mammals *Säugetier* derives from the verb *saugen*, to suck, and emphasizes the neonatal side of the interaction. Thus, both from the maternal and neonatal perspectives, mammals are noteworthy among vertebrates in their use of milk.

Lactation is a large part of a female's reproductive life. The energy and materials devoted to milk can be substantive. These demands are so large that lactation has been used as a model to understand metabolic ceilings (maximal energy use) in mammals. Lactation also takes time and may restrict movement, reducing opportunities for foraging while increasing opportunities for predation. But lactation and the mother-offspring bond have also led to social networks and cooperative breeding.

Lactation is central to the evolution of mammals. Mammals are defined by this multifaceted component of their reproduction, which has influenced their biochemistry, physiology, anatomy, behavior, sociality, ecology, and evolution. Just as gestation buffers young from the external environment, lactation buffers young from the need to forage and process adult foods. For juvenile females, the end of lactation is the start of their independence and leads to the beginning of their own reproductive lives. Those new beginnings are the subject of the next chapter.

CHAPTER TEN

Weaning and Beyond

Our mother spotted hyena is notably thinner than when we saw her last, even though she has lost one daughter. Her ribs jut from her thin torso, and she has a few small wounds that are healing poorly. Conversely, the growth of her surviving daughter seems to know no bounds. Her tiny belly is plump; yet, she is relentless in her demands for more milk from her haggard mother. Her fur has the distinctive spots for which the species was named. After almost 18 months, the mother is at her physiological limits, not only is she barely able to produce enough milk to provide for the increasing demands of her daughter, but also, as a subordinate of the group, she must grapple for every bite of food she converts to this precious milk. She is exhausted. Today is the day, and it is long overdue. Weaning day. The pup may not have noticed its mother has become more and more reluctant to allow nursing or even slightly hostile to such attempts, because the pup only notices it is hungry—always hungry. Weaning will not be a kind process for mom or for daughter. Each party wants things to go her way. Our mother would have benefited if this all had been done a month ago. The daughter, however, would not mind living another month or two in the lap of luxury that is maternal care. However, if she keeps providing milk, the mother may not reproduce again for another year, and her fitness, as well as the indirect fitness of her currently hungry daughter, will suffer. Once weaned, the young female hyena must fend for herself within the group, share the meat with her group-mates, and accept her meager portion, for she, the pup of a low-ranking female, is at the bottom of the female social hierarchy, although dominant to males. Times may be difficult, but if all goes well, the weanling will stay fed, disease free, and survive any catastrophic events so she may herself reproduce. But, first, she must go through puberty to reach sexual maturity. (Hill 1980; Holekamp, Smale 1993)

> Weaning is a singularly ill-defined term with many shades of meaning ranging from the specific to the general, the colloquial to the technical.
>
> —Martin 1984:1257

This book section (part 2) on the reproductive cycle ends by exploring the entrance to and the exit from a female's reproductive life. Unlike most of reproduction, the events discussed here are not cyclic but may only occur once in an individual female's lifetime. Weaning may be anomalous, because a female may have occasion to wean many offspring; however, she herself is only weaned once. In addition, she is only a juvenile once, a subadult once, and only goes through puberty once. Finally, a few females will go through menopause once to have a post-reproductive life. In this chapter, we cover the two major periods of a female's life when she is independent of reproduction: the ubiquitous pre-reproductive period from weaning to puberty and the much less common post-reproductive period that occurs primarily for cetaceans, elephants, and primates.

Weaning

Precise definitions for weaning are difficult, because weaning involves two perspectives: the mother's and the offspring's (Lee 1996). Weaning can be defined or assessed from either perspective, but the definitions do not always coincide. On the offspring side, weaning may mean first emergence from the natal den, first consumption of solid food, dispersal from the nest, ability to maintain weight after isolation from the mother, or the vague term *independence*. From the maternal perspective, weaning may mean cessation of lactation (i.e., milk production), reduction in the duration or frequency of suckling bouts, renewed ability to conceive another litter, or simply the time at which a female abandons her litter (Hayssen 1993). Depending on the definition, weaning may be sudden or prolonged. No one measure of weaning has consensus, because weaning is a process and not an event.

Whatever the exact definition, the event of weaning changes the lives of mother and offspring but in opposite ways. Energetically, after a period of intense investment, a mother's daily input into her offspring drops sharply, and the entire energetic burden falls on the offspring. The energetic transition is often difficult for newly weaned young, as evidenced by a decrease in growth rate at this time, e.g., deer mice (*Peromyscus*) and golden mice (*Ochrotomys*), pikas (*Ochotona*) and elephant seals (*Mirounga*; Bryden 1969, Linzey, Linzey 1967; Puget, Gouarderes, 1974), as well as by increased levels of stress hormones, e.g., free-ranging rhesus macaques (*Macaca*; Mandalaywala et al. 2014). For some species, such as ground squirrels (Hayssen 2008c), weaning occurs when environmental resources are abundant, thus diminishing the burden on newly weaned young. Other species may provision juveniles after weaning. Mothers may bring freshly killed prey (carnivorans), allow juveniles to take partially chewed food from their mouths (some hystricomorph rodents), spit out chewed food (platypus), regurgitate food (wolves), or provide special soft fecal material (koala, pika, *Ochotona*; Ewer 1973; Pond 1977). Newly weaned young may accompany their mothers when foraging, as with hedgehogs (subfamily Erinaceinae) and moose (*Alces*; Pond 1977). Such activities increase the survival rates of weanlings.

The shift of energetic burden from mother to offspring is a potential source of conflict between these two parties. From one theoretical perspective (Trivers 1974), a mother must choose either to provide resources to current offspring or to save her resources for future litters. This same perspective posits that offspring should demand as much from their mothers as possible. Any daughter clearly benefits from increased maternal attention, but she is also related to any future young her mother produces. Thus, kin selection (the genetic benefits of helping your relatives because you share genes with them) will mediate some parent-offspring conflict.

Weaning can also be a source of conflict among siblings. Within a litter, siblings may vary in their development, and a single weaning time may not benefit all offspring. As with the timing of birth, the timing of weaning may not suit all parties equally. In addition, if a female is simultaneously pregnant and lactating, the birth of the second litter may influence the weaning time of the first litter. Facultative delays caused by the suckling of young can delay the birth of the second litter (chapter 7). Even in species with singleton litters, if lactation is sufficiently long, birth may occur when an older sibling is present and nursing. For example, in Galápagos fur seals (*Arctocephalus galapagoensis*) and Galápagos sea lions (*Zalophus wollebaeki*) with 2-year lactations, "up to 23% of pups are born while the older sibling is still being nursed" (Trillmich, Wolf 2008:363). The younger pups are at a disadvantage when they compete with older siblings, and mothers may aggressively force weaning on the older sibling (Trillmich, Wolf 2008).

The idea of intergenerational conflict has been expanded to other aspects of reproduction and sparked an explosion of life-history theory related to who controls reproductive investment: parents or offspring. We explored some ramifications of this theory in our chapters on gestation (chapter 7) and lactation (chapter 9). Intergenerational cooperation appears later in this chapter when we investigate advantages of post-reproductive life. But before we move to the end of a female's reproductive life, we will dive a little deeper into weaning, for instance, just how large are weanlings?

Size at Weaning, Weaning Mass

Species vary a great deal in how large offspring are at the time of weaning. For one thing, the absolute size at weaning is related to adult size. That is, larger species tend to have larger young at weaning, for instance, an elephant at weaning is bigger than a weaned meerkat (*Suricata*). However, as a percentage of adult size, weaning mass tends to be larger for smaller mammals, e.g., 66–90% in some bats (*Eptesicus, Miniopterus, Nyctalus, Pipistrellus, Tylonycteris*), shrews (*Cryptotis, Neomys*), and rodents (*Peromyscus, Pseudomys, Nyctomys*; Hayssen 1985). Where do humans fit? While age at weaning in humans is culturally variable, toddlers at

30 months of age weigh 11–16 kg (25–35 lb.) and are, therefore, about 19–26% of a 60-kg (132 lb) woman (CDC 2000; Humphrey 2010).

Across 162 species of mammals from shrews to walruses, weaning mass is about 40% of maternal mass and ranges from 5% in manatees (*Trichechus*) to 100% (same size as mom) in a couple of bats (*Myotis nigricans* and *Megaderma spasma*; Hayssen 1985). Thus, other factors besides body mass are in play. For instance, weaning mass in squirrels is one-third of maternal mass, but, when examined by habitat, weaning mass is lower for ground squirrels, such as chipmunks and marmots (29%), than for arboreal squirrels (45%; Hayssen 2008b). Pikas occupy similar habitats to those of ground squirrels, and weaning mass in pikas (32% of maternal mass) is close to that of ground squirrels. But rabbits and hares also occupy similar habitats and have much smaller weaning mass (24% for the hares, *Lepus*; 14% for the rabbits and cottontails, *Brachylagus*, *Oryctolagus*, *Pronolagus*, *Sylvilagus*; Hayssen et al. 1993). Another example comes from the carnivorous, aquatic pinnipeds. Both eared (otariid) and earless (phocid) seals generally have weaning masses that are 20–30% of maternal mass, but that of the walrus (*Odobenus*) is 64%. Clearly, habitat or diet do not influence weaning mass across broad taxonomic categories (e.g., across carnivorans) but may be influential at smaller taxonomic levels, for instance, within a family as with squirrels.

Weaning mass by itself is not a key feature of reproduction, but the relationship of weaning mass to adult size is informative. That percentage indicates how much additional growth a female needs to do before she reaches adult size. Shrews and small rodents have less growing to do before they can reproduce than humans, squirrels, and rabbits. Put another way, shrews and small rodents spend less time growing and more time reproducing.

One problem with the difficulties of defining weaning is that deciding exactly when to weigh or to measure a weanling is an issue. Weaning can be a prolonged process. For Australian sea lions (*Neophoca cinerea*), the transition to independent foraging takes 3–6 months (Lowther, Goldsworthy 2016). Thus, data on weaning size, mass, or even extent of time are highly variable and are not often comparable across species.

Weaning is more than simply attaining a specific size. At birth, neonates lack teeth and their intestinal tract is primed for digesting milk rather than for an adult diet. Mammals rely on teeth with a precise match (occlusion) between the upper and lower jaws. As teeth do not grow in girth, the jaw must be nearly adult size to accommodate adult teeth and allow chewing of adult food. This is another reason why young animals have proportionally larger heads than adults (Pond 1977) and why many mammals have two sets of teeth (diphyodonty), a set of milk teeth without molars replaced by permanent teeth with molars. The digestive tract also must mature. This includes changes in biochemistry, size, and microflora to accommodate an adult diet (Hooper 2004). The accumulation of appropriate gut microflora may extend weaning in some hystricomorph rodents (Langer 2002). Overall, weaning will not be successful until a juvenile's body is ready for adult life.

Weaning is but one aspect of the life histories of mammals. In box 5.1 on page 87, we explored life-history strategies related to reproductive rate. Other theoretical simplifications are often used to explain reproductive patterns (box 10.1). For instance, small body size is often associated with a fast reproductive rate. Although mice and shrews fit this scenario, bats do not. Perhaps this is why many theorists restrict their analyses to terrestrial, non-volant mammals! More to the point, understanding the reproductive strategy of any species requires looking at many aspects of its reproduction, not just timing, or numbers of off-spring, or energetic input. One aspect is the time between weaning and repro-ductive maturity.

After Weaning: Juvenile, Pre-puberty, Subadult

After she is weaned, a female spends some amount of time on her own before she becomes a reproductive adult. This in-between time can be broken into juvenile and subadult phases, but the distinction between these two phases is fuzzy. Juveniles are younger than subadults, but no specific event, neither weaning nor puberty, marks the end of one stage and the beginning of another. For instance, young before wean-ing are sometimes called juveniles, but young still nursing would never be called subadults. Similarly, individuals after puberty, but before first reproduction, are sometimes called subadults but would not be called juveniles.

No matter the name, weaning is followed by a period of growth in which both physical tissues and behavioral responses mature. This growth phase slows dra-matically or stops with the onset of puberty. The duration and extent of the in-terim, non-reproductive period is highly variable across species and varies across individuals within a species, but it is not typically measured. It can be quite short as in the ermine (*Mustela erminea*) or the corn mouse (*Calomys musculinus*) both of which may mate or ovulate before or just after weaning, or it may last for years as in long-lived whales (Buzzio et al. 2002; Hayssen et al. 1993). One reason that the time between weaning and sexual maturity is rarely quantified is because con-ventional measures of the time to sexual maturity start from birth, not weaning, and thus include lactation. Information across a variety of species on the interval between weaning and sexual maturity is needed.

Regardless of the exact duration, the post-weaning period is challenging for individuals. After weaning, juveniles must find food and shelter, avoid illness, and evade predators. In social species, females may need to integrate themselves into the social system and establish a dominance rank. All of this is in addition to the basic task of growing and maturing into a functional reproductive adult (Fair-banks 2000).

BOX 10.1. Dichotomies

Physiologists and ecologists have devised a number of dichotomies to simplify our understanding of reproduction, such as r versus K, fast versus slow, capital versus income, altricial versus precocial (chapter 8), hider versus follower (chapter 8), and spontaneous versus induced (facultative) ovulation (chapter 6). All these dichotomies are useful because they simplify complex observations. They synthesize suites of characteristics into manageable units. However, the difficulty with useful dichotomies is that they can take on a life of their own, are taken as a given, and with time become dogma. We briefly describe three such dichotomies.

Some dichotomies have a theoretical framework. For instance, the r versus K classification rests on a theoretical sigmoidal curve of population growth over time with an exponential (r) phase of density-independent growth and a later asymptotic (K) phase of density-dependent growth. K-selected species (think elephant) are long-lived and large bodied, with small litter sizes, large neonates, slow development (late age to sexual maturity), and increased parental care. Conversely, r-selected species (think mice) are smaller with short lives filled with multiple litters of many small, rapidly developing neonates that are produced early and often. In reality, the traits that comprise r or K strategies need not be biologically connected. Bats are small in size but give birth to large singleton young, whereas pigs can be quite sizeable but have large litters.

The r versus K continuum generated much cogent theoretical dialogue. However, a newer dichotomy is the fast-slow continuum, which is based on mortality schedules, e.g., "living fast and dying young" (Promislow, Harvey 1990:417). Species with high mortality rates mature early and produce many small offspring after short gestations followed by short lactations. Thus, the fast-slow continuum still lumps similar suites of reproductive characteristics as does the r-K continuum, but body size is removed from the classification. As with the r-K framework, the speed-of-life continuum was instrumental in framing comparative studies of life-history evolution in mammals, but it, too, is ready for an update (Bielby et al. 2007). Reproductive traits in mammals do not arrange themselves along a single fast-slow continuum but instead vary over multiple complicated dimensions (Bielby et al. 2007).

A third dichotomy, capital versus income (Jönsson 1997), focuses on the allocation of resources using terms from economics. In many cases, it treats all aspects of reproduction (called breeding in the dichotomy) as though they had a single input or currency (often lumped as energy). Capital breeders use stored energy to fund a current reproduction, whereas income breeders use energy gained concurrently with that reproduction. Classic examples of capital breeders are females that go without food during lactation, such as hooded seals (*Cystophora cristata*) or polar bears (*Ursus maritimus*) that rely on fat stores to support both their own metabolic needs, as well as for the production of milk. Species with high metabolic rates, such as shrews might be income breeders, supporting reproduction only with the food recently obtained. However, because "the costs of accumulating capital can lead to pure income breeding, pure capital breeding, or a mixture of the 2 strategies" (Houston et al. 2007:241), the application of this dichotomy to specific species is difficult. Many species, such as Verreaux's sifakas (*Propithecus verreauxi*) use either stored or recently obtained resources to support reproduction, depending on external conditions (Lewis, Kappeler 2005). In addition, some elements needed for reproduction, such as calcium, may be stored, whereas others, such as water, may be obtained daily. Therefore, the specific currency important to reproduction will vary across species, across regions, and across seasons. Finally, capital and income are not the only options. Mothers can recycle nutrients or water by eating the placenta, urine, or feces of young; by resorbing the ejaculate; or by becoming more efficient in their use of resources. They may even cache food before conception and relinquish it to their offspring 10 months later as do American red squirrels (*Tamiasciurus hudsonicus*; Boutin et al. 2000). Chapter 9 on lactation reviews a few of the many complex ways females "pay" for reproduction.

While dichotomies provide a heuristic tool, the value of which cannot be understated, the reality is that natural selection operates on individuals and over long periods of time. It does not create dichotomous strategies. Dichotomies can be extremely helpful as long as we remember that they have human, not biological, origins and present a simplification of a complex world.

Many juveniles and subadults die before they reach sexual maturity. Survival, before weaning, is generally dependent on maternal care, but after weaning, survival is more closely linked to environmental conditions, as well as to predation and disease (Beauplet et al. 2005). The juvenile period is of "relatively high mortality in most undisturbed primate populations," especially due to predation (Fairbanks 2000:344). Regardless of the cause, mortality that happens before puberty may explain more variance in reproductive success than fertility or longevity (Fairbanks 2000).

One cause of mortality after weaning is dispersal. Dispersal often occurs after weaning but before the onset of fertility. Juvenile dispersal refers to migration away from the birth site or away from the maternal territory or range. Dispersal is a period of exploration often associated with high mortality, a lower familiarity with the new home area, and, in social species, a loss of the benefits of cooperation with relatives. Some benefits of dispersal include the potential for finding an area with more food or fewer predators, avoiding competition with kin, or reducing the potential for inbreeding (Handley, Perrin 2007).

Dispersal in mammals is generally regarded as a male trait with females tending toward philopatry (Handley, Perrin 2007). However, females are the primary dispersers for at least 20 species, from wombats (*Vombatus*) to the great apes (Handley, Perrin 2007). In some species, e.g., zebras (*Equus*) and gibbons (*Hylobates*), both sexes disperse, whereas in others, e.g., naked mole-rats (*Heterocephalus*) both sexes are usually philopatric (Braude 2000; Wolff 1993). Of course, irrespective of sex or general tendency, not all juveniles disperse. For instance, 77% of juvenile female (and 80% of juvenile male) kangaroo rats (*Dipodomys*) stayed within 50 meters of their natal burrows between weaning and reproductive maturity (Jones 1987). Some theoretical work connects sex-biased dispersal with mating systems, but the relationship is complex (Handley, Perrin 2007).

In social species, juveniles may remain within their social group and benefit from maternal support. Mothers defend their offspring against predators or from conspecifics (Andres et al. 2013; Brookshier, Fairbanks 2003). Between weaning and sexual maturity, in Old World monkeys, daughters stay with their mothers and, once mature, may remain in their natal troops the rest of their lives. Thus, daughters may benefit from maintaining proximity to their mothers long beyond weaning. Furthermore, juveniles and yearlings huddle with their mother at night and are groomed by their mothers. Daughters may obtain support from their mothers in aggressive encounters with other troop members, and injured juveniles will return to their mothers for aid. In general, juveniles, rather than mothers, maintain the relationship. The mother does not usually initiate the interactions but will respond if approached by offspring (Fairbanks 2000). Overall, sociality has many benefits for juveniles (Silk 2007). These and other benefits from conspecifics are discussed further in chapter 14.

Reproductive Maturity or Puberty

No one so far has been able to define this stage of development [puberty] in a way that will encompass adequately both sexes of all mammals, and simultaneously satisfy everybody concerned.—Bronson, Rissman 1986:157

Puberty is much more than just a set of interactions between brain cells, hormones and target tissues; it is a phenomenon of broad import in mammalian biology.
—Bronson, Rissman 1986:158

Sometime after weaning, a female becomes reproductively mature, i.e., able to produce offspring. This event can be assessed by a list of diverse firsts, such as first ovulation, first mating, first conception, or first birth. However, the processes that lead to any of those firsts often have gradual beginnings occurring well in advance of the specific event. Like weaning, reproductive maturity is a process, not an event.

Puberty has significance for animal breeders as well as for population ecologists. For efficient production, the physiological conditions that limit or accelerate reproductive development on individuals are important. But age-at-first-reproduction also influences population demographics. In this context, ecological, phylogenetic, and energetic factors are also relevant. Thus, reproductive maturity in mammals has been studied with both a mechanistic as well as an evolutionary framework.

From an evolutionary standpoint "the onset of fertility presages the risk of immense energy expenditure in the near future" (Bronson, Rissman 1986:163). Thus, puberty is likely to be timed so the chance of a female successfully raising her first litter is highest. Body size, taxonomy, habitat, dietary regimes, and thermoregulatory ability may be related to species-level differences in the timing of reproductive maturity. But no single factor has a commanding influence.

Body size is an example. Across 547 mammalian species, body size explains 56% of the variation in age at first reproduction (Wootton 1987). Not surprisingly, larger species generally take longer to achieve reproductive maturity (Charnov 1991). However, the huge baleen whales and smaller toothed whales have similar ages at first reproduction (5–10 years). Of course, humans are much smaller and take even longer to mature reproductively. Therefore, phylogenetic influences are also important.

What about ecology? Comparisons across large numbers of species suggest that ecological differences may have little effect (Wootton 1987). But in these large comparisons, much detail is lost and a fine-grained analysis would be more illuminating. To assess the effects of ecology on the evolution of reproductive maturity, we need comparisons within a taxonomic group with ecological variation, such as ground, tree, and flying squirrels, or talus versus rock dwelling pikas. One diffi-

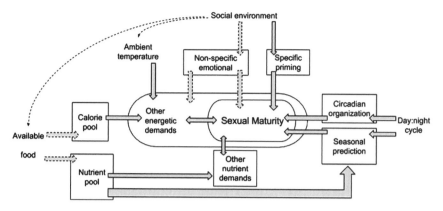

Figure 10.1. Influences on sexual maturation. The large ovoid represents the brain of a female in which stimuli are integrated to regulate sexual maturity (*small ovoid*). Factors operating within a female's body are shown in small rectangles. External stimuli are not boxed. Both abiotic (day/night, temperature) and biotic (conspecific interactions, food availability) variables influence the timing of puberty. Shaded arrows indicate pathways under strong selective pressures. Dashed lines indicate pathways that are more flexible in the face of random environmental variation. Modified by Abigail Michelson from Bronson, Rissman 1986.

culty is that data on first reproduction is gathered using inconsistent criteria, preventing quality comparisons at such a fine-grained level. In addition, ecological factors influence all of reproduction and not just age-at-first-reproduction. For instance, many ground squirrels produce only one litter per year, whereas tree squirrels produce more than one (Hayssen 2008a). Consequently, ground squirrels cannot reproduce the year of their birth, whereas tree squirrels can. Such constraints make understanding ecological effects on puberty across species difficult.

Natural selection will reinforce coordinating the timing of puberty to a variety of environmental factors, such as availability of food, water, and nutrients; favorable seasons and climate; low predator and pathogen abundance; and positive conspecific interactions. Any of these ecological variables could also provide individual females with external cues to trigger sexual maturity. She could use these cues to assess her environment and to coordinate the allocation of her resources to growth or to reproductive maturity (figure 10.1).

Cues from conspecifics may influence the timing of puberty. For instance, in laboratory rodents, puberty in females can be accelerated by the odor of adult male urine or inhibited by that from adult females (Bronson, Rissman 1983). In Arctic foxes (*Vulpes lagopus*), urinary pheromones of sexually active adults trigger puberty in juveniles and female pheromones have a stronger effect than those of adult males (Bartos et al. 1991).

Other biotic cues from the environment may trigger puberty. For instance, specific plant metabolites present in newly emerging grasses (chapter 11) may trigger puberty when ingested (Diedrich et al. 2014; Bronson, Rissman 1986). Abiotic cues, such as photoperiod, may also trigger puberty (Bronson, Rissman 1986). In females, cues that trigger seasonal reproduction (chapter 12) may also cue first reproduction. Seasonal reproduction might even be viewed as seasonally recurring puberty, except for the large differences in a female's ontogenetic profile (her past history) for her first compared with later reproductions. As with other aspects of reproduction, the timing of puberty is a complex interaction of multiple external and internal cues all within a female, who is constrained by her inherited genetic potential (e.g., phenotype).

Before we leave puberty, one fascinating case deserves mention. Female stoats (*Mustela ermine*) are unusual. They first mate at 17–75 days of age; however, the length of lactation is usually 5 weeks (35 days; Hayssen et al. 1993). Therefore, in a few cases, a female may mate before she is weaned, perhaps even before she opens her eyes (Amstislavsky, Ternovskaya 2000; King 1983; Weir, Rowlands 1973). In other words, puberty occurs before weaning. Gestation in female stoats includes a variable and facultative period of delayed implantation, so a juvenile female can mate (perhaps more than once) and put any conceptuses that result on hold while she completes her growth and development. At some future time, she can either abort the litter she conceived before she was weaned or allow her litter to develop without having to wait to find a male. For young females, gestation can last from 224 to 393 days (7.5–13 months), giving her ample opportunity to choose an appropriate time to complete reproduction (Amstislavsky, Ternovskaya 2000).

The reproductive pattern in *Mustela erminea* is even more interesting because females of a related species (*Mustela nivalis*), which often live in the same area, become sexually mature well after weaning. These females first mate at 3–4 months of age, and any resultant pregnancy proceeds without a delay (Weir, Rowlands 1973). The physiology of the early sexual maturity in *M. erminea* is not known, nor do we have a coherent explanation for why such divergent reproductive profiles evolved in related sympatric species.

Menopause, Reproductive Senescence, Post-Reproductive Life

Senescence refers to a decline in physiological condition with age. For many females, mortality preempts aging; nonetheless, some females do reach more advanced ages. Reproductive senescence highlights the decline with age in the ability to conceive and support offspring through gestation and lactation. Fertility varies with age. Many females have lower fertility just after puberty, and then fertility rises and, finally, drops off. For example, in meerkats, litter size, number of litters per year, and number of young emerging from the natal burrow all increase after puberty, peak at four years, and then steeply decline (Sharp, Clutton-Brock 2010).

Most female mammals reproduce until they die, because doing so is the most direct way to maximize their direct contribution to the next generation. Humans are an exception, because they experience reproductive senescence well before the end of their life span (see chapter 15).

Because life after the cessation of reproduction is so rare for most species, we do not have appropriate language for detailed discussion in a range of non-human mammals. Indeed, *menopause* is a human-centric term that identifies the time when menstrual periods stop. This is also the time when a woman can no longer conceive. Most other mammals do not menstruate, but any mammalian female may reach a time when she can no longer conceive. Rather than creating a new term, we follow Cohen (2004) and use "menopause" to refer to the time when the ability to conceive naturally stops because of age. Of course, the post-reproductive period is of greater interest than the event of menopause itself.

In mammals, how common is life after reproduction stops? Humans certainly can live well past menopause (Alberts et al. 2013; Cohen 2004). Many domestic (or semi-domestic) mammals have a post-reproductive life, such as cattle, red deer (*Cervus elaphus*), horses, dogs, cats, rabbits, lab mice, and Chinese hamsters (*Cricetulus griseus*; Cohen 2004). Life after reproduction has also been documented in at least 17 species of captive primates (Cohen 2004). However, one could argue that the lives of captive and domesticated mammals are anomalous relative to those in the wild.

Documenting a post-reproductive life in wild mammals is difficult and confounded by life span. Short-lived mammals, such as voles or shrews, almost by definition, cannot have much life after they stop reproducing, because their lives are so short to begin with. Longer-lived mammals have the potential to live after menopause, but long-term observations of individual females in the wild are limited. Among long-lived primates, menopause occurs, but only in a few individuals (Alberts et al. 2013). A few female lions (*Panthera leo*), polar bears (*Ursus maritimus*), and African elephants (*Loxodonta africana*) also live past reproduction in the wild (Cohen 2004). Among cetaceans, the very large baleen whales reproduce until death, but several of the toothed-whales (odontocetes) have post-reproductive lives (at least in some populations). These include killer whales (*Orcinus orca*), false killer whales (*Pseudorca crassidens*) and possibly sperm whales (*Physeter macrocephalus*), spotted dolphins (*Stenella attenuata*), spinner dolphins (*Stenella longirostris*), short-finned pilot whales (*Globicephala macrorhynchus*), and long-finned pilot whales (*Globicephala mclaena*; Marsh, Kasuya 1986). Bats and kangaroos are relatively long lived, but too few individual females have been studied to determine when females might stop reproducing relative to their life spans. Thus, although well documented for women, data on reproductive senescence are rare for other mammals.

Just because a few females manage to live after reproduction does not mean that selection has favored this aspect of a female's life. Direct selection on female post-reproductive life span is not possible, as no offspring are produced! However, if post-reproductive females can increase the reproductive output of their descendants, such as grandchildren, then selection may favor the extension of life after reproductive senescence. The odds of this happening are highest in taxa with cross-generational social systems. Such social systems provide an outlet for grandmothers to contribute to the health and well-being of their grandchildren. Data from captive vervet monkeys (*Chlorocebus*) suggest that the presence of older female relatives may increase the reproductive success of daughters by increasing fertility and lowering infant mortality. Older females may promote the survival of grand-offspring by actively defending them (Fairbanks 2000). Such cross-generational support requires a multi-level social system, such as present in elephants, orcas, and humans. A long natural life without post-reproductive senescence also allows inter-actions across generations.

Post-reproductive life may be a consequence of selection on males for in-creased longevity. Spermatogenesis does not stop completely as males age; there-fore, even older males can sire offspring. Females and males have nearly identical genomes. Any selection in favor of longevity in males that alters alleles on any but the Y-chromosome will be passed on to daughters. Those daughters could inherit any tendency toward longevity present in their fathers without altering the time when ovulation stops.

But let us back up a step. If females continued to create new gametes, then a post-reproductive life would be a moot point. So why is the number of oocytes restricted? Why do females stop creating new gametes? We do not know. In many animals, such as fish and frogs, oogenesis continues through adult life. These females retain stem cells that can create new oogonia. Thus, she can release new oocytes each time she wants to reproduce. This is not so in sharks, birds, and most mammals. For these females, a single, large population of oogonia is pro-duced early in life, and this population is the source of all subsequent oocytes (Rothchild 2002; chapter 6). Thus, most female mammals at birth have a limited supply of gametes, and this supply is often determined before a female is born. If a female survives long enough, she will eventually run out of gametes. But this is not a satisfying answer, because the number of potential gametes a female is born with is often much greater than the number of potential ovulations she may have. In addition, a few mammalian females produce hundreds of gametes at each ovulation, and some data suggest neo-oogenesis is also possible (chapter 6).

One hypothesis regarding menopause is that 40- or 50-year-old oogonia may be prone to errors in replication and may produce abnormal offspring. That aged oogonia are error prone is well documented (Jones, Lane 2013), but so, too, are mechanisms for the rejection of abnormal embryos (chapter 8). Also, baleen whales, which are reported to reproduce until they die, reach old ages (65–100 years; George et al. 1999; Hamilton et al. 1998; Marsh, Kasuya 1986) and Asian elephants (*Elephas*) have been known to calve at 62 years (Sukumar et al. 1997).

Direct answers may come by looking closer at molecular-based hypotheses regarding selection pressures on females to arrest gamete development before birth (Mira 1998). These hypotheses explore the roles of mutation, genetic recombination, or competition among polar bodies in the creation of successful gametes. Perhaps, questions about post-reproductive life revolve around events happening before a female is even born. We continue this discussion in the last chapter.

Weaning and Beyond, Revisited

For an individual female, the stages of her reproductive life are oogenesis, conception, implantation, placentation, birth, nursing, weaning, juvenile, subadult, puberty, and then a number of cycles of gestation, birth, and lactation until (if present) menopause stops her reproductive life. Once a female becomes a reproductive adult, she produces offspring fairly constantly, with interspersed pauses and/or non-reproductive periods, depending on her condition and that of the environment. Most adult females die sometime during that reproductive period, but a few survive to enjoy a post-reproductive life.

The primary focus of this chapter has been on those parts of a female's life in which she is independent of both her mother and her offspring: the intervals from weaning until puberty and from menopause until death. These "non-reproductive" periods nonetheless influence her reproductive success, and the age at first reproduction is important for assessing population demographics and life-history theory.

Regardless of the importance of a female's independent life, the primary coverage of this third section of the book has been those parts of a female's life in which she directly interacts with mates and offspring. Reproductive females are inherently social. A female's connections with mates and offspring occupy nearly all of her reproductive life even in otherwise "solitary" mammal species. Females carry out their reproductive activities in an environment that may aid or injure them. Weather, predators, parasites, and disease vectors may influence the success of a specific, reproductive attempt. Food must be obtained and a microbial community supported. Females that are part of social groups may also interact with conspecifics. These conspecifics may help a female by protecting or provisioning a female's young, but they may also hurt offspring by infanticide. The next section explores how females respond to the outside world to successfully raise their offspring.

PART THREE

REPRODUCTION IN CONTEXT

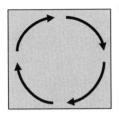 The reproducing female is no island. Part 3 investigates female reproduction in the context of the environment, both non-living (abiotic) and living (biotic). Each reproductive female must meet the challenges and bounties of the abiotic world, which includes changing weather, air or water pressure, temperature, and water availability (chapter 11). Her life is also interwoven with those of other organisms. In this biotic world, she encounters heterospecific organisms: predators, parasites, and prey alike (chapter 12). Finally, females are also sisters, aunts, mothers, mates, competitors, or allies as part of complex, social conspecific interactions (chapter 13). To symbolize the connectedness of the female with her environment, the schematic for this section (see figure) nests the reproductive cycle in a box depicting the environment. This section explores the "big picture" constraints in which females reproduce.

CHAPTER ELEVEN

Abiotic Influences on Mammalian Reproduction

She is an equatorial mammal (latitude 28° S to 17° N) but her environment is not in equilibrium. Seasonal rains drive much of the biology around her. Drought may occur from July to October; flooding in other months. But rains are unpredictable. Certainly not predictable enough to provide a reliable cue for either conceptions or births. Because she cannot predict the rains, she must deal with the consequences. A prior drought limited her milk production and her cubs starved. The last heavy rain flooded her birthing den. She dragged her drowning daughter from certain death, and both emerged caked in mud. When the rain stopped, she salvaged the den by carefully scraping out the mud inflow. Although the weather may create challenges, death of her cub at the tooth and claw of her clan mates is also likely. (East et al. 1989; Holekamp et al. 1999a; Lindeque, Skinner 1982a; White 2005)

> Live in each season as it passes; breathe the air, drink the drink, taste the fruit, and resign yourself to the influence of each.
> —Henry David Thoreau, journal entry, August 23, 1853; Shepard 1927:119

The abiotic world is the non-living world, a world that changes but does not adapt. As humans, we can easily recognize major features of the abiotic world, such as meteors, weather, climate, seasons, and pollution, as well as how they clearly influence reproduction. Features with a less obvious connection to reproduction, such as minerals or moonlight, can be just as important. For example, females need calcium and salt both for themselves and for the development of their young (figure 11.1). Consequently, natural salt licks may be sought out during gestation or lactation (Atwood, Weeks 2003). Or, moonlight may expose a female or her young to increased predation (Griffin et al. 2005; Prugh, Golden 2014). Even more subtly, changes in pressure, as with increasing ocean depth, or changes in oxygen content, as with increasing altitude, can alter placental nutrient and gas exchange (Zamudio 2003).

Even after gestation, the abiotic environment is important, especially for offspring as they suddenly shift from in utero warmth and protection to the cold, hard world. Offspring may not have the capacity to deal with external conditions as well as an adult. Adjusting to these limitations of offspring may have

Figure 11.1. Mountain goat (*Oreamnos americanus*) nanny and kid licking minerals, Glacier National Park, Montana. Photograph by Sumio Harada; used with permission from Minden Pictures.

far-reaching consequences. For sperm whales (*Physeter*), abiotic effects on calves may have influenced the social system of the adults. Because calves of sperm whales cannot make the deep, foraging dives that are routine for adults, the young whales must remain in surface waters where they are exposed to predators. Consequently, when calves are present, females stagger their dives such that an adult is always present in the vicinity of the calves (Gowans et al. 2001). Kin selection would reinforce any tendency for related females to work together to protect the calves. Thus, the matrilineal social organization of sperm whales may be a result of the inability of young whales to do deep-water foraging. Clearly, the abiotic environment can influence reproduction in obvious and subtle ways.

Most of the well-known adaptations to the abiotic environment concern survival. For example, kidney tubules lengthen to retain water (kangaroo rats, *Dipodomys*); dense fur retains heat (fur seal, *Callorhinus*); fat metabolism is modified to increase lipid stores for periods of fasting (camel, *Camelus*; chipmunks, *Tamias*); elongated ears cool blood (kit fox, *Vulpes macrotis*; antelope jackrabbit, *Lepus alleni*); modifications of the circulation allow for deep-sea diving (elephant seal, *Mirounga*; sperm whale,); and hemoglobin has a high altitude form (llama, *Lama*; deer mice, *Peromyscus maniculatus*). These are just a few of the many adaptations mammals have evolved to meet environmental challenges. But what changes in reproduction have occurred to meet these challenges?

Abiotic factors that influence reproduction are many and varied: too many and too varied to explore in detail. Thus, we will focus on just a few to showcase the myriad ways in which the abiotic environment affects reproduction. Here, we think about influences of the abiotic environment on reproduction in three ways and present an example from each way of thinking.

First, we explore the effects of individual abiotic variables on reproduction. Oxygen pressure, temperature, abundance of water or trace elements, and general weather patterns influence reproduction. We chose to explore water because it is both a substrate for life, as well as a component of life. Living in water puts demands on marine mammals that land animals do not face, especially with respect to heat loss. On land, water can limit reproduction, especially in combination with very hot or very cold temperatures. Marine and desert mammals have adapted their reproduction to abundant or limiting water.

Second, multiple abiotic factors co-vary to create a climatic region. Regional variation often occurs because abiotic features can be correlated. For example, at high altitudes, low temperature, low oxygen, and low pressure are the norm. Similarly, in deep areas of the ocean high pressure, constant but cold temperature, and darkness are normal. Other examples of regional variation include climate, ocean currents (think El Niño), and other weather patterns. Climatic regions are called biomes when the living component is incorporated with the abiotic environment.

Geographically, different areas may have the same constellation of abiotic features. These areas can be widespread or widely separated. For instance, in oceans, the zone to which light can descend (e.g., the photic or epipelagic zone) differs from the dark, constant temperature of the deep ocean (e.g., the bathypelagic zone). Both are widespread. On land, water is limited in desert regions, whether the Gobi in Mongolia or the Sonoran in the United States and Mexico, and oxygen is limited in the high altitudes of mountains, such as the Andes in South America or the Himalayas in Asia. Although deserts, mountaintops, and tropical jungles all have specific similar abiotic characteristics, their plant and mammal communities differ by geography. These biotic communities can be rich and complex.

Our focus for a regional example is the Arctic. We selected the Arctic for several reasons. First, the Arctic is a single geographic region rather than a fragmented region across the globe, as are mountaintops. Of course, Antarctica is also a geographic unit, but it has much more topographic variation (think mountains) and, more importantly, no land mammals. Second, relative to most other choices (e.g., jungles or wetlands), the Arctic has relatively few mammalian species, <1% of the world's total. This smaller number of species allows us to cover a greater proportion of the Arctic mammals in limited space. Finally, although the number of Arctic mammals is small, the adaptations females have made in their reproduction are highly variable. Thus, we can explore a greater variety of adaptations across just a few species. Having examined marine mammals under "water," we focus on a few iconic Arctic land mammals, such as lemmings (*Lemmus*), reindeer (*Rangifer tarandus*), and polar bears (*Ursus maritimus*).

Our third way of thinking about the abiotic environment focuses on a more abstract abiotic characteristic, that is, time or, more importantly, timing. Natural disasters such as lightning, earthquakes, hurricanes, and volcanic eruptions undoubtedly impede reproduction, but the timing of these erratic phenomena is unpredictable; thus, it is unlikely that natural selection will lead to adaptations specific to each disaster. However, females can evolve adaptations to abiotic changes that occur at predictable intervals, such as seasonal changes, tides, and the rising and setting of the sun. Plants take advantage of this predictability to initiate growth and flowering. Female mammals can indirectly use abiotic cues by depending on plant growth (a biotic cue), but, more importantly to our focus in this chapter, female mammals also use abiotic cues directly to time their reproduction. Here, we will focus on seasonality as our example of the temporal aspect of the abiotic environment, because timing reproduction to periods of relative abundance is critical to successfully rearing offspring.

Thus, water, the Arctic, and seasonality are our choices for exploring abiotic influences on reproduction. Of course, a major feature of the Arctic is the seasonal ebb and flow of frozen water; thus, our examples also illustrate the interconnected nature of the abiotic environment.

Water: A Global Factor with Diverse Influences on Reproduction

Water is essential for mammalian physiological processes but is also a place to live. And as a place to live, water challenges mammalian physiology. Water conducts heat about 25 times faster than air; thus, staying warm requires extra insulation either through modifying fur or by adding internal layers of fat (Dalton et al. 2014). Mammals have used both strategies. Polar bears, have both fur and fat. Otariid seals, sea otters (*Enhydra lutris*), and beavers (*Castor*) have dense pelage and thin layers of fat, whereas cetaceans, phocid seals, and walruses (*Odobenus*) have thick subcutaneous layers of fat (blubber). Whales have dispensed with fur almost entirely. Both toothed (odontocete) and baleen (mysticete) whales retain only a smattering of whiskers (vibrissae) on their heads, although for some species these hairs are only present before birth (Berta et al. 2015; Drake et al. 2015). What are the reproductive consequences of using either fur or fat for insulation?

One main difference between fur and fat is that blubber allows storage of energy from food for later use. Females with blubber can store energy while simultaneously keeping themselves warm. Once they have accumulated sufficient reserves, they can move to warmer waters or haul out onto land where they fast while nursing their young. Baleen whales and phocid seals use this strategy. For instance, gray whales (*Eschrichtius*) rely almost entirely on stored fats and nutrients to support the last stages of a 13-month pregnancy and most of a 7-month lactation. They feed primarily in the cold, nutrient-rich Arctic waters of the Bering and Chukchi Seas and travel south to warm, but nutrient-poor, nursery lagoons in Baja California, Mexico, where they fast as they give birth and provide milk, which is up to 53% fat, to their calves (Oftedal 1997, Perryman et al. 2002).

The warm southern waters allow rapid development of calves because the young can spend less energy on thermoregulation and more on growth. Phocid seals, such as elephant seals, are similar. They forage in open water for months then leave the ocean for isolated haul-out areas to give birth and suckle their pup for about 23 days, before returning to the ocean (Carlini et al. 2000). We discussed lactation of phocid seals in detail in chapter 9.

Otarids (fur seals and sea lions) differ from phocids. Otariids use a thick, dense, fur coat to keep warm rather than blubber. The storage of nutrients (for example, omega-3 fatty acids or vitamin D; Kuhnlein et al. 2006) in blubber is thus not a readily available option. Without fat reserves, fur seals must alternate foraging and fasting during lactation (Gentry, Kooyman 1986). For example, female northern fur seals (*Callorhinus*), in the Pribilof Islands, have thick fur and little subcutaneous fat. These females congregate in dense colonies (up to a million in one colony in the Pribilof Islands), where they each give birth to a single pup. Over the 4-month lactation, females alternate 1–3 days on shore with pups with 4–8 days foraging at sea (Nordstrom et al. 2013). Consequently, the pups must fast when their mothers forage, just as mothers fast while their pups nurse. Indeed, periods without food are a normal part of terrestrial life for seals.

For both whales and phocid seals, fasting while lactating provides the additional benefit of deterring predators that might attack an unattended neonate. However, many seal species, both otariid and phocid, give birth in large numbers in the same location and over a short interval. Thus, the sheer number of animals lowers the risk of predation to any individual neonate.

Both whales and seals have representative species that primarily fast during most of lactation, as well as species that combine foraging and fasting. In both groups, the strategies have a phylogenetic component. Whales are divided into baleen (Mysticeti) and toothed (Ondontoceti) whales, with the baleen whales generally much larger and capable of storing huge amounts of blubber. Baleen whales first sieve and then ingest microscopic plankton in great quantities. Baleen whales tend to fast during their 5–7 month lactations. The much smaller toothed whales, such as dolphins and porpoises, feed on nimble prey, such as fish. Their body forms are relatively trim and shaped for speed and maneuverability with less room for blubber. With less space for fat storage, toothed whales forage during their 1- to 3-year lactations.

What about seals? Seals, too, belong to different families: earless (Phocidae) and eared (Otariidae). Although seal families lack the body size or dietary divisions seen in whales, phocids have large amounts of blubber, whereas otariids do not. Again lactational strategies are divided by phylogeny, phocids tend to fast during their 4- to 60-day lactations, whereas otariids tend to forage intermittently during their 3- to 12-month lactations. Thus, both seals and whales have evolved the strategy of fasting during lactation but from different selection pressures. For whales, diet and body-size differences may be key features driving these differences, whereas, in seals, the mode of thermoregulation (fur vs. blubber) may be the critical factor.

At this point, the prescient reader may be asking: If both phocid seals and toothed whales eat fish and have similar body forms, why can one store enough blubber to fast during lactation, whereas the other cannot? One possible answer is that the duration of lactation is much longer in toothed whales, 1–3 years versus the 4–60 days for phocid seals. Storing enough blubber for a 1- to 3-year fast is probably not possible. A related issue is that the milk fat of toothed whales, 10–30%, is about half of that in phocid seals, 30–60% (Oftedal 1997). Theoretically, toothed whales could double their milk fat and cut their lactation in half, but that would still mean fasting for 6–18 months. Overall, toothed whales provide lower fat milk but for a longer period of time. Longer lactations keep mother and calf together for longer and allow for more socialization and more opportunities for learning such things as migration routes and foraging strategies. In contrast, phocid seals pump high-fat milk into their young over a short period of time and then abandon those young to fend for themselves. Thus, even with the same diet, body form, and habitat, species differ in reproduction. Even closely related species have different reproductive patterns (as we illustrated with *Arctocephalus* in chapter 9). Mammalian reproduction provides infinite diversity in infinite combinations.

Before we leave the oceans, let's very briefly consider one consequence of locomotion in water versus air. Whales, with their giant size, streamlined shape, and large amounts of buoyant, subcutaneous fat, may spend less energy not only on thermoregulation but also on locomotion. If the energetic cost of locomotion and physiological maintenance is lower in aquatic mammals, these mammals could use that extra energy on reproduction (Bartholomew 1972). This may account for the extremely fast growth rates observed in whales (Case 1978). At the opposite extreme, the tiny denizens of the air, the only truly volant mammals, bats, spend large amounts of energy on locomotion and may be expected to have long gestations or lactations, which they do (Hayssen, Kunz 1996). Now, what about water in more terrestrial environments?

> A large number of studies have examined the energy costs of pregnancy and lactation; but few have measured water requirements during these physiological states. (Degen 1997:248)

Water is essential for normal physiological function. Not only is water the major component of our bodies, but it is also a major part of the food we eat. However, teasing apart the role of water availability from that of nutrition, and in turn the overall condition of females, is nearly impossible. Some females use green vegetation more than water to cue reproduction (Degen 1997). Much of the scientific literature concerning water and reproduction on land is focused either on energetics or on the availability of water in arid environments relative to timing

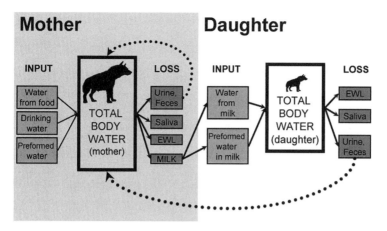

Figure 11.2. Paths of water loss. Female mammals have several routes for water loss during reproduction. Maternal total body water derives from metabolic breakdown of food, drinking water, and water in food (preformed water). Water is lost from urine, feces, saliva, evaporative water loss (EWL), and when females lactate via milk. Some water may be reintroduced into the system via coprophagy and/or ingestion of urine. Pups meanwhile are dependent on their mothers for water balance during lactation. They receive water in milk and can obtain metabolic water when they metabolize milk products. Their paths for loss are the same as those for their mothers (urine, feces, saliva, EWL); however, these materials are recycled, not by the pups but by their mother via ingestion of neonate fecal materials, as well as urine. Modified by Teri Orr from Degen 1997.

reproduction relative to precipitation. Here we will give a few examples of how water is used in reproduction.

The physiological influence of water on reproduction in terrestrial environments hinges on water balance, i.e., gain and loss. Water may be gained directly through drinking, indirectly through free water in food, or metabolically through digestion of food. Water is lost through urine, feces, saliva, or through evaporation via sweat or from respiratory surfaces during exhalations. After birth, water may be regained through placentophagy as well as recycling of pup urine and fecal water (Degen 1997; figure 11.2). As water accounts for much of our bodies, water is important in building babies but water also surrounds and cushions developing embryos.

Gestation requires significant water: water for metabolism, for the developing embryo(s), for the placenta, and for amniotic fluid. Toward the end of gestation, fluids surround the fetus, amounting to up to 20 liters in cows, although in arid-adapted camels, which are close to the same size as cows, only 12 liters accompany the fetus near birth, a fluid savings of 40% (ElWishy 1987). In general, "longer gestation develops drier animals" (Adolph, Heggeness 1971:59). The data that prompted this quote are from only nine species, including humans, but they suggest that longer gestations produce leaner neonates with less water and more dense tissues, such as muscle and bone. Thus, although water requirements

increase during gestation as embryos grow, on a per gram basis, embryos become denser as they develop (Olsson 1986). We don't know about changes in
water content of the placenta over the course of gestation or near birth.

Diet may also influence water balance in gestation. An example is frugivorous
bats for which a diet of only fruit may put females in a poor water balance. We
think of fruit as being water rich. It is. But, consequently, it also has fewer minerals, such as calcium, and fewer vitamins per unit mass. To process enough fruit
to obtain the extra nutrients required by developing embryos, females must actually ingest water to enable digestive processing of food materials to eventually
(with increased fruit consumption) meet the demand. Thus, during pregnancy,
the Jamaican fruit bat (*Artibeus jamaicensis*) may be especially water stressed
(Morrison 1978; Orr et al. 2016). Similar changes in water demand occur in Egyptian fruit bats (*Rousettus aegyptiacus*; Korine et al. 2004).

For most females, water balance is especially critical during lactation. Although milks vary in fat, protein, carbohydrates, and minerals, all milks have a
large water component—water that must come from the mother either through
direct ingestion or via metabolic pathways. When freshwater is abundant, such as
in mesic environments like tropical forests and wetlands, water will not limit
reproduction. However, in dry environments (e.g., deserts, tundra), reproduction,
in particular, lactation, may need to be carefully timed to periods of water availability, such as increased rain or snow melt. In some areas, the amount of rain
falling in the rainy season can be massive and thus not a problem for water balance, but let's turn our attention to situations where that is not the case.

Milk production is a route for water loss that poses challenges in arid regions
(Degen 1997). Water can be recycled. Australian desert rodents, dingos (*Canis*),
and kangaroos (*Macropus*) all consume the feces and urine of their offspring and,
hence, re-ingest up to a third of the water lost in milk (Baverstock, Green 1975). Is
the milk of xeric mammals special? The poster-mammal for adaptations to arid
regions may well be the camel. Camel milk is normally 86% water, but, paradoxically, when water is restricted, water content of the milk rises to 91% (Yagil,
Etzion 1980). Fat content is similar to that of cow milk 3–4% (Farah 1993) but
declines to 1–2.4% when water is restricted (Yagil, Etzion 1980). Thus, camels
adjust to drought by providing calves with more water and less fat. In contrast,
milks from two desert rodents (*Acomys cahirinus*), the common spiny mouse,
from Asia and African arid regions, and the rock cavy (*Kerodon rupestris*), from
the Brazilian Caatinga, are high in solids and thus their milks conserve water
(Derrickson et al. 1996). Clearly, no single change in milk composition occurs for
desert-dwelling females.

Overall, water exemplifies a single feature of the abiotic world that presents
challenges to mammals at many levels. Water surrounds all mammalian embryos until birth and continues to be a substrate for the lives of marine mammals
even after birth. Being surrounded by water challenges thermoregulation and

thereby energy transfer. Finally, because bodies are mostly made up of water, sufficient water is required for successful reproduction. Water balance and water demands change with reproductive stage in many mammals (Degen 1997).

In the rest of the chapter, water will also appear, frozen in the Arctic and as rainfall for influencing reproduction in areas with wet and dry seasons. In the Arctic, water is present in its solid form (ice) and provides a substrate for mating, birth, and lactation. To explore this aspect of water, we move to the Arctic and to an investigation of abiotic factors that have regional homogeneity rather than global heterogeneity.

The Arctic: A Regional Constellation of Abiotic Factors

Polar environments are among the most extreme on Earth. They are cold and dry, with long periods of constant darkness, as well as long periods of constant light. Although both the Arctic and Antarctic have similar polar climates, they differ greatly in their mammalian diversity. Except for a few human scientists (we recommend a viewing of Werner Herzog's *Encounters at the End of the World* documentary of this strange mammalian life), no terrestrial mammals currently live on Antarctica, but the Arctic has a diverse and multi-leveled mammalian fauna. Why?

About 200 million years ago (MYA), Gondwana, a supercontinent, sat on top of the South Pole. Gondwana was composed of most of what is now South American, Africa, Australia, and Antarctica. Over the next 150 MY, the southern continents broke off from Gondwana and moved north: Africa first, then South America, and finally Australia. That left the southern polar region, Antarctica, isolated. Antarctica began to freeze over. Ice replaced forests. Glaciers replaced rivers. Plants became fossils. With the elimination of plants, the base of the terrestrial food chain for mammals was gone. Once the mammals were gone, the isolation of Antarctica from boreal or other high-latitude terrestrial biomes prevented colonization. Hence, no new land mammals could migrate to Antarctica. Antarctica, and its penguins, remain free from mammalian terrestrial predators.

Whereas Antarctica is isolated and has been so for some time, the northernmost polar regions have been repeatedly connected to the large continents of both Asia and North America by the ebb and flow of pack ice. These recent ice age migrations and retreats allowed repeated speciation in the north, with the result that the Arctic has a set of lineages from several orders (Artiodactyla, Carnivora, Lagomorpha, Rodentia), all of which have adapted to polar climes. The smaller-bodied mammals from these lineages have multiple species across the region (e.g., hares, *Lepus arcticus*/*Lepus timidus*; or lemmings, *Lemmus sibericus*/*Lemmus trimucronatus*), whereas specialized lineages of larger-bodied mammals are circumpolar (reindeer aka caribou, or polar bears). Thus, ironically, Antarctica has land but no land mammals, whereas the Arctic has little land but plenty of land mammals.

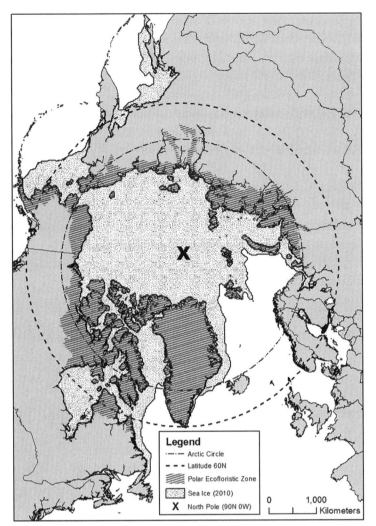

Figure 11.3. Definitions of the Arctic. The Arctic can be defined as the region north of the Arctic Circle (*long-dashed line*), the region north of latitude 60° N (*short-dashed line*), the greatest extent of sea ice (*stippled region*), or by the extent of polar vegetation (*hatched area*). The legitimacy of political claims on Arctic resources depends on these definitions. Emily Fusco synthesized the image from the following sources: "World Latitude and Longitude Grids," Esri; "World GeoReference Lines," Esri; "Global ecofloristic zones" mapped by the United Nations Food and Agricultural Organization FAO, 2000, adapted by A. Ruesch and H.K. Gibbs, 2008; "World Boundaries and Places Alternate," Esri, DeLorme, HERE, MapmyIndia © OpenStreetMap contributors, and the GIS User Community; "Arctic Ice Cover," National Snow and Ice Data Center, *nsidc@nsidc.org*.

Antarctica is easy to define because it is a continent surrounded by ocean. The Arctic, however, is not so obviously defined, because the northern polar region is an ocean surrounded by continents. Because defining the extent of an ocean is fluid, abiotic definitions of the Arctic vary, especially regarding how much peripheral land should be included. The four major definitions differ in the amount of area included, especially land as opposed to ocean (figure 11.3). From least to most inclusive they are (a) the area north of the Arctic Circle (above latitude ~63° N), (b) the area above latitude 60° N, (c) the area of Arctic vegetation (a terrestrial definition), or (d) the area of the maximum extent of pack ice. This last definition changes from year to year and can include areas as far south as the southern area of Hudson Bay (think Churchhill, Manitoba, and polar bears) and the northern parts of Japan. Antarctica is a named continent, but the Arctic is a region of variable extent and not a continent at all. Although the extent of the Arctic varies with each definition, its abiotic character does not.

Like all habitats, the Arctic experiences a set of abiotic conditions that influence, regulate, and constrain life. The Arctic is cold and dry with strong winds and little precipitation (<50 centimeters annually). Much of the region is in either constant light or constant darkness for much of the year. Thus, normal circadian cues, such as sunrise and sunset, are either lacking or restricted to brief periods in the spring and fall. Surface temperatures in winter are well below zero (down to −50°C), but water temperatures are never below −2°C (28°F). Coastal areas, around continents and islands, are somewhat intermediate as the ocean warms the land. Ocean currents bring warm waters to the north and move polar waters to the south. These currents are relatively stable geographically but may have seasonal or multi-year cycles. The currents create mixing and upwelling of water and nutrients, resulting in an abundance of microscopic life (plankton, krill) that constitutes the base of the aquatic food chain. Currents in the form of wind are also important on land. Without major vegetation for protection, wind is a constant. In general, Arctic winds are light, although gales reaching hurricane strength can last for days. For both terrestrial and marine mammals, wind chill is an issue. Not only does the body cool off faster, but both land and aquatic mammals must breathe the cold air, and their respiratory surfaces have adapted to the freezing temperatures (Pielou 1994). Arctic air is also low in water vapor because cold air holds less water than hot air. The Arctic can be as dry as the Sahara (National Snow & Ice Data Center, nsidc.org). In addition, precipitation has a different character in the two regions. For instance, unlike rain in the Sahara, snow in the Arctic may last for years (Pielou 1994; sidebar on pages 202–203).

With respect to food on land, the Arctic is a boom-or-bust region. The summer growth season is very short (a few weeks at most). Perennial plants can't afford to put energy into defensive alkaloids. To do so would be at the expense of either producing seeds or storing nutrients underground to survive the winter. Thus, the available leaves are highly nutritious. Annual plants survive winter in the form of seeds. Seed production is high and a source of food under the snow for many small mammals. The relative abundance of food for small mammals in the summer results in numerous prey to feed an abundance of avian predators, such as snowy owls, gulls, and jaegers. For small mammals, the summer skies are full of aerial marauders. In winter, weasels (*Mustela nivalis*) and Arctic foxes (*Vulpes lagopus*) can pounce on lemmings under the snow and can burrow in the snow to find subnivian nests. For mammals, just as plankton is the base of the aquatic food chain, lemmings are the base of the terrestrial food chain.

The Arctic is above treeline and, thus, has no arboreal or volant niches available for mammals (no bats, no tree squirrels). Even birds are either aquatic (e.g., gulls), roost on the ground (e.g., snowy owls), or nest on cliffs (e.g., dovekies). Permafrost restricts options for burrowing (fossorial) mammals. Although the

Statistics and Shaking Conventions
Evelyn C. Pielou (1924–2016)

As an amateur I was beholden to nobody and could follow my inclinations and make my own decisions without the need to justify them to granting agencies, senior academics, or anybody else. —Personal communication, Langenheim 1996

Although the name Pielou may not immediately come to mind when considering mammalian reproductive biology, one would be hard-pressed to find an ecologist's bookshelf devoid of at least one of her statistical books. Such is her contribution that her name is used for a measure of species evenness! Known more commonly as E.C. or Chris Pielou, Evelyn Chrystalla Pielou was a mathematical and statistical ecologist who made enormous contributions to quantification and understanding of ecology, as well as modeling natural systems both past and present. Equally fascinating is Pielou's non-traditional life path. Given the huge influence of Pielou on the field of ecology and her uniqueness as a scientist, she is in-

cluded here. Pielou not only helped forge a new discipline but did so as a part-time, self-taught, stay-at-home biologist!

Pielou's contributions are prodigious. She pioneered the field of multivariate statistics in ecological research. On the scholarly end, her books include *Introduction to Mathematical Ecology* (1969), *Population and Community Ecology: Principles and Methods* (1974), *Ecological Diversity* (1975), *Mathematical Ecology* (1977), *Biogeography* (1979), and the iconic *Interpretation of Ecological Data: A Primer on Classification and Ordination* (1984). Her writing also includes books for a general audience such as *World of Northern Evergreens* (1984), *After the Ice Age: The*

Return of Life to Glaciated North America (1991), *A Naturalist's Guide to the Arctic* (1994), *Fresh Water* (2000), and *The Energy of Nature* (2001).

Pielou was born in England but spent much of her career in Canada. As a young woman, she entered the sciences with a certificate in radio-physics (at age 18) and subsequently completed 3 years of war service as a navy technical assistant. Eventually, she achieved her bachelor of science in botany from the University of London. She published her first paper 2 years later and raised three children while continuing to publish on statistical ecology. In 1962, 12 years after her first publication and without an adviser, she concatenated several of her excellent publica-

Arctic has a much-reduced mammalian biota compared to more southern climes, many iconic mammals call the far north home: lemmings, Arctic hares, reindeer, muskox (*Ovibos moschatus*), ermine (*Mustela erminea*), Arctic foxes, polar bears, walrus, and harp seals (*Pagophilus groenlandicus*), as well as the marine narwhals (*Monodon monoceros*), bowheads (*Balaena mysticetus*), and beluga whales (*Delphinapterus leucas*).

Not just trees but also time is limited in the Arctic. Summer is exceptionally short, and terrestrial females must rapidly use this ephemeral season for two key purposes: (1) reproducing and (2) depositing enough fat to survive the winter. How do Arctic mammals adjust reproduction to fit these competing demands? No single set of reproductive adaptations characterizes Arctic mammals; instead, a

tions into a dissertation and was awarded a PhD from the University of London (a mere formality given her obvious scholarly achievements by this point). She obtained a second PhD in mathematical ecology (Maingon 2016). In 2001, she was also awarded an honorary PhD from the University of British Columbia. In her acceptance address she remarked "a person who blocks out math is a mental couch potato" (Maingon 2016). Pielou did not suffer fools gladly.

Impressively, Pielou invented the field of mathematical ecology (Gill 2012). Her *Introduction to Mathematical Ecology* "literally changed the direction of ecological research" (Bentley 1986:30). She also contemplated macroecology before it was broadly studied. When she was nearly 40, she took a job as a research statistician for the Canadian Departments of Forestry and Agriculture and began her paid academic career as a full professor at

Queen's University a few years later (1968–1971). Subsequently, she was employed at Dalhousie University in Halifax, Nova Scotia (1974–1981), and worked until retirement at the University of Lethbridge, Alberta (1981–1986).

In 1986, she was the second woman to be awarded the Eminent Ecologist Award from the Ecological Society of America (ESA) for forging the field of mathematical ecology. The first was in 1972 given to Ruth Patrick, a freshwater wetlands specialist. Pielou did not hesitate to correct mathematical errors, even those of eminent ecologists, as exemplified by her published correction of an error by Robert MacArthur (Pielou, Arnason 1966). He did not seem to appreciate the correction (1966).

Pielou retired to Comox Valley, British Columbia, but retirement did not slow her productivity as she continued to write. These later books focused on making complicated subjects accessi-

ble to general audiences and included lovely drawings done by Pielou herself. Her books on the Arctic and freshwater delve into the workings of the abiotic world. During retirement, Pielou actively worked for conservations issues and was instrumental in having the Clayoquot Sound designated as a United Nations Biosphere Reserve. She also participated in Arctic expeditions and served as a scientific adviser for ecotourism. In her honor, ESA has a graduate student award for Statistical Ecology.

(Photo is from a 2011 walk at the Goose Spit at the mouth of the Comox Bay during which Evelyn Pielou expounded on the post-glacial history of the estuary; Loys Maingon, personal communication. Used with permission of the Comox Valley Naturalists.)

diverse array of specializations have evolved, related, in part, to how big the mammal is, where it lives, and what it eats.

Body size is an important aspect of life in the treeless, terrestrial Arctic. With large expanses of snow and ice and little shelter of appropriate size, large mammals are constantly exposed to the elements, but small mammals can burrow under the snow and escape the blistering wind chills. Many lemmings may spend up to 9 months under the snow, and females can give birth and raise litters during the winter (Dechesne et al. 2011; MacLean et al. 1974; Millar 2001). Although large mammals cannot hide under the snow, they can move great distances to find food, mates, or shelter for neonates. This vagility (freedom of movement) has evolutionary consequences. Because polar bears and caribou/

reindeer are circumpolar, all polar bears are members of the same species, *Ursus maritimus*, just as all reindeer (called caribou in North America) are *Rangifer tarandus*. The reproductive biology of each is the same across their wide distributions because their genetic population is circumpolar.

On the flip side, populations of smaller mammals are geographically isolated from one another with minimal genetic exchange. As a result, these lineages have multiple species across the Arctic. For instance, four species of hares call the Arctic home: *Lepus americanus*, *L. arcticus*, *L. othus*, *L. timidus*. Hares are the only medium-sized herbivores in the Arctic, and each species has a specific geographic range. Even smaller than hares are the lemmings. These 40-g herbivores are even more diverse than the hares and include two genera (*Dicrostonyx* and *Lemmus*), not just one. By recent count, the Arctic is home to eight species of *Dicrostonyx* and five species of *Lemmus* (Musser, Carleton 2005). Over the majority of the Arctic, one species of *Dicrostonyx* co-occurs with one species of *Lemmus*. The only exception is the most northern Arctic, where only *Dicrostonyx* is present.

Hares and lemmings are major prey for a multitude of avian and terrestrial predators, and correspondingly, both lemmings and hares are known for their rapid reproduction and resulting cycles of abundance. For example, in a 3- to 4-year period the number of lemmings per hectare may vary from <1 to well over 100 (Batzli, Jung 1980). The larger hares have a longer, 8- to 10-year cycle. When numbers are high, a multitude of predators are attracted to the area and, as a consequence, decimate prey populations in short order. As populations decline, so too does predator abundance, thus increasing the survival of pups and their mothers. Both lemmings and hares must have high reproductive output to compensate for heavy predation. Let's look at lemmings first.

From a colloquial human perspective all "lemmings" are the same, but the mammals we lump together as lemmings really belong to two genera, *Dicrostonyx* and *Lemmus*. The genera share many attributes. Females of both genera are nearly the same size and live in the same geographic area. They are both thick, sausage-shaped creatures covered by dense fur and have reduced extremities (Stenseth, Ims 1993). Both genera are leaf-eating herbivores, but *Lemmus* prefers grasses (monocots), whereas *Dicrostonyx* prefers shrubs (dicots), especially willows (*Salix*). The food that each prefers is deleterious to the growth and reproduction of the other (Batzli, Jung 1980). Plant metabolites produced early in the growing season in grasses can trigger reproduction in *Lemmus* but have no effect on *Dicrostonyx* (Negus, Berger 1998).

Most other aspects of *Lemmus* and *Dicrostonyx* reproduction are similar, but some differ. Although, for both, the usual number of young or embryos in the wild is 3–4, the maximum recorded litter size is 11 embryos in wild *Dicrostonyx* and 16 neonates in captive *Lemmus* (Manning 1954; Semb-Johansson 1993). The large difference between maximum and observed litter size in both genera suggests considerable prenatal mortality during the 19–21 days of gestation. For both genera, young are 4–5 g at birth for a total of 12–15 g for a litter with the average litter size. Given that mom normally weighs about 40 g in either genus (Haim et al. 2004; Jensen, Gustafsson 1984; Nagy et al. 1995), her litter is huge, about 30–40% of her mass. For comparison, that would be equivalent to a 63-kg (140-lb.) woman giving birth to triplets weighing a total of 18–23 kg (40–50 lb.), or a per-baby weight of 6–8 kg (14–17 lb.). But a human mom would do this over 9 months, while a lemming does it in only 20 days. Imagine eating enough leafy greens to create 20 kg (45 lb.) of babies in just 3 weeks. Then, too, remember the large maximum litter sizes: a litter of 10 young could be the same weight as mom. Thankfully, neonatal mass decreases with litter size; unfortunately, so does neonatal survival (Semb-Johansson 1993).

Even though neonates are large, they are naked, toothless, and blind at birth. Phenomenally, after just 2 weeks of nursing, these neonates become independent individuals with a full coat of fur, a full set of teeth, eyes that can see, and ears that can hear. Their weight is about three times what it was at birth, all of which came from milk (Batzli et al. 1974; Hansen 1957). Not only that, but while mom was suckling that litter she may also be pregnant with the next because females can conceive immediately after parturition. Thus, lemmings can potentially give birth to a litter every 21 days. One captive pair of *Dicrostonyx* produced 17 litters in succession (Hasler, Banks 1985). In the wild, such productivity is unlikely (Miller 2001). No wonder.

Let's get back to the lives of female lemmings. Lemmings stay active under the snow throughout the winter, scavenging plant material, while shielded from the surface wind. When the snow cover is deepest and offers the most insulation, often in March, females make spherical nests of grass and fur under the snow and start the year's reproduction (MacLean et al. 1974). Snow cover is key to pulling off winter litters (Duchesne et al. 2011), and winter litters may be critical for survival of the population, because predation during the summer is intense, up to 3.4% of the female population each day (Gilg 2002). That amount of predation can exceed the summer reproductive rate.

Female lemmings are adapted for rapid reproduction. Ovulation is not impromptu in *Dicrostonyx* and *Lemmus*; instead, females keep their ova on hold ready for ovulation so that they can mate and start a litter as soon as a potential mate appears, which may be quite often in years of high population density (Coopersmith, Banks 1983; Hasler et al. 1974; Mullen 1968). Thus, females are ready to start a reproductive effort whenever they find a mate and do not have to wait for the next hormonal cycle.

Dicrostonyx has another trick for rapid reproduction: subadult or juvenile females are able to advance to sexual maturity within only 2 weeks if they sense the presence of males (Hasler, Banks 1975). Not surprisingly, photoperiod has little effect on either sexual maturity or other aspects of reproduction. Females are often living with either 24 hours of light or 24 hours of dark, and photoperiod has an entirely different meaning to these mammals compared to those in more temperate climates (Hasler et al. 1976; Nagy et al. 1995; Weil et al. 2006).

Lemmus uses a different strategy. Females can conceive just after weaning, which occurs at 2 weeks of age, although first conceptions at 3 weeks are more usual (Semb-Johansson et al.1993). Unfortunately, these early/first conceptions have higher prenatal mortality (Jensen, Gustafsson 1984). Birth and weaning are also times of high mortality. Even in captivity, 30% of neonates die immediately after birth or at weaning (Semb-Johansson et al. 1993). However, one behavior of *Lemmus* females may increase weanling survival. Just before birth of the next litter, females may shift their home ranges about 30 m (Heske, Jensen 1993). This movement may expose pregnant females to increased predation but provides juveniles of the first litter the protection of familiar territory. Clearly, transition periods are difficult for females and offspring.

For the most part, although lemmings are members of different genera with distinct ancestries, the reproductive adaptations to living in the Arctic are remarkably similar, not so with hares. Hares belong to the genus *Lepus*, with four different species in the far north. Unlike lemmings with similar reproduction across genera, the four *Lepus* species exhibit two different reproductive patterns related to body size and distribution rather than to specific features of the Arctic. Large hares have more young per litter and only one litter per year. They also have restricted distributions. The small and medium-sized hares have fewer young per litter but more litters per year, and these hares range widely, extending much farther south. Details follow or you can skip the next paragraph and move to another Arctic mammal, the reindeer!

The four hares come in three sizes: two large (4–5 kg), one medium (3 kg), and one small (1.5 kg). The medium-sized hare, *Lepus timidus*, has sole possession of the Old World from northern Europe through northern Asia. The other three species divvy up the New World Arctic, with the two large species flanking the smaller one. Starting in western North America, the Alaskan hare (*Lepus othus*) has a restricted distribution in western coastal Alaska. It is a very large hare (5 kg) with a large litter size, six, but only one litter per year. Farther east, covering the rest of Alaska and much of Canada and extending south into the mountains of the United States, is the much smaller (1.5 kg) snowshoe hare (*Lepus americanus*), with a small litter size (3.5) and perhaps three litters per year. North of the snowshoe hare lives the second large hare, the Arctic hare (*Lepus arcticus*), at 4 kg. It ranges through far northern Canada east into Greenland. Like the large Alaskan hare, the Arctic hare has large litter size, 4.5, and only a single litter per year. Finally, in the Old World is the tundra hare (*Lepus timidus*). At 3 kg, the tundra hare is larger than the snowshoe hare but smaller than the Alaskan and Arctic hares. The tundra hare's reproduction is similar to the much smaller snowshoe hare, having a litter size of two to three and three litters per year, generating six to nine young per year. In sum, each of the two species with large bodies and restricted distributions has only one litter per summer of four to six leverets, or four to six young per year. However, each of the two species with smaller size and wider distributions has three litters of two to four young each summer, resulting in 6–12 young per year (Hayssen, in prep). Although lemmings represent two distinct lineages, they have similar reproductive strategies, but hares belong to a single genus, with different strategies. Clearly, for small herbivores, no one reproductive strategy is best living in the Arctic. What about large herbivores?

The only large herbivores in the Arctic are muskox (*Ovibos moschatus*) and reindeer (aka caribou). Similar to hares, the smaller of these species, reindeer, has the wider distribution: reindeer are circumpolar, muskox are limited to North American and Greenland. Both are ruminants; that is, instead of a simple stomach, they have a complex series of chambers filled with microbes that digest whatever grass, twigs, lichen, and moss are available on the barren Arctic tundra. This means both have less room internally to fit a large neonate.

Muskox and reindeer are not closely related. Muskox are more closely related to cattle, goats, and sheep, whereas reindeer are related to moose, elk, and deer. These differences in ancestry are reflected in differences in morphology and in reproduction. Female muskox (200 kg) are twice as big as female reindeer (100 kg), but reindeer have relatively larger neonates, 8% of mom's mass versus 6% for muskox (Parker et al. 1990). They both give birth to singletons in the spring, but muskox suckle their young well into the winter, whereas reindeer wean their young in early autumn. Another difference is that growth rates of neonates in the first week of life are nearly five times higher in reindeer, than muskox (571 vs. 121 g per day; Parker et al. 1990). Reindeer mothers put so much milk into their babies the first week after birth that the mothers lose weight. However, over the course of the summer, growth rates of young muskox and reindeer even out, and female reindeer regain the weight they lost early on. In contrast, during late winter, muskox are still lactating and may lose 242 g of fat and 55 g of protein each day (Adamczewski et al. 1997), whereas female reindeer stopped producing milk months earlier. Milk fat for both species is 10–15% but generally higher for reindeer (Baker et al. 1970; Chaplin Follenbensbee 1993; Gjøstein et al. 2004). Female muskox have an additional way to negotiate their energetic demands. Both reindeer and muskox conceive in the fall, but muskox may terminate their pregnancies during the winter if their body fat falls (Adamczewski et al. 1998). Overall, the timing of mating and birth are similar in muskox and reindeer, but reindeer have larger neonates that are weaned earlier than those of muskox. The larger muskox is able to store enough fat to have concurrent gestation and lactation at least in good years; in contrast, without excess weight, reindeer are more mobile and can search for food over a wider area unimpeded by a nursing calf. No one solution works better. Lemmings, hares, reindeer, and muskox all illustrate the complex ways in which herbivores successfully negotiate the demands of the Arctic.

We cannot leave our discussion of the Arctic without at least some mention of polar bears, the icons of the Arctic. Interestingly, the reproductive challenge for polar bears is not the Arctic, but rather the bear reproductive patterns they inherited from their ancestors.

Polar bears descended from temperate zone bears and have inherited the same reproductive profile (see box 9.1). In the spring, female polar bears advertise for males and usually mate with more than one. After mating, they conceive, but instead of continuing with pregnancy, they put the embryos on hold for most of the summer. Those embryos sit, unattached, in the uterus (see discussion about delayed implantation in chapter 7) while mom fattens up as much as possible. In the fall, females find a snow den in which to hibernate. After denning, implantation occurs and females resume their short pregnancy to give birth to several (1–4 usually 2), very small (700 g, 0.3% of mom's mass), uninsulated, blind, and relatively undeveloped cubs (Blix, Lentfer 1979). While denned, the female provides milk with a high-fat content (~33%) and the cubs gain weight (Jenness et al. 1972). Mom fasts the entire time, supporting both herself and her cubs with stored fat. In the spring, the cubs and mother emerge and immediately search for food. After emergence, females modify their milk composition by lowering fat content as the cubs become more independent and thermally stable. This change in composition also allows females to rebuild their own fat stores (Derocher et al. 1993). Nevertheless, nursing may continue for over a year. Amazingly, pregnant females must accumulate fat stores to accommodate months of fasting and to provide high-fat milk for their cubs. Not surprisingly, pregnant females (234 kg) in the autumn weigh almost 50% more than females (159 kg) with new cubs in spring. Thus, females lose about 75 kg over the winter (Ramsay, Stirling 1988).

This pattern of using hibernation to raise vulnerable cubs works well as an evolutionary strategy for temperate zone bears but is out of synch with polar bear food availability. For many bears, winter is a time of low food availability, but this is not the case for polar bears. Phocid seals are a major food for polar bears, and seals do not hibernate. Seals are easier to find in winter because they rely on breathing holes in the ice that are fewer in winter. Consequently, food may be more available to polar bears in the winter than in the summer. Non-reproductive polar bears do not hibernate, although they may find shelter and sleep through long winter storms. Only pregnant polar bears den up and fast for the months the cubs need to develop enough fat, fur, and coordination to live in the Arctic. Although polar bears are morphologically adapted to the Arctic (e.g., hair color, proportions), their reproductive patterns are stuck in their no-longer-beneficial ancestral past.

We have used the Arctic as an example of how a specific regional constellation of abiotic features leads to an extraordinary diversity of reproductive adaptations. The mammalian diversity of the Arctic is small compared to that of many other habitats. Nonetheless, Arctic females have found diverse solutions to reproducing in polar regions, ranging from raising litters even during the winter (lemmings) to having only one litter with an increased litter size (Alaskan hare). Or a species such as the polar bear simply makes do with its ancestral reproductive pattern. In all cases, timing reproduction to the extreme seasonality of the Arctic is key. In the Arctic, time is of the essence.

Seasonality: Adaptations to Regular Abiotic Change

Having explored both a single abiotic factor—water—and a regional constellation of abiotic factors—the Arctic—we now turn to a more abstract component of the nonliving world, which is time, or, phrased another way, timing. As far as reproduction is concerned some aspects of the environment are predictable, that is, have a regular timing, and other features are unpredictable. Only predictable features are worth attending to and only predictable, recurring features are a target for natural selection.

In many environments, some times of the year have more abundant resources, such as food, than others. For instance, in temperate forests, cold winters generally result in lower food availability relative to warmer summers, and in equatorial grasslands, wet seasons result in abundant food but usually alternate with dry seasons when food is scarce. If survival of young is generally poor in one season, say, winter, compared to the other, then limiting winter reproduction could leave more energy to put into reproducing in the summer. What we call seasonality is the tendency for reproduction to occur in particular seasons but not others. Definitions of seasonality vary and often are biased by body size.

For large mammals, reproduction often takes an entire year or even longer. Thus, for large mammals, what we call seasonality is the occurrence of one reproductive event, say, mating, during only one part of the year, such as in autumn. For large mammals, one of the most general adaptations to their environment is that females time their reproductive periods of greatest energetic demand, usually lactation, with periods of environmental abundances, usually of food or water. Because gestation and lactation are often of long durations, females may instead use environmental cues to time short duration events, such as mating, conception, implantation, or birth to the appropriate period. If all females are cued to the same environmental signal, reproductive events, such as mating or birth, will occur over short and specific times of year and, as a result, are often said to be synchronized.

Unlike many large mammals, numerous small mammals reproduce more than once in a year (temperate bats and hibernating ground squirrels are notable exceptions). Small mammals, too, may have seasonal reproduction but not by means of timing a specific event to a specific time of year. Instead, small mammals may use some environmental signal to start a series of reproductive cycles at the beginning of a season and may use a second environmental trigger to stop reproduction at the end of the season.

The timing of reproduction to seasonal resource abundances is a matter of responding to cues well in advance of when critical demands are made on a female's reproduction. Regular abiotic features of the environment may be important proximate triggers for reproduction. For instance, implantation in many southern hemisphere fur seals (*Arctocephalus*) occurs near the March (autumn) equinox (Boyd 1991). But are all cues created equal?

For any cue to be effective, two criteria must be met. First, the cue must be present only at the appropriate time. For instance, at the equator, 12 hours of light occurs many times and thus would not be useful for equatorial mammals, but 12 hours of light only occurs twice in temperate zones so light might be useful there. Second, the cue must be sufficiently distinct from other cues present at the same time (McAllen et al. 2006). Among abiotic features of the environment, photoperiod has received the most study and has been linked to reproductive cycles in hamsters (*Mesocricetus, Phodopus*) and voles (*Microtus*; Diedrich et al. 2014; Król et al. 2012), as well as to embryonic diapause in wallabies (*Macropus*), mustelids (*Meles, Mustela*), spotted skunks (*Spilogale*), and fur seals (*Arctocephalus*; Boyd 1991). For this reason, we will explore photoperiodic cues in some detail and then more briefly examine rainfall, temperature, and one biotic factor as additional reproductive cues.

Photoperiod

Photoperiod presents two potential cues that females could use to time reproduction: (1) the absolute photoperiod, that is, the actual number of hours of light and dark in 24 hours and (2) the rate at which photoperiod changes, that is, the number of minutes of increase or decrease in light (or dark) each day compared to the day before. Over most of Earth, day length increases from the winter solstice to the summer solstice and then decreases until the following winter solstice (figure 11.4). The absolute length of a given day is related to latitude. The authors of this book, at 42° N, experience 9 hours and 5 minutes of light at the winter solstice and 15 hours and 17 minutes of light at the summer solstice (U.S. Navy, http://aa.usno.navy.mil/data/docs/ Dur_OneYear.php). Areas near the poles have a greater range of day lengths than locations closer to the equator. Above the Arctic or Antarctic Circles, a 24-hour day or night occurs near the solstices, and the number of 24-hour days or nights increases as one approaches the poles. Thus, maximal day length can occur over more than just one single day.

Changes in photoperiod follow a different pattern. At the North Pole (90° N), the switch from maximum day length to maximum darkness, that is, from 24 hours of light to 24 hours of darkness, occurs over a single day. One day (24-hour period) is in constant light, and the next day is in constant dark. Just 1° farther south (89° N) the switch takes 5 days. Move another degree south (88° N) the change takes 9 days, and jumping to 80° N the switch takes 53 days. Most of our readers live below the Arctic Circle and do not experience constant

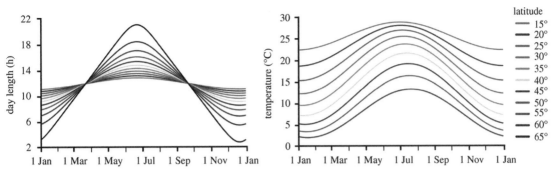

Figure 11.4. Photoperiod, temperature variation. Variation in day length from 15° N to 65° N (*left*) and the concomitant variation in average daily temperatures from 15° N to 65° N (*right*). Twelve hours of light occurs everywhere at the March and September equinoxes (the nexus points on the left figure). The higher the latitude, the steeper the change in day length (steeper sine wave on left figure). Variations in daily temperature are also larger over the year at higher latitudes (lower lines on right figure). In addition, the peak temperature is later at higher latitudes (lower lines on right figure) as is the nadir. From Wilczek et al. 2010; used with permission.

light or constant dark. For most of us, the switch from maximum day length to maximum darkness takes six months. Thus, the rate, as well as the amount, that day length increases or decreases also varies with latitude. The rate is faster closer to the poles because the amount of change that has to occur is larger. The speed at which day length changes is related not only to latitude but also to time of year.

Changes in day length are faster near the equinoxes and slower near the solstices. From the winter solstice until the spring equinox, daylight increases at a faster and faster rate. Starting at the spring solstice, those increases slow down until the summer solstice. Starting at the summer solstice, the decreases in day length accelerate until the autumn equinox when the rate of decrease begins to slow down. Thus, the rate at which the photoperiod changes is fast in March and September and slow in June and December. In contrast, absolute day length is generally maximal or minimal at the solstices. So, the rate of change is maximal at the equinoxes, whereas the absolute length of dark or light is maximal at the solstices.

What does all this mean for females? Higher latitudes have a reduced time span over which a particular rate of change will be available as a cue to start or stop reproduction. McAllan et al. (2006) figured out the details for a small Australian marsupial (*Antechinus*). They calculated that, if the reproductive trigger was a 35- to 90-second increase in day length per day, *Antechinus* in Tasmania would have a photoperiodic window of 2–3 days, while those closer to the equator in New South Wales would have about a 2-week window (McAllan et al. 2006). Thus, females farther from the equator could synchronize their reproduction more precisely than those closer. The equator is not a good place to use any photoperiodic cue, as day length changes little over the course of the year. Overall, photoperiod is probably a useful cue only from high latitudes to those in the mid tropics.

Use of photoperiod for timing is generally common in females with relatively long life spans, especially those in regions with more variable photoperiods (Zerbe et al. 2012). However, for females with shorter life spans, use of photoperiod is less common at any latitude (Bronson 2009). Why the life span difference? Except for bats, life span is often connected to body size. Elephants and whales outlive shrews and gerbils. Species with short lives do not have as many seasons within which to reproduce as those living multiple years. Thus, small mammals cannot wait until next year to breed, because they have a high probability of dying before that can happen. In addition, photoperiodic cues occur on a yearly basis, and again, short-lived mammals won't be around to experience the suite of those cues. Thus, small mammals and equatorial, tropical mammals generally use other cues to time reproduction. Photoperiod is also not useful for a third group of mammals, those that live in desert and dry grasslands with unpredictable rainfall and thus unpredictable resources.

However, things are not always simple as illustrated by the California vole (*Microtus californicus*). This small rodent lives in areas where, if photoperiod cues were used, females would reproduce in the spring, but in reality, they produce offspring in the fall. Experimental manipulation of both water and food availability demonstrated that photoperiod is not the main driver of reproduction in California voles (Lidicker 1973; Nelson 1987). Fall corresponds to a period of rainfall and associated abundances of their food (grasses). Voles ultimately prepare for mating 2 weeks after the rainfall rather than in the spring, as photoperiod would dictate. Thus, both rainfall and food abundance, but not photoperiod, cue reproduction (Nelson 1987). Why rainfall?

Rainfall

Annual changes in day length have less variability and are more predictable than other abiotic factors, such as temperature or precipitation. However, changes in rainfall may be more closely and directly linked to resource availability (e.g., vegetation). In predictable environments, selection favors females relying on predictable cues to time reproduction, such as photoperiod. In less predictable environments, relying on day length to time reproduction may not be advantageous. Here is an example.

Northeastern Brazil has areas where climate is regular and predictable (e.g., Feira de Santana) and areas where climate is irregular (e.g., the Caatinga). When climate is irregular, rain is the chief ecological factor cuing reproduction for female cricetid rodents (Cerqueira, Lara 1991). However, an interesting point (made by these same authors) is that rain cannot be the only factor because prolonged droughts may last up to five years, which exceeds the typical female's life expectancy (Cerqueira, Lara 1991). In very long periods without a rain cue, females with sufficient body reserves may initiate reproduction but with a smaller litter size or reduced milk output.

In areas where variability is unpredictable, plasticity will evolve. When the rains come, females that respond to them will be selected, but if the rains do not come, females that respond to some other cue will pass those genes to the next generation. Thus, the population will include females that can respond to multiple factors. A hierarchy might exist, such as responses to rains over-riding responses to some other cue. But does the trigger prevent or permit reproduction? In other words, does drought prevent females from reproducing or do rains cue reproduction? We don't know, and the answer may differ for different species.

Another example of a reaction to rain comes from the Namaqua dune mole-rat (*Bathyergus janetta*) that lives primarily belowground in very arid regions of southern Africa. Much of the time not enough water is available to support gestation and lactation. Not surprisingly, reproduction is correlated with the onset of the seasonal winter rainfall. When the rains come, they soften the soil and allow individuals to tunnel, extend burrow systems, and search for mates. Thus, by softening the soil such that females and males can meet, rainfall plays an indirect role in the seasonal reproduction of this subterranean rodent (Herbst et al. 2004). Here, the rains are permitting reproduction but do not physiologically initiate the process.

Temperature

Although winters are generally cold and summers warm, daily temperature fluctuations can be quite variable. Thus, as a cue, air temperature would be difficult to use for timing reproduction because much of the annual variability may be experienced by an individual over a single 24-hour period. However, water temperatures, especially those in the open oceans, are much less variable than air temperatures and could provide a better seasonal cue. For example, minimum sea-surface temperature is highly correlated with birth dates in gray seals (*Halichoerus*; Boyd 1991). These seals may use some aspect of water temperature to time reproduction. The effect could be indirect via an influence of water temperature on their demersal (bottom-dwelling) prey.

Biotic Triggers

Although the focus of this chapter is on the ways non-living features of the environment influence reproduction, biotic aspects of the environment may cue seasonal reproduction and may be based on abiotic cues, such as rain. For many small mammals, the availability of food determines whether reproduction will occur. For instance, edible dormice (*Glis*) do not reproduce if the seed crop of the beeches and oaks in their area is poor (Lebl et al. 2011). But simply because a female does not reproduce without sufficient food does not indicate that some aspect of that food is cuing reproduction. Reproduction could be physiologically curtailed below a certain weight or when body condition is poor without the necessity of a single external cue.

In the previous discussion of water, we mentioned that desert mammals time their reproduction with patterns in rainfall; however, the cue for this timing may not rain itself but rather the availability of green vegetation (Degen 1997). This has been demonstrated by providing females with ample food out of synch with rainfall. The result is that food wins out over rainfall (Degen 1997). How does this happen?

As with abiotic cues, any biotic cue must be present only at a specific time of year and must be distinctly recognized. One possibility is that plant secondary compounds are synthesized at specific times of the year. For example, some newly sprouted monocots (e.g., grasses or corn) contain the metabolite MBOA (6-methoxybenzoxazolinone), which deters insect herbivory but also promotes gonadal growth in some rodents, e.g., *Microtus montanus* (Negus, Berger 1977). In addition, MBOA has a structural similarity to melatonin, a hormone involved with entraining circadian rhythms (Diedrich et al. 2014). Thus, MBOA has been proposed as a cue to seasonal reproduction.

At present, MBOA is known to stimulate reproduction in three voles (*Microtus montanus, M. pinetorum, M. townsendii*), white-footed mice (*Peromyscus leucopus*), Ord's kangaroo rats (*Dipodomys ordii*), Harwood's gerbils (*Dipodillus harwoodi*), as well as house mice (*Mus musculus*) and Norway rats (*Rattus norvegicus*; Diedrich et al. 2014). In contrast, MBOA has no effect on reproduction in Merriam's kangaroo rats (*Dipodomys merriami*), Djungarian hamsters (*Phodopus sungorus*), and at least two voles (*Microtus arvalis, M. ochrogaster*; Diedrich et al. 2014; Król et al. 2012). Some genera, e.g., *Dipodomys, Microtus*, contain species that respond to MBOA as well as species that do not. No clear criteria separate responders from non-responders (Dietrich et al. 2014), although clearly if they don't eat plants with MBOA, they aren't going to respond.

Before we leave seasonality, we would like to mention the unique case of the Mexican deer mouse (*Peromyscus nudipes*) a denizen of the cloud forests of Costa Rica (10° N) with a novel mechanism for achieving seasonal reproduction. The cloud forests these *Peromyscus* inhabit have distinct wet and dry seasons. Females conceive in both seasons but only bear young in the wet season. Females ovulate, mate, and conceive year-round, but during the dry season, the embryos either do not implant or are reabsorbed early in gestation. Captive females subjected to mild restrictions of food or water reject their embryos just as wild females do (Heideman, Bronson 1992). Thus, these cloud forest *Peromyscus* achieve a seasonal reproduction by the early rejection of embryos when environmental resources are insufficient to bear young. The energetic cost to the mother is probably small, especially compared to energy saved by stopping the pregnancy. Overall, this strategy works not only because females are "ready" to reproduce at a moment's notice (when conditions are right) but also because females avoid the energetic pitfalls associated with maintaining a litter destined to die during lactation.

The Abiotic World: A Non-evolving Foundation

The non-living environment underlies all of mammalian reproduction. Mothers must obtain oxygen, water, and minerals and convert them into embryos and milk. Changes in light, temperature, and pressure can challenge the basic survival of mothers, as well as of their offspring. We explored the influences of the abiotic environment in three ways. First, we looked at the myriad ways in which a single abiotic factor, in our case, water, affects reproduction, from bodies surrounded by water to bodies made up mostly of water. Temperature is another variable of equal import we could have discussed. Second, we looked at an example of how a regional constellation of abiotic factors can influence reproduction. Here, we choose the Arctic because of its limited mammalian fauna, but many other habitats, such as deserts, oceans, or the tropics, have mammals whose reproduction is specifically adapted for the combination of abiotic factors in each of those environments. The take-home message for this second section is that differences in body size, diet, and ancestry all impose additional constraints on adaptations to specific abiotic challenges. Finally, we explored the most abstract aspect of the abiotic environment, time. We choose to look at the effects of time somewhat indirectly by examining what abiotic cues females can use to time their reproduction to the temporal availability of resources.

The fact that the abiotic world is not alive means that it does not, except in a metaphorical sense, evolve or adapt to changing conditions as the biotic, living world does. Why is this important? Let's take an example. One adaptation of Arctic hares to winter is to molt from their thin, brown summer pelage to a dense, white winter fur coat. The hair is white because it is filled with air pockets. These air pockets provide insulation and keep body heat from escaping to the air (Stegmaier et al. 2009). Keeping heat in is more important than absorbing sunlight during short or non-existent days. So, white hair is as much, or more, an adaptation to abiotic cues as to avoiding predation, a fact made especially clear as polar bears are white and have few predators. A side benefit is that the white coat may hide the bears while hunting seals, but anyone who has seen polar bears in the Arctic realizes that the yellow tinge of their fur is visible against the snow from great distances. Returning to the main point, the fact that hares change to white coats in winter has no effect on the abiotic environment of the Arctic. The abiotic environment provides a relatively constant set of challenges. But the biotic environment is constantly changing. Let's look at Arctic hares again. Arctic hares have adaptations to their biotic environment; for instance, they can restrict their litters to only that time of year when food is available. Predators can adapt to this change in the Arctic hares' biology. Predators can migrate to areas where Arctic hares are breeding only when Arctic hares have progeny. This change in predator behavior can then select for changes in Arctic hare reproduction and lead to further changes in the predators. Thus, biotic systems can respond over time to one another, whereas the abiotic component of the environment does not do so. Adapting to the biotic environment is much more complicated, as we will explore in the next chapters.

CHAPTER TWELVE

Interactions with Other Species

Fleas and ticks are a part of everyday life for our female hyena. She acquired these ectoparasites as a pup while nursing from her mother and will pass them on to her daughter. She may also harbor endoparasites, such as hookworms or a cestode or 2 (or 3 or 12 or 500). Each day she loses a bit of herself to these co-inhabitants, and these losses affect her ability to provide for her young. However, not all of her co-inhabitants are harmful. She also houses a commensal microbiome, a host of bacteria in her gut and on her skin that improve her life by aiding digestion or competing with more harmful microbes. Her skin bacteria contribute to her social interactions because her pheromones are, in part, derived from her symbionts! The fermentation processes from these microbes are the source of her unique smell. Outside her body, her heterospecific interactions extend beyond microbes to other species, such as prey and predators. Annually, she is able to eat her fill when thousands of zebra and hundreds of thousands of wildebeest migrate through her territory, but her food supply is reduced at other times of the year. Without prey, she stands no chance for survival, least of all successful reproduction. If she starves, she may become prey to lions. Thankfully, she can scavenge and feed from carcasses. Her cubs may not be so lucky. (Engh et al. 2003; Gombe 1985; Holekamp et al. 1996, 1999a; Lindeque, Skinner 1982a; Theis et al. 2013)

In nature nothing exists alone.—Rachel Carson, *Silent Spring*, 1962:51

Females live in a complicated environment. In addition to reproducing in the face of numerous abiotic challenges, females interact with many other organisms. Positive interactions range from microbial digestion of plant cellulose in a sloth's gut, to oxpeckers feeding on a Cape buffalo's (*Syncerus*) ticks, to hyraxes of different species responding to one another's alarm calls. Neutral interactions include dispersing seeds from fur or feces, transferring soil microbes from one location to another, or sharing a roost site in a vast cave. Negative interactions receive the most attention. Predators, parasites, and disease vectors curtail reproduction in diverse ways. They can alter energetic balances, lower survival, prevent ovulation, induce abortion, or reduce milk production.

In this chapter, we explore how other organisms influence a female's reproduction. We start with the easily observed and long studied predator-prey interactions and then move to smaller, but no less significant, threats, such as parasites and disease vectors. Although many organisms cause mammals harm, the reverse is also true. The nature of mammals is to eat other organisms. Herbivores eat plants; carnivores eat animals. Thus, other organisms directly support reproduction, but that food may resist. Plants produce fatal toxins or thorny deterrents. Animals repel predators with noxious sprays and sharp teeth or claws. We review some deterrents directly connected to reproduction. Finally, we conclude with mutually positive interactions, such as cooperation among species and our extensive, commensal microbiome.

Eat or Be Eaten: Reproduction in the Context of Predators

Lions eat baby antelope; shrews eat baby voles. To successfully reproduce, females must not only obtain food but also evade predators. Unfortunately, reproduction can increase the risk of predation. A heavily pregnant female may have difficulty darting from a swift predator. Giving birth increases vulnerability and often leaves debris that attracts predators. Even after birth, a female may carry or escort her offspring as she forages, thus reducing her ability to escape. These are explicit examples of a direct increase in predation risk due to reproduction. Reproduction can also increase the risk of predation indirectly. The energetic demands of reproduction may increase exposure to predators if a female must increase the frequency or duration of her foraging bouts to support additional mouths to feed. In addition, if dependent young are in a burrow or nest, a female's foraging radius may be restricted and her behavior may become more predictable, again increasing her risk of predation. In spite of the odds, females must protect themselves as well as their offspring. In addition, offspring must do their part to avoid attracting predators. Protection can come from direct action when an attack is imminent or from modifications to decrease the risk of an attack.

Three options are open for evading predation: (1) run to safety, (2) fight back, or (3) avoid encounters. Flight or fight reactions occur when a predator is present, whereas actions to reduce predation risk occur in the absence of a predator. Most documented evidence of individual females successfully defending young comes from large ungulates with hooves and teeth, such as zebras (*Equus*), rhinos (*Diceros*), moose (*Alces*), pronghorns (*Antilcapra*), buffalos (*Syncerus*), or mountain goats (*Oreamnos*; Caro 2005). Unfortunately, not all efforts are successful (Creel, Creel 2002). Fighting a predator is usually an option of last resort, especially without help. In social groups, joint efforts may deter predators, such as the cohesive line of defense muskox (*Ovibos*) use to separate their calves from Arctic wolves (*Canis lupus*). In general, escaping is a safer option than attacking a predator. Alarm calls and signals from mothers or group members may encourage offspring to flee or to escape from a nearby predator. Meerkats (*Suricata suricatta*), yellow-bellied marmots (*Marmota flaviventris*), and vervet monkeys (*Chlorocebus pygerythrus*) make good use of this anti-predator communication (Caro 2005). So do Belding's ground squirrels (*Urocitellus beldingi*), who are more likely to call if relatives are at risk of attack rather than unrelated individuals (Sherman 1985).

Females may modify their behavior when they know predators are in the area. For instance, when wolves are present, female *Cervus elaphus* (called elk in the United States and red deer in the United Kingdom) spend less time foraging and more time looking for wolves. They also move to forested areas with more cover but less nutritious forage. Progesterone levels and birthrates are lower in these females, documenting the indirect effect of predation on reproduction (Creel et al. 2009).

Predator odors can suppress female reproduction. For example, female field voles (*Microtus agrestis*), when exposed to the smell of their mustelid predators, suppress reproduction. The suppression could be either because females did not mate or because they reduced foraging and lost enough weight that ovulation was inhibited (Koskela, Ylönen 1995). Besides suppression of mating, sexual maturity may be delayed or litter mass and size (number of pups) may be smaller (Caro 2005). Most of these observations are experimental and use fresh predator cues; however, aged predator cues might signal the absence of predators and either have no effect or a positive effect on reproduction.

The stress caused by encountering a high density of predators can be passed from mothers to offspring and, subsequently, alter the reproductive output of offspring. Such intergenerational effects of predation on population cycles occur with snowshoe hares (*Lepus americanus*). The hormonal profiles of mothers fluctuate with predator density, and the profiles are echoed in their offspring. Essentially daughters have the stress-hormone levels of their mothers even though the predation levels the daughters experience differ from those of their mothers (Sheriff et al. 2010). This intergenerational, maternally inherited effect creates a lag in population response to predation.

In addition to direct action when a predator is threatening or in the vicinity, mothers and their offspring may have specialized behaviors or morphologies to reduce predation risk. Avoiding encounters with predators is the most common anti-predator mechanism. Unfortunately, "attempts to categorize adaptations to avoid detection are still crude and somewhat arbitrary" (Caro 2005:35).

Mechanisms to decrease the risk of predation have both spatial and temporal components. Mating and birth are especially vulnerable periods and, probably due to strong selective forces, these events often occur at times or in locations where the risk of predation is low, such as at night or in burrows. These events are also short in duration and generally inconspicuous. Mating calls and birthing sounds are rare, and, if possible, a female will leave the area after the event, or if she is tied to the area, she will often clean up any cues that could attract predators.

The choice of birthing sites may be greatly influenced by predation risk. For instance, hooded seals (*Cystophora*) give birth on transient ice floes that are inaccessible to polar bears and Arctic wolves. Subordinate hyenas give birth in dens some distance away from the communal den to avoid infanticide by dominant females. Many ungulates give birth away from the herd perhaps to avoid trampling, but also to keep the young away from the much more obvious movement of the herd.

The risk of predation is also influenced by how vulnerable the young are. When a female produces blind and naked young, she risks losing her entire litter should a predator find the nest or burrow while she is away foraging. Consequently, these females choose nest sites carefully and behave in ways to reduce discovery by predators. What are their options? Option 1: limit the time near the nest. Bank vole mothers (*Myodes glareolus*) limit their time near the natal burrow, thus reducing the olfactory cues available to predators (Liesenjohann et al. 2015). Option 2: rather than leave the nest, guard it. For this a female needs help. Help can come from mates in a monogamous pair, from family groups such as wolf packs, or from social groups as with hyenas, meerkats, and ground squirrels. Option 3: make the nest inaccessible to predators. For example, cave ceilings and the narrow rock crevices used as maternity roosts in some bats are difficult for predators to access. Option 4: swamp the predators or at least decrease your risk on a per-pup basis. Synchronizing births with those of other females in your population, such as with some bats and many ungulates, may yield more food than a predator can use. Option 5: carry your young with you as do kangaroos (*Macropus*), koalas (*Phascolarctos*), marmosets (*Callithrix*), baboons (*Papio*), giant anteaters (*Myrmecophaga*), and sloths (*Bradypus, Choloepus*).

A pervasive anti-predator adaptation is increased vigilance during lactation. In general, "mothers with young are more vigilant than females without young" (Caro 2005:168). This effect is known for many open-country species from elk to elephants to kangaroos and may vary, depending on the characteristics of their young. For instance, moose mothers are more vigilant when their progeny are active than when they are resting, or, as with cheetahs (*Acinonyx jubatus*) mothers with larger litters may be more vigilant than those with smaller litters (Caro 2005).

Nursing behaviors, too, can be influenced by predator avoidance. Female European hares (*Lepus europaeus*) give birth in relatively unprotected areas rather than in a protected burrow, but the ambulatory young (leverets) disperse away from the birthing site within a day. Just after sunset, the leverets congregate and await the doe's return. When she arrives, nursing occurs for 6 minutes or less after which the leverets and the mother disperse in different directions (Broekhuizen, Maaskamp 1980). Thus, not only is the vulnerable time limited but even if a predator found a single leveret during the day the other members of the litter would still be hidden.

Anti-predator adaptations can be a coordinated effort among both parents and offspring. In tree shrews (*Tupaia*), the male builds the natal nest 1–5 days before birth (Martin 1966). Thus, the nest has male rather than female odors associated with it. As males do not nest with females nor their pups, male odors do not indi-cate the presence of vulnerable young. After birth, the female suckles her young until their stomachs are distended, at which point each neonate weighs about 15 g, of which 6 g is milk. The female next removes the embryonic membranes and umbilical cord and leaves the nest for about 48 hours (Martin 1966). Nursing continues at 2-day intervals throughout lactation. Nestlings contribute to their own safety by suppressing vocalizations during lactation (Benson et al. 1992). Together, these coordinated activities reduce the chance predators will find the vulnerable nestlings.

Not just mothers but also neonates and juveniles have diverse adaptations to avoid predation. Two were previously described: leverets scattering after they have been born, and tree shrew nestlings suppressing vocalizations. Such changes in behavior are important because predators can be attracted to the sight, smell, or sound of young, as well as to their movements. To reduce visual cues, offspring may have a coat color that more closely matches the neonatal en-vironment than the adult pelage would (Caro 2005). As mentioned in chapter 8, tapir (*Tapirus*) calves and deer (Cervidae) fawns have striped or spotted coats that allow some measure of concealment when the young are lying in the open or under brush. Maned wolves (*Chrysocyon*) and spotted hyenas (*Crocuta*) are born underground and have a solid dark pelage (as illustrated by our cover photo), which matches the interior of their dens. Small mammals also have distinct age-specific coats; for instance, juvenile deer mice (*Peromyscus*) have a gray coat rather than the red-brown pelage of their mothers, which may more closely match their dark nest. If young are carried for most of their early lives, their pelage may more closely match that of their mothers' (chapter 8). For instance, the markings of giant anteater pups can blend in with the striping pattern of their mothers when the pups ride above their mothers' shoulders (figure 12.1).

Because coat color serves other purposes, such as thermoregulation or social recognition, selection on neonatal fur may not be related to avoidance of preda-tors. For instance, vervet monkey neonates have dark fur and pink faces in sharp contrast to the black-faced, light-colored bodies of their mothers. Consequently, they are conspicuous both to predators but also (perhaps to their salvation) to other group members. The adaptive benefit of the natal coat has been investigated in primates and may be due in equal parts to infant defense (against infanticide via signals for cooperative defense) and paternal cloaking (obscuring paternity of pups to fathers), but eliciting alloparental care is not a factor (Treves 1997). Neo-nate coat color presents an interesting area for investigation in a broader range of mammalian orders.

Not just visual signals but also odor and sound may attract predators. Our understanding of the odors pups emit is limited, and we found no studies of how the smell of pups may match their environment. Mothers could choose specific nesting materials that might mask pup odors. Waste materials have odors, and

Figure 12.1. Pelage as camouflage. Mother and juvenile giant anteater (*Myrmecophaga tridactyla*) foraging in the dry Cerrado grassland of Brazil, illustrating the interaction of juvenile pelage (fur) and behavior in anti-predator adaptations. Image by Tui De Roy; used with permission from Minden Pictures.

mothers may eat or remove material to reduce those odors (as done by tree shrews). Perhaps because humans have a limited olfactory repertoire, olfactory investigations are also limited. More work is available regarding sound. Neonates may have distinctive vocalizations that can be difficult to localize if they are soft in volume, or out of the normal hearing range of their predators, such as the ultrasonic vocalizations of voles, *Microtus* (Blake 2012).

Behaviors may help pups evade predation. Many offspring use distinctive postures, such as freezing or scattering. In chapter 8, we described a well-known behavioral dichotomy of young ungulates (hoofed mammals) as hiders or followers. The dichotomy also works for marsupials. For macropod marsupials, juveniles of most genera are solely hiders (13 species in *Dorcopsis, Onychogalea, Petrogale, Thylogale, Wallabia*) or followers (3 species in *Lagorchestes, Setonix*). The diverse genus *Macropus* has both hider (4 species) and follower (7 species) juveniles (Fisher et al. 2002). Different adaptations are expected of hider versus follower neonates. The differences include fur color and behavior, as well as physiology and development. For instance, neonates who follow might have muscle mass and limb morphologies suited to very early locomotion.

Eat or Be Eaten: Reproduction in the Context of Microbes and Parasites

Predators are not always other vertebrates and instead can be tiny and live on or in a female host. Traditionally, parasites are of two types, external (ectoparasites) or internal (endoparasites). Some ectoparasites, e.g., fleas or chewing lice, live in the fur or sit on the skin and take occasional bites of skin or debris. Others attach and drink for longer periods as do ticks, leeches, and sucking lice. Endoparasites live inside a female's body and usually have more long-lasting interactions with her. Such endoparasites include various worms, such as roundworms, flukes, and tapeworms but also single-celled protozoa, such as trypanosomes that cause Chagas disease or *Plasmodium* that causes malaria.

Not all parasites fit neatly into the bins we assign them. For instance, fleas are considered parasites, but female mosquitoes (females and not males feed on blood) and other biting insects are not, even though biting insects may cause as much or more irritation and subsequent diversion of resources as fleas do. Also, for mammals, only invertebrates and protozoa are considered parasites. Vampire bats (*Desmodus*), fungi, molds, and bacteria are not called parasites. Adding to the confusion, ringworm is not a worm at all but a fungus. Meanwhile, viruses are neither alive (according to some) nor cellular but share many of the tricks that protists play and in many ways may be considered intracellular parasites.

For mammalogists, a parasite is any invertebrate attached to (ectoparasite) or found within (endoparasite) a female once she is trapped and processed. Therefore, to be called a parasite, the invader must remain with its host throughout processing in the field. Thus, a flea is a parasite but not a mosquito, because a mosquito flies from its host after feeding rather than staying with it. Vertebrates are not considered parasites, although special terms such as *brood parasite* or *cleptoparasitism* describe behaviors resembling parasitism. The Centers for Disease Control and Prevention defines a parasite as "an organism that lives on or in a host and gets its food from or at the expense of its host" (http://www.cdc.gov /parasites/).

No matter the name, parasites, infections, and diseases can all have negative effects on reproduction. They all cause females harm in small doses rather than killing outright as a predator might. Another key difference is the time frame over which parasites versus predators take their toll on a female. Predators often kill quickly. They take offspring or tear off a limb in a relatively short time frame. Parasites and their ecological congeners, infectious and disease-causing entities, may have effects over much longer periods, perhaps over the life span of their host. Parasites have hosts; predators have prey.

What are the negative effects of parasites and their ilk on reproduction? Parasites can alter a current reproductive attempt in diverse ways. A blood meal may take resources from a female that she would otherwise put toward her offspring, but that meal will have different consequences if it is given to a mosquito, a leech, or a vampire bat. Koalas with reproductive tract infections (*Chlamydia*) may have reduced fertility (Phillips 2000). Female bank voles (*Myodes*), infected with the coccidian protozoan *Eimeria*, have a lower post-partum body condition (Laakkonen et al. 1998). Infections may have even more extreme effects. For instance, a 30-year decline in bison numbers from 10,000 to 2,200 may be related to infection with the tuberculosis and brucellosis bacteria (*Mycobacterium*, *Brucella*) that subsequently depressed both winter survival and pregnancy rates (Joly, Messier 2005).

Does the presence of parasites always lead to reduced reproduction? Certainly if a parasite kills a young mother or weakens her such that she falls prey to a predator, she loses her current litter and her lifetime reproductive output is diminished compared with mothers without parasites. However, what if the parasite doesn't kill her immediately but shortens her life? Can a young mother with a parasitic infection compensate for a shorter life span by increasing the energy she puts into a current litter? In some cases, she can. For example, female deer mice (*Peromyscus*) were allowed to reproduce after being experimentally infected with a blood fluke (*Schistosomatium*) related to the parasite that causes schistosomiasis in humans. On average, infected females delayed their first reproduction but produced litters that were 6% heavier than females without infections even though litter size was the same (Schwanz 2008). Thus, although reproduction over a lifetime may be reduced, input into a current litter may be increased. This is consistent with the theoretical supposition that, if the future prospects for reproduction (residual reproductive value, see chapter 5) are low or risky, a female should put more energy into her current reproduction.

A usual assumption about parasites is that the higher the host's population density, the more often the parasite will be transmitted, leading to more parasites in more individuals. This assumption may not always hold. Transmission rates may differ for different parasites even in the same host population. For instance, in central Colorado, montane voles (*Microtus montanus*) have two intestinal parasites, a cestode (flatworm) and a protozoan (*Eimeria*, the cause of coccidiosis). When vole populations were large, so too were the numbers of cestodes, but the numbers of protozoans did not vary with vole denisty. As a corollary, voles with either or both parasites did not have poorer body condition, suggesting reproduction may not be affected (Winternitz et al. 2012). Females were infected by fewer intestinal cestodes relative to males. However, of the infected females, the cestodes in reproductive females produced more eggs than cestodes in non-reproductive females (Winternitz et al. 2012). Clearly, the dynamics of parasite influences on reproduction are complicated. The vole study also highlights that females may have multiple species of parasites simultaneously, even in a single anatomical region (in this case the gut). A female may host her own community of parasites that will have interactions both with each other and with her.

Parasites sometimes tie their life cycles to a female host's reproductive patterns to infect the new young before they disperse from the natal nest. A classic example is the rabbit flea (*Spilophsyllus cuniculi*) and its host the domestic rabbit. Adult fleas apparently crawl from mom to baby at birth, and then, 12 days later after feeding and laying eggs, the fleas return to mom to wait for her next litter. The fleas respond to their host's reproductive state using hormonal cues (Rothschild, Ford 1964).

Some parasites may target organs associated with reproduction itself. A key and rather shocking example comes from *Placentonema gigantissima*, a nematode known only from the uterus and placenta of sperm whales (*Physeter macrocephalus*; Gubanov 1951). Although relatively common, little is known about the life cycles of these worms (Dailey 1985). One thought is that when the placenta is expelled at birth with the female worm inside the female worm dies, but her eggs are released as she decomposes. As a final note, these worms, as the name implies, are quite large, the largest of any known nematodes, with female worms reaching up to 8.4 m in length.

While some parasites might be transferred through the placenta, as may be the case with *P. gigantissima*, milk serves another easy path from mother to pup. Infectious parasites (roundworms, *Strongyloides stercoralis*) occur in the milk of dogs (Shoop et al. 2002). In another case, nematodes of the genus *Crassicauda* infect the muscle and ducts associated with mammary glands in cetaceans. Rate of infection may be as high as 47% in mammary glands of female Atlantic white-sided dolphins (*Lagenorhynchus acutus*; Dailey 1985). The transmission of *Crassicauda* is unknown but may be via milk. Presumably these metazoans diminish a female's ability to produce and/or transmit milk (Dailey 1985).

Pathogens may alter a female's reproduction either directly, e.g., via abortion, or indirectly, by diverting energetic resources to combat the pathogen (Pioz et al. 2008). Both scenarios have been well-documented in mammals. The list of pathogens that cause abortion or infertility in domestic mammals includes many bacteria, fungi, protists, and viruses (Givens, Marley 2008). For example, for more than 20 years, embryo rejection due to bacterial infection altered the overall annual reproductive output of alpine chamois (*Rupicapra rupicapra*), more than variations in the weather (Pioz et al. 2008). An example of indirect effects is in Columbian ground squirrels (*Urocitellus columbianus*). Females from which ectoparasitic fleas were experimentally removed (at the time of mating) increased their body mass during lactation. They also had a larger number of pups at den emergence, 5.25 versus 3.6 (Neuhaus 2003). The litter size differences could be due to an increase in ovulation, a decrease in post-implantation loss, increased milk production, or some combination.

Disease and parasites may have other indirect effects. Behavioral changes associated with infections, including those that indicate sickness, may result in lower chances of mating and thereby have important fitness consequences. Many parasites manipulate their hosts, for example, by altering the host's behavior. *Toxoplasma gondii* is a particularly fascinating and well-studied example. This parasite and its disease, toxoplasmosis, alters host-female mate-choice. The protozoan makes infected males more attractive as potential mates (Vyas 2013), presumably by increasing testosterone and thereby enhancing sexually selected traits in males. Sex-ratio changes due to toxoplasmosis infections occurred in experimentally infected house mice (*Mus musculus*), which gave birth to more daughters, but the mechanism is unclear (Kaňková et al. 2007). Finally, parasites may also drive a divergence from an even sex ratio by influencing female condition and thereby the trade-offs associated with bearing sons rather than daughters. This is not to say that all parasites have negative impacts on females, as many may be neutral, especially if they have co-evolved with their hosts for extended periods.

Worms, fleas, and protists are not the only organisms that curtail a female's reproduction. Bacteria and viruses infect mammals and can have major impacts on female biology, ranging from altering her background physiology to death of herself or her offspring. Some bacteria and pathogens become so linked to their host's physiological system and ecology that the relationship can gradually become altered on an evolutionary timescale. In some cases, the relationship may become a symbiosis. Other parasite-host pairs may remain (or become) neutral. However, before these infections can reach either stage, they must first interact with a female's immune system.

Reproduction and Immunity in Reproductive Females

The sterile womb paradigm is an enduring premise in biology . . . [and] remains dogma, as any bacterial presence in the uterus is assumed to be dangerous for the infant.—Funkhouser, Bordenstein 2013:1

Reproduction is tightly linked with immunity. Often pregnancy or lactation suppresses immunity. Significant changes in the maternal immune system occur in association with pregnancy, including cell-mediated and humoral immunity. These changes, especially immune suppression, could have consequences for the resistance of pregnant females to various diseases (Jamieson et al. 2006). Susceptibility to infection, and the ultimate severity of the disease once a female is infected, are two sides of the immunological coin (Jamieson et al. 2006). However, the general view of immune suppression during gestation is likely an over-simplification (Mor, Cardenes 2010; Racicot et al. 2014) because pregnant females can mount full immunological responses to infections during pregnancy (Racicot et al. 2014).

The placenta plays an important role in immune regulation (Mor, Cardenes 2010; Robbins, Bakardjiev 2012). The placenta protects the developing fetus from blood-borne pathogens and may have been a factor in the evolution of this fetal organ (Robbins, Bakardjiev 2012). For instance, when pregnant mice are infected with *Salmonella*, the uterus and the trophoblast become infected but not the developing embryo (Robbins, Bakardjiev 2012). In some cases, pathogens cross the placental barrier and infect the fetus. When this happens, a mother would benefit by rejecting that young. In this case, the pathogen also benefits as the aborted tissues are likely to be consumed by scavengers and allow transmission of the pathogen. Thus, inflammation-mediated abortion may benefit both mother and pathogen (Robbins, Bakardjiev 2012).

Of course, pathogens could still infect developing embryos via the reproductive tract. As a consequence, the cervix and its secretions help prevent infection of the developing fetus (Racicot et al. 2014). In this case, mother and placenta are joined against the pathogen and in aid of the developing embryo. One more complication is present in species with more than one young. In species with large litters, such as pigs, pathogens may not infect all litter-mates equally. Some may live and some may die. Sows routinely have large litters but, only rarely, give birth to fewer than four young. Sows usually terminate any pregnancy that will result in fewer than four offspring. Thus, even healthy offspring may be rejected (Givens, Marley 2008). Overall, interactions between pathogens, embryos, and mothers are complex and natural selection operates on all the participants with myriad results.

Eat or Be Eaten: Reproduction in the Context of Obtaining Food

Although all mammals are prey to some predator, all mammals are predators themselves. Up to this point, we have explored the diverse ways in which other species interfere with a female's reproduction; however, when she forages or hunts, she is the aggressor. What a female eats influences her reproduction, and her reproduction influences what she eats.

Gestation and lactation are energetically demanding and females must increase their intake of food and water to meet the demands of developing young. A mother's specific resource needs may also change. For instance, calcium is crucial for the development of bones and teeth but is limited in some diets, such as those composed largely of fruit or insects. Thus, females may need to secure higher amounts of calcium during reproduction. Consequently, female foraging patterns and sometimes dietary choices change during reproduction. For example, females of several rodents (*Dasyprocta leporina, Hylaeamys megacephalus, Proechimys cuvieri*) shift from frugivory to insectivory or granivory (seed eating) during pregnancy and lactation (Henry 1997). Similar dietary changes may occur in frugivorous bats (Orr et al. 2016). In addition, diets may shift to include more energy or water (Barclay 1994). Changes in diet require changes in foraging, thus females may change their behavior and activity patterns during reproduction. For instance, pregnancy and lactation improve the hunting performance of female rats perhaps through augmentation of the visual system (Kinsley et al. 2014).

Dietary requirements of offspring may alter other aspects of reproduction. For instance, Barclay (1994) proposed that the small (generally singleton) litter size of most bats was due to the large amounts of calcium needed by young to develop wings strong enough to fly.

Food abundance is rarely constant. Thus, changes in the availability, accessibility, and composition of food will affect a female's reproduction. For instance, prey reproductive cycles and life histories may drive the timing and duration of the reproduction of their predators. A classic example is the reproductive cycle of the Canadian lynx (*Lynx canadensis*). Lynx are specialist predators on snowshoe hares (*Lepus americanus*). When snowshoe hares are abundant, lynx have sufficient food to raise full litters, but when hare populations are low, malnourished lynx stop reproducing (Stenseth et al. 1997).

Similar elements are in play with herbivores. For instance, the reproductive patterns of frugivorous bats mirror periods of fruit ability (Fleming 1971; Racey, Entwistle 2000). In some cases, a change in food quality may instigate reproduction as described in chapter 11 for grass-eating rodents and the plant metabolite MBOA (6-methoxybenzoxazolinone).

As a food resource, plants are probably far more diverse in composition than animals. Plants produce multiple secondary metabolites, such MBOA, to deter herbivory. Some of those metabolites are intended for specific insect herbivores, but any mammal may benefit or be harmed by these compounds if they are medicinal or toxic. In addition, molds, mildews, smuts, and rusts may invade plants, and these fungi may have their own cornucopia of metabolites.

Anti-fertility and abortifacient effects of herbs are a substantive portion of human medical ethnography (Shah et al. 2009). Similarly, the animal science literature has documented the abortifacient effects of plants, such as locoweed (*Astragalus, Oxytropis*) and broom snakeweed (*Guttierrezia*). As for the fungi, mycotoxins from moldy grain will alter reproduction. For instance, zearalenone, a mycoestrogen, causes infertility in livestock, whereas aflatoxins, a carcinogen in rodents, can be transferred to milk after ingestion. Other mycotoxins reduce milk production, cause abortion, or reduce litter size (Atanda et al. 2012). Most of the negative effects of plant compounds on reproduction are well known in humans and mammals of concern to humans, but their influence is less well understood in mammals overall.

Natural selection on female behavior and physiology during reproduction may be particularly intense because not only does her own survival depend on successful escape from predators but the survival of her young pivots on her ability to provide for them.

Interactions of a Cooperative Nature

Although many of the interactions females have with other organisms are negative, in the sense that one species benefits at the expense of the other, other interactions are neutral or positive. Unfortunately, the emphasis is usually on competition and conflict, not on cooperation and integration. Consequently, commensalism, among species of mammals, is not well studied. Fortunately, we know something about the vast array of microbes we carry with us, although our understanding of the influence of our microbiome on reproduction is in its infancy. First, we explore cooperation among mammalian species.

Among mammals, the potential for mutual benefit exists in foraging, in avoiding predation, and in shared maternal care (Stensland et al. 2003). In mixed-species herds, increased foraging efficiency and predator avoidance will indirectly aid successful reproduction; however, we will focus on direct reproductive benefits of living with other mammalian species. Synchronized birth among individuals, even of different species, may be beneficial for the survival of offspring if the sheer numbers of offspring are more than predators can consume. For monogamous species, a mixed-species group could achieve the benefit of large group size without the costs of competition between mated pairs or the possible benefits of extra-pair copulations (Stensland et al. 2003).

Care of juveniles could be shared in mixed-species groups. A clear example is the shared nursery groups of rock hyrax (*Procavia capensis*) and bush hyrax (*Heterohyrax brucei*). In Tanzania and Zimbabwe, the two species occupy nearly identical niches, produce offspring synchronously, and have their young share the same protective areas. When rock and bush hyraxes share the same territory, they also share living holes, have similar activity patterns, huddle together, and use similar communication (Hoeck 1989). Juveniles of both species play together, and adults may simultaneously attend to juveniles of both species (Barry, Mundy 2002). To our knowledge, no other mammalian species have such clear evidence of shared parental care.

Mixed-species foraging troops of primates and African savannah ungulates are well known, but the reproductive consequences either do not exist or have not been observed (Stensland et al. 2003). Some anecdotal reports of shared care of juveniles in mixed-species groups of dolphins are known: one with common dolphins (*Delphinus delphis*) and bottlenose dolphins (*Tursiops truncatus*; Stensland et al. 2003); a second with Atlantic spotted dolphins (*Stenella frontalis*) and bottlenose dolphins (Herzing, Johnson 1997); and a third with Indo-Pacific humpback dolphins (*Sousa chinensis*) and finless porpoises (*Neophocaena phocaenoides*; Wang et al. 2013). Finally, human facilitation of reproduction in companion animals, farm animals, and zoo animals might be an example of mixed-species care, but the direction of care-giving is not mutual. These few anecdotal accounts suggest that either mixed-species cooperation is rare or at least rarely observed. However, we agree with Stensland et al. (2003:219) who concluded "mixed species groups occur in antipredator, foraging and social-reproductive contexts. This interesting phenomenon warrants further studies."

We Are Not Alone: The Microbiome and Reproduction

... many so-called "sterile" niches—notably in and among the female reproductive tract (such as the placenta)—may function as active low biomass ecological niches that harbor unique microbiomes.—Prince et al. 2015:2

The complex role of microbial communities in reproduction overturns two assumptions: first, that microbes are necessarily harmful and, second, that the reproductive tract and reproductive processes evolved to limit potential microbial contact. Some aspects of reproduction may promote the transfer of beneficial microbes between the female and her various partners, both mates and offspring. Research on a variety of taxa clearly illustrates the beneficial roles of our microbiomes. But how do we get our microbiome? Is it acquired early on and, if so, what role does it play during our development and during reproductive efforts?

The community of microorganisms and microbiota that inhabits the body is collectively termed the microbiome. The composition of the microbiome varies across time, between individuals, and within a single individual in different body regions. The microbiome is increasingly considered and discussed as another organ of the body (Mueller et al. 2015). Currently, our appreciation of the role and importance of this biota has resulted in a particularly active area of research and National Institutes of Health established the Human Microbiome Project to fund this research and coordinate databases (Prince et al. 2014, 2015). This funding focuses on humans, and little is understood about endemic microbial relationships in non-human mammals. However, the Earth Microbiome Project used crowdsourcing to document microbial diversity across the globe and has made every effort to keep these data available to other researchers online (Gilbert et al. 2014). What have these studies told us? How do microbes relate to reproducing females?

Transfer of microbes occurs throughout reproduction, before, during, and after birth. Microbes are transferred between mates during courtship and mating. They move between mother and offspring during gestation, birth, and nursing through amniotic fluid, meconium, colostrum, milk, regurgitated food, coprophagy, and allogrooming (Funkhouser, Bordenstein 2013).

Inter-species interactions within microbial communities may be an important component of mammalian reproduction. Commensal microbes might cooperate or compete for transmission to the next generation via incorporation into ova or milk, or via transport across the uterus into the placenta. Maternal and fetal physiologies may have evolved to facilitate or reduce microbial interactions. In fact, "microbes that promote host fitness, especially in females, will simultaneously increase their odds of being transferred to the next generation" (Funkhouser, Bordenstein 2013:6). Thus, females may benefit by providing environments that promote the survival and reproduction of beneficial microbes.

The maternal microbiome changes during pregnancy and influences the microbiome of the offspring (Prince et al. 2014). Changes in the vaginal microbiota likely prevent different bacteria and microbes from entering the uterus and/or pathogens from infecting the reproductive organs (Mueller et al. 2015). The placenta may have a complex microbiome that differs from its mother's reproductive tract (Aagaard et al. 2014). In rhesus macaques (*Macaca mulatta*), more than 300 species of microbes have been found in the placenta (Prince et al. 2015). The biota of the placenta and uterus are the initial origin of the developing offspring's own microbiota, and the vagina is a key place for exposure during birth (Prince et al. 2015).

Even after birth, a mother may provide her offspring with beneficial microbes. A study of human newborns to toddlers found a gradual increase in microbial diversity after birth (Prince et al. 2015). Where do these new microbes come from? Milk is one possible source. Mouse milk contains antibodies that interact with the intestinal microbiome (Prince et al. 2015). In humans, breast-feeding has a long-term impact on the composition of offspring intestinal microbes (Prince et al. 2015). Overall, the microbiome probably has wide-ranging effects on reproduction, but we have yet to understand the full extent of its influence.

The Biotic Environment: Heterospecific Interactions

Cooperative interactions among species are a hallmark of life on Earth. The mitochondria that power our cells and the nucleus that holds our DNA were once independent microbes (Pennisi 2004). Not only within our cells but also within ourselves we carry a vast microbiome. In these respects, we are not alone; we are a village.

Conflict and competition also characterize life on Earth. Females avoid or evade predators, scratch resident parasites, send T-cells to attack invading viruses, kill prey, and graze grasses. Aspects of mating, pregnancy, birth, lactation, and weaning are all influenced by predation, parasitism, and disease.

The reproducing female and her offspring are in constant interaction with heterospecifics from the large predator threatening survival to the tiny microbes enabling a mother to digest foods and thereby produce milk for her babies. These interactions have shaped the anatomy, physiology, and behavior of reproduction.

Historically, negative interactions among species received the lion's share of scientific effort, but in recent years, more scientists are exploring positive interactions. Our understanding that heterospecifics are not always detrimental to females is a new and exciting area of research.

The abiotic environment does not interact with us. We can only respond to it. Although microbes created an oxygen atmosphere billions of years ago and humans have accelerated the accumulation of carbon dioxide in recent times, the abiotic environment changes but does not evolve. Not so the biota that surrounds us. Co-evolution is a fact of life. A female's reproduction puts her in contact with a multitude of other life-forms. She alters their evolution, and they alter hers. But a female also interacts with members of her own species apart from her offspring and mates. A female's interactions with the biotic environment include those with conspecifics, and those social interactions are the subject of the next chapter.

CHAPTER THIRTEEN

Social Life

Help and Harm from Conspecifics

The hyena's world is one of social interactions. Our female, with her daughter, inhabits a communal den where she conducts much of her reproductive life. Her social rank influences everything from access to food to the safety of her daughter. Were she an alpha female, life would be easier for her and for her daughter. However, as a subordinate, she is often the recipient of aggressive behaviors. Within the matriline, her social rank dictates much of her perspective, her social role, and her obligations. As a low-ranking female, unless resources are abundant, her reproductive success will be low, whereas her high-ranking counterparts maintain high fertility in good times and in bad. Her offspring depend on her for their own rank; thus, they too will be low-ranking adults. Even so, they will outrank males. Living with other females is better than living alone. Unlike the Serengeti in Tanzania, where multiple clans inhabit the same area, her group in the Masai Mara in Kenya has fewer clans. Her matriline cooperates with others to defend specific territories against other clans. The group also defends their prey against other hyenas and even lions. Being part of a complex social network of matrilines and clans with diverse allegiances and rank relations sharpens her intelligence overall. Thus, even the lowest member of the group benefits from membership. (East, Hofer 2001; Engh et al. 2005; Frank et al. 1989; Henschel, Skinner 1991; Holekamp, Dlaniak 2011; Holekamp et al. 1993, 1997, 2007; Jenks et al. 1995; Smith et al. 2010; van Horn et al. 2004)

> Our knowledge about the effects of the relationship between group size, group composition and social relationships on the fitness of females is incomplete for even the best-studied mammal species.—Silk 2007a:553

People often refer to other mammals as solitary or social. Bears are solitary; lions are social. Dogs are social; cats are solitary. What do these terms mean? Social behavior is generally understood as the interactions among members of the same species. Are pregnant females social? Most mammalogists would say no, but these females certainly exchange information with their in utero offspring (chapter 7). Taking the female perspective calls into question our definition of *social* because reproductive females are seldom solitary. For instance, female black bears (*Ursus americanus*) have cubs at roughly 2-year intervals and stay with those cubs for most of that interval. Why, then, are black bears considered solitary? Is this another instance of a male bias? Male bears are generally alone while roaming large areas searching for females. Females, however, are generally not solitary and instead associate with their cubs and advertise for a mate when the time is right. In this case, the solitary (aka asocial) categorization correctly describes males but not females.

Tree squirrels, such as the gray tree squirrel (*Sciurus carolinensis*) and the American red squirrel (*Tamiasciurus hudsonicus*) are frequently considered asocial. In summer, during the reproductive months, females live with their offspring while males are often alone. But, in winter, females (and males) may nest in single-sex groups of 2–10 individuals. Female groups are of related individuals, whereas male groups are not (Koprowski 1996; Williams et al. 2013). Thus, female gray tree squirrels are with female relatives during the long winter nights and with their offspring during most of the summer. Again, what we mean by social is unclear from a female perspective.

Taking a female-centric perspective, all reproduction is social. From mating to weaning and often beyond, females interact with their offspring and mates. Females may also interact with other members of their species (conspecifics) who are not part of the current reproductive effort. Through such conspecific interactions, a reproducing female may take a broad range of roles: mother, daughter, sister, aunt, niece, cousin, granddaughter, grandmother, mate, babysitter, and neighbor. Although the names of these roles are of human design, many have biological relevance. For instance, in Belding's ground squirrels (*Urocitellus beldingi*), mothers, daughters, litter-mate sisters, and non-litter-mate half-sisters cooperate, whereas grandmothers, granddaughters, aunts, nieces, and first cousins do not (Sherman 1981). The DNA shared between close relatives provides a mechanism (kin selection) for natural selection to favor cooperative behavior. However, not all interactions among conspecifics, even closely related ones, are beneficial.

Some social groups are as small as a single female and her current offspring. But a number of mammalian species form social networks with several females. Many social groups also include past and current offspring, as well as adult males. Often females in social groups have daily interactions with one another. In fact, this version of sociality is the norm for many mammals. For example, 81% of diurnal primates are gregarious (Sterck et al. 1997). The interactions may be affiliative or confrontational. Cooperative interactions in larger social groups may result in cooperative hunting, maintaining territories, removing parasites, avoiding harassment by males, or communally rearing young. Competitive interactions may result in injury, reduced access to food, or infanticide. Both cooperative and competitive interactions take time and energy, but, given the persistence of complex social groups in many species, the advantages may outweigh the disadvantages. Furthermore, the presence of kin generally enhances a female's reproductive output (Silk 2007a).

This chapter explores several conspecific social interactions that influence female reproduction. The literature on social behavior is vast and diverse. We limit our brief foray to one mechanism, one benefit, and one cost. First, we examine what forces might drive females to form social bonds. Then, although the benefits of sociality are numerous (Silk 2007a), we explore one, alloparental care, as an example. At the opposite extreme, the costs of sociality are also numerous, but we choose infanticide as perhaps the most costly conspecific interaction.

Formation of Female Bonds

Female mammals exhibit a huge range of social behaviors from minimal short-term interactions with both offspring and mates to involved, repeated, and even continuous interactions with many individuals over long periods. For gregarious females, social bonds provide a context within which females may mature, mate, undergo gestation and lactation, and wean their young. At each of these stages, fellow group members may aid in the care of young or help with vigilance against predators, but they may also compete for limiting resources, such as water, food, nest sites, or mates.

To illustrate how these costs and benefits might drive the evolution of social groups, we examine one model (Sterck et al. 1997) hypothesized for primates. It starts with food availability. A clumped food distribution allows females to band together with little competition. When a favorable food distribution occurs in an area with high predation, the many eyes and ears of group members allow females to benefit from one another if alarm calls or actions alert other females to danger. Thus, food availability and predator risk influence the evolution of groups. For females, an additional variable is possible male harassment. If males impose costs on females in the form of harassment, injury, or death, females may form coalitions to reduce those costs (Sterck et al. 1997). Female coalitions among relatives will strengthen the evolution of female social groups. Social bonds may also form among unrelated females, and these bonds also have benefits. When related females begin to form social groups, inbreeding avoidance may cause males to disperse. Female philopatry results. When females are grouped closely together, rank relationships can reduce injurious, competitive interactions with resulting dominance hierarchies. This is a simple scenario, but one that involves ecological, demographic, and social factors in the evolution of female groups (figure 13.1). Different initial conditions or alternative physiological requirements will change the outcome. Changes in one parameter, such as food distribution, might quickly alter female gregariousness and ultimately social relationships. Not surprisingly, social groups are highly variable among mammals.

Three major forms of female social groups are matrilines, alliances, and coalitions. Matrilines are cross-generational groups of female relatives. All species have matrilines, but in social groups with dominance hierarchies, social rank as well as genetics may be inherited. Even if rank is not inherited, social groups may

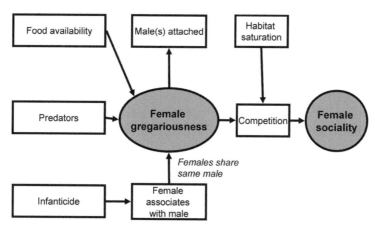

Figure 13.1. A female sociality model. A female's sociality can be influenced by many different aspects of her environment. Only a few are illustrated here. Both food availability and predators can alter female gregariousness. Indirectly, infanticide can lead to different patterns of female associations with males and thereby gregariousness. Females that share the same male also alter gregariousness. Habitat saturation may change the competitive regime and directly feed into female sociality. Further details regarding the nuances of female sociality are in the text. Modified by Teri Orr from Stereck et al. 1997.

be organized by matrilines. Individuals within a matriline may travel together, share resources, or have closer spatial relations than unrelated females.

Matrilines are a prominent aspect of the social organization of primates and cetaceans, ground squirrels and hyenas. Among primates, baboons (*Papio*), vervets (*Chlorocebus*), and macaques (*Macaca*) have strong matrilineal associations that organize their social relationships and activity (Silk 2007b). In cetaceans, matrilines are prominent in several odontocetes, such as killer whales (*Orcinus*), sperm whales (*Physeter*), and at least two pilot whale species (*Globicephala*; Kasuya, Marsh 1984). In particular, short-finned pilot whales (*G. macrorhynchus*) live in long-lasting matrilineal groups perhaps, in part, because they have an extensive post-reproductive life. Females bear their last calf before age 40 but may live until 63, providing 20 years for interactions between mothers, children, and grandchildren (Kasuya, Marsh 1984). Killer whales and false killer whales (*Pseudorca*) both live in matrilineal groups and may also have post-reproductive females (Marsh Kasuya 1986). The other group of cetaceans, baleen whales (Mysticeti), rarely has post-reproductive females (Marsh, Kasuya 1986).

Toothed whales (Odontoceti) have fission-fusion associations similar to those of spotted hyenas (Rendell, Whitehead 2001; Smith et al. 2008). Fission-fusion societies are composed of subgroups that frequently split or merge into groups of variable compositions. Matrilines are often the nexus of larger scale fission-fusion societies. In addition to hyenas and cetaceans, matrilines are central to the larger social order of such diverse taxa as bats (e.g., *Myotis* but not *Eptesicus*) and elephants (Archie et al. 2006; Kerth et al. 2011; Metheny et al. 2008). With the continued ease of genetic sampling, understanding the details between kinship and social organization is increasing.

Alliances and coalitions are social groupings in which two or more individuals join efforts for mutual benefit (Chapais 1995). The terms are used interchangeably and inconsistently across authors (Mesterton-Gibbons et al. 2011). When distinctions are made, they may refer to time frame (short versus long-term associations) or exclusivity (overlap across groups versus mutually exclusive groups; Mesterton-Gibbons et al. 2011). We will use the term *partnership* for both. Long-term and reciprocal partnerships may require nuanced social decisions about who to ally with or against and in what contexts (Smith et al. 2010). These complex social relationships have been used as indicators (proxies) of intelligence (Holekamp et al. 2007; Silk 2007a; sidebar on page 237).

Complex social interactions are a hallmark of primate social behavior but occur in other taxa. Coatis (*Nasua*), bottlenose dolphins (*Tursiops truncatus*), wild dogs (*Lycaon pictus*), wolves (*Canis lupus*), and hyenas (*Crocuta*) all form partnerships that function in disputes among group members (Romero, Aureli 2008). Not surprisingly, primate partnerships have received the most attention. Most often, primate female-partnerships are between related individuals, such as sisters or mothers and daughters, but in gelada baboons, *Theropithecus gelada*, they also occur between unrelated individuals (Dunbar 1980).

In the well-studied yellow baboon (*Papio cynocephalis*), key factors for the origins of female partnerships have been carefully evaluated (Silk et al. 2004). This species is particularly interesting, because partnerships are only present in some parts of its range. Silk et al. (2004) investigated several hypotheses for the formations of partnerships, including kin selection, reciprocal altruism, and individual benefits. The philopatry of females, linear dominance hierarchies, and the associated acquisition of maternal rank were all key factors in the formation and existence of female alliances (Silk et al. 2004). Dominant females were more likely to take part in coalitionary aggression, perhaps because they derive the most benefits from group membership or incurred fewer costs.

Partnerships (which can include males) may be used to defend territories that are key for the successful rearing of young, especially in environments with limited resources. Thus, these behaviors may vary with the season, whereby territories are only defended during periods of food scarcity or during periods of reproductive importance, such as lactation (e.g., northern red-backed voles, *Myodes rutilus*; West 1982). Coalitions present an interesting case in which cooperation allows for the possible resolution of conflicts between groups.

Overall, to support reproduction, females participate in diverse associations from long-term, cross-generational connections to short-term alliances. Support is sometimes more direct and takes the form of alloparental care.

The Hyena Watcher
Kay E. Holekamp (1951–)

A Smith College alumna, with a PhD from University of California, Berkeley, Kay E. Holekamp has studied hyenas in Kenya since early 1988. As a result of her work, we now have a much-improved understanding of hyena biology, in particular female reproduction and behavior. Her research focus is on the social and hormonal aspects of group dynamics in both spotted and striped hyenas. She has also made substantial contributions through her work on other mammals. Indeed, despite her current reputation as the world's expert on hyena biology, her early work was rodent-centric; in addition to examining sex-specific natal dispersal, she also worked extensively on the topic of sexual size dimorphism. Her dissertation on the proximal mechanisms of natal dispersal in Belding's ground squirrels produced a series of papers that examined dispersal in ground squirrels. In these, she showed that males disperse probably when they reach a certain body mass (the proximate cue for dispersal), whereas females nearly always remain on their natal territory. The ulti-

mate explanations for sex-biased dispersal may lie in avoiding familial incest or increasing access to mates. Holekamp also worked to understand the hormonal profiles of free-living ground squirrels when she addressed these interesting questions: What is the role of prolactin particularly in free-living animals? What is the typical seasonal variation in hormones in such animals?

Her research interests shifted to maternal rank, and this appears to coincide with her interest in hyenas (in particular spotted hyenas). In an early article, she demonstrated that, in many mammals with maternal rank inheritance, both kin and non-kin coalitions may form to help juveniles secure their ranks. Other early hyena publications focused on provisioning and lactation in hyenas and illustrated that provisioning is minimal at dens. Returning to studies of maternal rank, Holekamp used long-term data on spotted hyenas (*Crocuta crocuta*) from her field sites in Kenya to document the enormous reproductive benefits to high-ranked females (e.g., more surviving offspring, earlier age at

first reproduction) relative to low-ranking females. Her work now often includes striped as well as spotted hyena behavioral endocrinology.

More recently, her research on the group dynamics of hyenas allowed her to address questions relating to the evolution of mammalian intelligence. In recent comparative studies of Carnivora, she demonstrated that brain size, but not complexity of social structure, may correlate with solving puzzle boxes (Benson-Amram et al. 2016; Benson-Amram, Holekamp 2012).

Kay Holekamp is a C. Hart Merriam Award recipient from the American Society of Mammalogists and a fellow of the Animal Behaviour Society. In 2012, she was admitted to the American Association for the Advancement of the Sciences, and in 2015, she was elected a member of the American Associate of Arts and Sciences.

(Photo courtesy of Kay E. Holekamp.)

Caring for Offspring That Are Not Your Own—Alloparental Care

Despite the obvious, required, and biological link between reproduction and motherhood, once young are born, the role of caregiver may be filled by others. Social living can take advantage of this fact. We now move to discuss care of offspring beyond one's own.

Alloparental care occurs when individuals other than the genetic parents take care of the offspring. Cooperative care can be from unrelated individuals but is usually from siblings, aunts, uncles, or grandparents (Reidman 1982). Such help allows a mother to devote her energetic resources to lactation or reduces predation. Adoption is also a form of alloparental care. Cases of adoption are known for at least 120 mammalian species from diverse taxa, including marsupials, shrews, bats, primates, rodents, carnivorans, ungulates, elephants, hyraxes, and cetaceans (Reidman 1982).

Alloparental care takes many forms. Muskox (*Ovibos*) form defensive circles around their young when wolves threaten. Bottlenose dolphins work together to bring dead calves to the surface (Cockcroft, Sauer 1990) and presumably do the same for injured or ill calves. Female roost mates of the Rodrigues fruit bat (*Pteropus rodricensis*) assist mothers with the births of their neonates (Kunz et al. 1994). This direct assistance is dramatic but rare. More often, group members give alarm calls, or other warning behaviors, allowing mothers time to move their young to safety.

In some cases, females care for each other's young. Giraffe mothers deposit their calves in a crèche (nursery) some distance from the main group. When danger threatens, one mother will lead the young from the crèche to safety (Bercovitch, Berry 2012; Pratt, Anderson 1979). Nubian ibex (*Capra ibex*) mothers also leave their kids together in crèches. Mothers may forage at some distance from the crèche but stay with them overnight (Levy, Bernadsky 1991). Crèches are not always small. Thousands of Mexican free-tailed bat pups (*Tadarida brasiliensis*) are left in crèches. For bat pups, the chief benefit is energetic. Sharing warmth with other pups decreases the cost per pup of thermoregulation. Meanwhile, mothers can forage outside the cave while their young are safe within that dark, constant temperature environment (McCracken, Gustin 1991).

As with bat maternal colonies, sharing parental care can be passive. The key requirement is that females give birth synchronously over short time frames. Synchronous births allow for shared thermoregulation of developing young, but they also provide a strategy against predation. If many young are born at one time, predation on any single neonate may be reduced (Rutberg 1987). In the long term, reducing the per capita risk for each neonate passively benefits all females. Wildebeest (*Connochaetes taurinus*) are the textbook example. Wildebeest newborns are a major prey for spotted hyenas (*Crocuta*), because a single wildebeest mother cannot defend her young from a group of hyenas. Pups born during the peak period of calving are much more likely to survive relative to those born outside the peak, but only if they are in larger social groups (Rutberg 1987). Of course, the major cause of birth synchrony is the seasonal abundance of resources (Rutberg 1987). Predator swamping may reinforce the benefits of seasonal births and may tighten the birthing window.

A special and more extreme type of alloparental care occurs when the entire social group reproduces as a collective. In other words, a single female produces all the young that are subsequently cared for by the entire social group. Cooperative breeding is limited to a few of the about 5% of all mammalian species that are socially monogamous (Lukas, Clutton-Brock 2012). Wolves and naked mole-rats (*Heterocephalus*) are the classic examples. Cooperative breeding evolves primarily in groups of closely related individuals. Monogamous mating may be a precursor to the development of communal breeding (Lukas, Clutton-Brock 2012). In groups with high levels of social monogamy, the kinship between members is likely to be much higher and cooperative breeding (e.g., reproductive altruism) may evolve via kin selection (Lukas, Clutton-Brock 2012).

In mammals, the most extreme form of cooperative breeding is seen in naked mole-rats and Damaraland mole-rats (*Fukomys damarensis*). Naked mole-rats live in extended familial colonies of up to 290 individuals, with an average group size of 75–80. Reproduction is extremely restricted, usually a single female. The dominant queen suppresses reproduction in subordinate females (Jarvis, Sherman 2002). This matriarch solicits copulations from up to three males, generating litters of up to 27 young, with multiple paternity, maximally every 76–85 days (Jarvis, Sherman 2002). With average litter sizes of 11 and a life span of 30 years without a decrease in fertility, the productivity of a single queen is substantial (Buffenstein 2008; Sherman, Jarvis 2002). Such extreme productivity reflects the fact that reproductive females do nothing but mate, gestate, and lactate. They are completely cared for, fed, and protected by their nonbreeding kin. Nonreproductive colony members (workers) also groom and protect the young (Jarvis, Sherman 2002). After 5 weeks of suckling from the queen, pups beg for specialized feces (caecotrophes) from workers (Hood et al. 2014). Although naked mole-rat queens nurse their own young, females from other mammalian species may share even this burden.

Another costly way of helping is through co-lactation. In mammals, the opportunity for co-lactation is limited for several reasons. Some nursing strategies prevent cross-nursing, e.g., those marsupials that nurse young in pouches. Also, many mammals nurse in the confines of individual, dispersed nests or dens and, hence, do not have the opportunity to exchange young. A third factor is timing. To be effective, mothers need to synchronize lactation to ensure that the milk they produce is suitable for all young. Another aspect of timing is that lactation for many mammals is short; in a few cases, only a few days. Finally, lactation is enormously demanding; therefore, a female might better place her resources into her own young.

Nonetheless, nursing offspring of other females has been recorded for more than 60 species from diverse taxa, such as bats, primates, carnivorans, rodents, and artiodactyls (Packer et al. 1992). In species with larger litters, co-lactation is more common when mothers live in small groups (Packer et al. 1992). Closer to home: many different groups of humans have co-lactated, including the Samoans, Dakota, Alor, Bororo, and Arrenta (Shaw, Darling 1985).

Sharing milk with the young of relatives can evolve by kin selection, but females may also provide milk to unrelated young. This unusual situation is called allonursing (MacLeod, Clutton-Brock 2015) and can occur sporadically or as a common part of reproduction. Sporadic milk-sharing can occur in ungulates with singleton young. These females have two or more teats but only one calf. Thus, an unrelated calf can steal milk from one teat, while the mother's own young is nursing from an adjacent teat. Benefits to the young are clear. They obtain increased nutrition and energy, as well as any compounds in the milk that enable immunity, all at no cost to their own mothers (MacLeod, Lukas 2014). For shared lactation to evolve as a regular occurrence, donors should also benefit. But do any benefits accrue to the donors? Or is allonursing simply accidental?

Perhaps surprisingly, allonurses may reap several benefits. Over the long run, they may gain reciprocal care. If they are young mothers, they may gain experience. Finally, if mothers have lost all or some of their own pups, their unused milk may impede locomotion or be painful (MacLeod, Lukas 2014). Alternatively, allonursing may be an accidental and unavoidable result of living in close-knit burrows and nests, an inability to recognize one's own young, or an inability to simply get away. These explanations may apply to communally nesting species as well as to incidents of milk-theft (MacLeod, Clutton-Brock 2014).

The elements that go into co-lactation may be complicated. In a longitudinal study of wild house mice (*Mus musculus*), co-lactation occurred in 33% of cases. Females may selectively choose nursing partners or nurse alone (Weidt et al. 2014). With more young in a combined litter, thermoregulatory costs decrease. In addition, leaving pups in a nest allows individual females to forage unimpeded. For house mice, allonursing may be mutualistic with direct benefits to both mothers.

Reproductive Suppression and Infanticide

Although females may support the reproductive efforts of other females, they may also hinder them. Increased transmission of disease or parasites and female-female competition for limited resources are two major avenues by which females impede the reproduction of each other. Competition among females is not always obvious, but the intensity of female-female competition may be "as great or greater than in males" (Clutton-Brock, Huchard 2013). For example, when a single female reproduces for the entire group, as in naked and Damaraland mole-rats or meerkats (*Suricata*), competition when that female dies may be intense and physical. Subsequently, the physical abuse received by the losers may contribute to their reproductive suppression (Clarke, Faulkes 2001; Faulkes, Bennett 2001; Young et al. 2006). Meanwhile, in groups in which several females reproduce, fighting occurs over access to resources, such as food or burrows, as well as to establish dominance hierarchies (Clutton-Brock, Huchard 2013). In short, female-female interactions can result in the suppression of reproduction or the abortion of current reproduction in conspecifics of lower social status (Clutton-Brock, Huchard 2013). This type of reproductive suppression is observed in highly social rodents, e.g., Damaraland mole-rats, and in primates, such as olive baboons (*Papio anubis*; Clutton-Brock, Huchard 2013). Chronic stress can also hinder later stages of reproduction, such as lactation in chimpanzees (*Pan troglodytes*; Markham et al. 2014). These are also the same females who experience the most aggression from males (relative to females of higher rank). Psychosocial stress may, thus, result in physiological changes. Social stress may have additional repercussions (Markham et al. 2014). Group living is dangerous, especially in the context of matrilines and other social dynamics that may place one's pups at risk of infanticide.

Infanticide is the non-accidental killing of conspecifics during the interval between their conception and weaning. It occurs in many vertebrates, and mammals are no exception. Infanticide can take many forms and may be perpetrated by females or males. Infanticide allows individuals to alter the reproductive success of others. Males who commit infanticide may alter the short-term timing of a female's reproductive cycle and increase male opportunities for mating. Females may kill one another's young to reduce competition. Maternally induced infanticide, such as embryo rejection, may allow apportionment of reproductive effort into future, rather than current, offspring during environmental stress, while siblicide may allow survivors to obtain additional resources. Despite these potential benefits, infanticide is clearly a loss to the recipient. Documented cases of infanticide are generally most dramatic during lactation; yet, the evolutionary effects of infanticide at earlier stages may be more pronounced (Hayssen 1984).

Conflict between mothers and offspring can arise concerning the duration of either gestation or lactation. During gestation, a fetus is warm, wet, and protected. It receives water, minerals, vitamins, and other predigested nutrients. All these benefits are lost at birth. Thus, the fetus may benefit by extending gestation whenever it is able to do so (Hayssen 1984). However, the larger a fetus gets, the more dangerous the subsequent birth. Although a difficult birth (dystocia) may kill both the mother and the offspring, it is generally not considered infanticide.

Infanticide during Lactation

Although gestation is energetically more efficient than lactation, long gestations with continued fetal growth may decrease a female's ability to evade predators quickly and may increase the potential for a hazardous birth once it occurs. Partitioning resources into lactation rather than gestation allows a female to terminate reproduction with less risk (Hayssen 1984).

Infanticide during lactation is more easily detected by an observer and is less dangerous to a female. Another interesting aspect is that during gestation, fetal tissues may alter the maternal hormonal status. However, during lactation, young can only physiologically alter their mother's metabolism via the stimulus and hormonal correlates of suckling. The mother, however, can terminate this form of manipulation easily.

The marsupial mode of reproduction puts emphasis on lactation rather than on gestation. These mothers also produce extremely undeveloped neonates. These two conditions allow a female to abort reproduction throughout almost the entire reproductive period. Unlike the embryo rejection of eutherians, marsupial rejection of "embryos" incurs little or no risk to the mother's survival (Hayssen et al. 1985). She can cut her losses almost as soon as conditions deteriorate and renew reproduction when conditions improve.

Apart from embryo rejection, most mothers do not kill their own young, but other females may. However, which females commit infanticide varies. The variation can occur even within the same taxonomic family under similar ecological conditions. For instance, infanticide by females occurs in several ground squirrel species. In black-tailed prairie dogs (*Cynomys ludovicianus*), resident lactating females kill and eat the offspring of close kin. In these populations, infanticide may partially reduce or totally eliminate up to 40% of all litters (Blumstein 2000; Hoogland 1985). However, in the smaller Belding's ground squirrels, females do not kill close relatives but only unrelated pups (Blumstein 2000; Sherman 1981).

Infanticide by Males

In at least 114 species of mammals (2 bats, 31 carnivorans, 1 ungulate, 58 primates, 2 lagomorphs, 20 rodents), males are known to commit infanticide. Why do males kill infants? Of these species, infanticide is more likely when social groups have a female-biased sex ratio, when a larger percentage of offspring are sired by dominant males, and when the length of time dominant males are associated with the social group is short (Lukas, Huchard 2014). A different study, looked only at the roughly 230 species of primates and determined that 56 species have high levels of infanticide, 120 have low levels, and 54 species have not been studied well enough to be able to determine the presence or extent of male infanticide (Opie et al. 2013). Analysis of just primates suggests that the evolution of social monogamy may be tied to male infanticide (Opie et al. 2013); however, looking at mammals overall, social monogamy is correlated with breeding females who are intolerant of one another and not male infanticide (Lukas, Clutton-Brock 2013). To add another wrinkle, infanticide could drive the formation of female coalitions to defend against infant-killing males (figure 13.1). These are very different and, in fact, conflicting consequences of infanticide on the evolution of social groups. In addition, at least nine adaptive explanations exist for the causes of infanticide (Dixon 2013). Why such a muddle?

Both infanticide and social groups are the result of complex selective pressures that vary by species, environment, and even time period. An added complication is that scientists usually define and measure phenomena of interest before they make conclusions. For instance, defining infanticide in a specific instance may be obvious, a male kills an infant, but defining a species as infanticidal is more difficult. How many times must infanticide be observed? What percentage of males must commit infanticide? How many populations must have infanticidal males? For social systems, finding consistent operational definitions is even more difficult (Kappeler 2014). Thus, not only will the variation present in the natural world make simple answers difficult, our human methods for simplifying the natural world generate additional variation. For these reasons, we should not expect to have a single explanation for the evolutionary causes or consequences of infanticide and we do not.

Some infanticide by males is unrelated to mating opportunities. For instance, subordinate male southern sea lions (*Otaria byronia*), abduct and sometimes kill pups even though that behavior does not increase mating opportunities (Compagna et al. 1988). This sort of infanticide by males also occurs in other otariids (*Callorhinus, Neophoca, Phocarctos*) and at least one phocid (*Mirounga*; Campagna et al. 1988). Similarly, among terrestrial mammals, male golden marmots (*Marmota caudata*) kill pups, including those that may be potential mates in succeeding years (Blumstein 1997).

In highly predictable, seasonal environments, infanticide by males is less likely. In these conditions, female and offspring survivorship strongly correlates with the timing of birth. Natural selection will operate such that females give birth synchronously at that time. Because intraspecific variation in gestation is small, copulation and conception must occur in all females a set time in advance of birth. When females synchronously choose their mates over a very short period of time, a male must compete with other males for access to as many females as possible. Thus, male-male competition will be high, and a male's major energetic demand will be concentrated into a short period just prior to gestation. Because female receptivity, or lack thereof, is due to environmental cues, not conspecific ones, infanticide, in these cases, will not shorten a male's wait until the next opportunity for mating and will not increase his reproductive success (Hayssen 1984).

Infanticide by Siblings (Siblicide)

One of the hallmarks of the mammalian radiation is the extensive provisioning of maternally refined nutrients to immature offspring. These nutrients, however, are not infinite, and nonidentical siblings may be in conflict regarding the allocation of maternal resources. Even if maternal energetic resources are not limiting, mammary glands or implantation sites may be restricted. Thus, siblings may not have equal access to maternal resources. Siblicide may allow surviving offspring to receive a larger proportion of available nutrients or even to obtain exclusive access to them (Hayssen 1984). We described the murderous blastocysts of the pronghorn antelope (*Antilocapra*) in chapter 7 but less obvious forms exist.

Many small to medium-sized marsupial mothers give birth to more offspring than they have teats (Hayssen et al. 1993). Consequently, not all young who leave the uterus find a teat on which to attach. These extra young die. Thus, these marsupials have an obligate form of infanticide and sibling competition at birth.

For many mammals, the entire uterus can support embryonic development. However, in some mammals, such as elephant shrews (*Elephantulus*), glossophagine bats, viscachas (*Lagostomus*), and pangolins (Pholidota, *Manis*), only a restricted portion of the uterus is available for implantation and subsequent embryonic development. If blastocysts outnumber implantation sites, in utero sibling competition can occur. An extreme example of this is the eastern rock elephant shrew (*Elephantulus myurus*) in which two embryos are brought to term out of the 100 ova produced. Although some ova may never fuse with sperm, the presence of only two implantation sites (one per horn) forces sibling conflict and eventual death. The plains viscacha (*Lagostomus maximus*) also illustrates competition for implantation sites. Up to seven zygotes implant; however, all but the two farthest from the ovaries are reabsorbed during gestation (Weir 1971a, 1971b).

Even without morphologically distinct implantation sites, not all portions of the uterus will support development equally. For instance, in pigs (*Sus*), rabbits (*Oryctolagus*), mice (*Mus*), and guinea pigs (*Cavia*), fetal weight is correlated with uterine position; smaller fetuses often occupy the middle segments of the uterine horns. In addition, fetal size decreases as litter size increases. Thus, competition between blastocysts exists before implantation for the optimal position in the uterus. Rabbit zygotes spend about 3 days in the uterus before implanting, allowing many hours for in utero competition (Hayssen 1984).

Competition for parental resources also occurs during lactation. The finite number of mammary glands may result in conflict among siblings if litter size exceeds the number of teats. If a phase of semi-permanent teat attachment occurs during lactation, as in the woodrat (*Neotoma*), as well as dasyurid and didelphid marsupials, only the first young to find teats will have a chance for survival. For mammals without teat attachment, sharing of teats is possible. However, milk composition changes during the course of a nursing session (e.g., rabbits, pigs), and subsequent young may not obtain equal nourishment. In addition, young that suckle later may not be able to suckle as long, especially if females control the duration of nursing. Aggressive interactions between siblings for teat ownership have been observed in piglets and kittens and result in the establishment of a dominance hierarchy among siblings (Hayssen 1984).

Most of the previous examples are of obligate sibling competition, but sibling conflict can be facultative and occur only when resources are limited (Morandini, Ferrer 2015). Perhaps best understood is the case of siblicide in domestic pigs, in which mothers produce the maximum numbers of piglets and allow siblicide to occur if resources (i.e., milk) are sufficiently limiting (Andersen et al. 2011).

The Tuco Tuco Whisperer
Eileen A. Lacey (1961–)

When Eileen Lacey took first year high school biology from a woman named Laine Gurley, she realized that biology—and in particular animal behavior—could be a career path for women. As an undergraduate at Cornell University, she was introduced to naked mole-rats, which set the stage for a career devoted to studying subterranean rodents. Early papers examined cooperative breeding in these fascinating rodents, and she still publishes with her undergraduate mentor on topics, such as "what is eusociality?" By the time she completed her PhD at the University of Michigan, she was focused on understanding the social biology of burrowing rodents.

Now a professor at the University of California, Berkeley, and a curator in the Museum of Vertebrate Zoology, she trains future scientists while exploring the lives of tuco-tucos, tojos, cururos, and other subterranean rodents in their natural environments in South America. Using comparative studies, she seeks to disentangle the impacts of ecology, phylogeny, and physiology on patterns of social behavior, including patterns of reproductive success and suppression, within and among species. Many of her ideas on this topic are covered in the edited volume *Life Underground* (Lacey et al. 2000). Her contributions to our understanding of trade-offs between sociality and repro-

duction include papers that explore how reproductive success is related to group size, alloparental care, and territoriality.

Recent work in her lab has expanded to encompass mechanistic aspects of social behavior. Current research topics include the effects of social environment on stress physiology and reproductive success, as well as the role of the maternal environment in shaping those interactions. Her numerous contributions and administrative acumen led to her election as president of the American Society of Mammalogists in 2012.

(Photo courtesy of Eileen A. Lacey.)

Mothers enhance sibling conflict via staggered ovulations and multiple matings, thus, creating embryos of different ages in the same pregnancy. Sibling competition has a morphological component in piglets, which are born with specialized teeth. The third incisor and the canine are angled such that sideways biting during competition for teats results in facial lacerations of siblings. A reduction in competition occurs after the first week of nursing when individual piglets consistently use the same teat at successive nursings (Drake et al. 2008).

Another example of facultative siblicide comes from our focal species: the spotted hyena. In spotted hyenas, littermates may kill one another during times of resource shortages, but litter-mates may survive during periods of plenty (Golla et al. 1999; Smale et al. 1999). In contrast to swine, hyena mothers intervene when sibling aggression occurs. The intervention may even include temporarily housing the fighting pair in separate dens and providing private nursing bouts to the subordinate cub (White 2008).

Conspecific Interactions Have Multiple Facets

Reproduction requires the interaction of conspecifics who may be in conflict regarding the exact timing of, or energetic investment into, the various reproductive stages. Infanticide is only one manifestation of that conflict (Lacey 2004; Lacey, Sherman 1997; sidebar on page 245). The occurrence of female-female conflict, sibling conflict, and maternal-offspring conflict are all common but so, too, is cooperation.

Females play various roles throughout their lives. As infants, juveniles, aunts, mates, or mothers, females have numerous opportunities to take part in social interactions. Many of these interactions support reproduction. From shared crèches to shared defense, females may help one another.

In this chapter, we discussed many cases of conspecific interactions as they relate to the reproductive female mammal, but numerous additional interactions are important and beyond the scope of this book. For example, social cues can suppress reproduction or advance sexual maturation. Sociality is under selective pressure from abiotic influences, such as seasonality and topography, as well as from biotic influences, such as food availability and parasite abundance. Key is that aspects of the biotic environment, in particular relationships with conspecifics, can influence mammalian reproduction either to the benefit or to the detriment of the individual female.

PART FOUR

THE HUMAN SIDE

 Just as all female mammals reproduce in the context of their abiotic and biotic environments, so, too, do we, as a species, alter and are altered by other species and our environment. In this section, we explore the human side of female reproduction. Schematically, we put mammalian reproduction inside the human community to represent our impact on the environment, our own physiologies, and our essential nature as mammals (see figure). We first focus on our environmental impact and our attempts to rectify what has largely been a detrimental association (i.e., destruction and consumption of natural resources; chapter 14). However, we also investigate how reproductive biology might help conservation efforts. In our final and most comprehensive chapter, we cover humans from genes to ecosystems. We examine ourselves as a species and ask whether, as the "naked ape," we are really all that different from other female mammals (chapter 15).

CHAPTER FOURTEEN

Conservation and Female Reproduction

Shot, snared, trapped, speared, and poisoned, her life is not an easy one. Despite being one of the most abundant mammalian predators in Africa, her relatives in the Serengeti are in decline. Aside from these direct threats, hyenas are also subject to the indirect effects of habitat destruction. Noise and other forms of pollution from human activities may reduce fertility or alter milk quality. All these negative influences have left populations at risk.

Humans try to help our female and her species. Conservation efforts aim to preserve hyenas in the wild. Eco-tourism provides funds for habitat preservation while putting her and her cubs on display. Tourists and their cars provide new and unusual camouflage: hiding behind vehicles in order to steal prey from cheetahs. Humans also attempt to preserve the genetic variation of her species. To do so, a captive, sedated female may be artificially inseminated with sperm from an unrelated male at some distant zoo. Human assistance has mixed results. (Haysmith, Hunt 1995)

> Females "encounter dangerous situations almost daily but encounter predators far
> more rarely."—Caro 2005:111

Humans have altered and destroyed the living space of nearly all mammals, including themselves, and continue to do so. We as a species have hunted many other species to extinction and continue to do so. More recently, some individuals are trying to rectify these patterns and conserve biodiversity. Conservation biology is a relatively new, and welcome, human endeavor. As its professional society (the Society for Conservation Biology) was founded only in 1985, the science of maintaining and restoring biological diversity is in its infancy. Understanding the reproduction of species is key to maintaining them.

Successful reproduction is essential for population growth and stability. Species-specific reproductive patterns provide key information for both wildlife management and conservation initiatives. Perhaps the most obvious link between reproductive biology and conservation research is the shared aim to understand how mammalian life histories dictate recruitment, growth, and other population-level parameters. Estimates of maximum population growth rates use many reproductive variables, such as age at first reproduction, litters per year, and interbirth intervals (Bowler et al. 2014). But what other aspects of a mammal's life history and reproductive strategies are important for conservation efforts? Furthermore, how might the timing of reproduction be altered by our changing world?

This chapter is divided in two parts. Part one explores how anthropogenic changes to the environment have affected female reproduction. Climate change and environmental pollutants are the main players. The focus of the second part is how humans are now trying to ameliorate problems associated with environmental perturbations. Here, the importance of reproduction to conservation efforts is key. Captive breeding and assisted reproduction are the focal elements of this section. We highlight areas in which we see promise in relation to conservation efforts.

Anthropogenic Influences on Reproduction

Humans have caused the extinction of many mammals from the Steller's sea cow (*Hydrodamalis gigas*) to the Tasmanian wolf (*Thylacinus cynocephalus*). Of course, all extinctions and population declines can be connected to reproduction, but in some cases, the links are more direct. Two major disturbances are climate change and environmental contaminants, but a host of smaller disruptions also impede reproduction. The exciting new field of conservation physiology explores how the physiology of organisms, including their reproduction, is altered by anthropogenic environmental change (Seebacher, Franklin 2012). This section is part of that effort. Knowing how mammalian reproduction is altered may elicit ways to combat the damage. Avid readers looking for a thorough investigation of the topics covered in this chapter are referred to *Reproductive Science and Integrative Conservation* (Holt et al. 2003) or *Marine Mammal Research: Conservation beyond Crisis* (Reynolds et al. 2005).

Effects of Climate Change on Reproduction

The predicted changes in climate include more extreme temperatures and weather events, as well as an increased randomness of previously regular weather patterns. If these changes are realized, the very cues used by mammals to time reproduction may become less reliable, resulting in reproduction at sub-optimal periods, as already happens with birds (Visser et al. 2004). The problem may be particularly severe in species with long reproductive cycles (Bradshaw, Holzapfel 2006).

Again, biases in our knowledge limit conservation efforts. For instance, we know much about the timing of reproduction in temperate regions but know far less about the timing of tropical reproduction (Bronson 2009). This gap is especially problematic given the huge tropical biodiversity combined with extremely high levels of habitat destruction and human population growth in the area (Laurance et al. 2014).

In Australia, evidence of the impact of climate change became apparent in January 2002 when temperatures over 42°C resulted in the deaths of at least 3,200 flying foxes, primarily females and juveniles (Welbergen et al. 2008). Extreme temperature events are increasingly common and likely to result in large die-offs when individuals are unable to negotiate the severe challenges. Temperature changes may also alter water chemistry and water availability with subsequent effects on wildlife biology.

Climate change leads to habitat change and the loss of breeding grounds. The ice floes on which Arctic mammals depend for reproduction (often parturition) are melting. Alpine habitats on mountaintops essential for pika reproduction are shrinking.

The impacts of climate change may not be as obvious as shrinking ice floes. Instead, climate change may gradually and cryptically impede reproduction via the physiological consequences of stress. Physiological reactions to stress alter the hypothalamic-pituitary gonadal axis (chapter 5, physiology; Toufexis et al. 2014). In addition, environmental stress may alter behavior and change male-female or mother-offspring interactions (Anthony, Blumstein 2000).

Effects of Contaminants on Reproduction

Additional environmental challenges faced by the reproducing female include the presence of chemicals in the air, water, soil, and thereby food. Agricultural runoff, pesticides, industrial waste, toxic waste, medicinal antibiotic or hormonal prescriptions, and oil spills are among the potential contaminants that may disrupt hormonal action, cause fetal abnormalities or stillbirths, or be passed through milk to contaminant offspring (Fair, Becker 2000).

Agricultural fertilizers add phosphorus and nitrogen to soil to increase production. Excess nitrates and phosphates run off fields into rivers and lakes where they can augment the growth of blue-green algae (cyanobacteria) causing algal blooms. Cyanobacteria produce toxic microcystins, which can contaminate drinking water. Although liver damage is a major effect, these toxins also decrease ovarian size (Wu et al. 2014).

Organochlorines (e.g., DDT, chlorinated benzenes, polychlorinated biphenyls) are highly toxic compounds used as pesticides, solvents, coolants, dielectric fluids, plastics, and as insulating material in electronics. Many countries in North America and Europe have restricted their use, but their extreme longevity and employment in other countries means they remain a source of pollution. Organochlorines accumulate in top predators.

The effects of DDT have been noted in ringed seals (*Pusa hispida*), harbor seals (*Phoca vitulina*), and gray seals (*Halichoerus grypus*; Fair, Becker 2000). From 1900 to the 1970s, ringed seal populations in the Baltic Sea sizes dropped from ~200,000 to ~5,000 individuals and gray seal populations dropped from 100,000 to 4,000 individuals. Much of this decrease was due to hunting. However, declines between the 1950s and 1960s were due to reproductive anomalies, such as uterine stenosis and occlusions, and many aborted pups were noted during this time.

Polycholorinated biphenyls (PCBs) are lipophilic, meaning they are able to bind fats and as a result remain within the body for extended periods of time, especially in blubber. PCBs can also cross the placenta (Tanabe et al. 1982). The U.S. Department of Defense maintains a large "force" of bottlenose dolphins (*Tursiops truncatus*) and has monitored their organochlorines. Females with higher blubber organochlorines also had more stillborn calves and higher neonatal mortality (Reddy et al. 2001). Milk can also be a source of organochlorines in neonates (Tanabe et al. 1982). As a result of reproductive transfer via placenta and milk, female bottlenose dolphins and killer whales actually offload more of their pollutants to their offspring than they take in and these reproductive females actually have lower concentrations than males (Ross et al. 2000; Wells et al. 2005; Yordy et al. 2010). The sex and age differences in the concentrations of persistent organic pollutants is another reason understanding reproductive biology is a critical component of conservation efforts.

Public awareness of the effects of environmental contaminants from crop production and industry is increasing, but pharmaceutical waste may also be problematic and is less commonly discussed. Hormonal birth control and hormone replacement are two ways in which women regulate their lives. Meanwhile, steroids are used extensively for livestock. The use of antibiotics to treat the perceived or real risk of infection is rising both for humans and for livestock. Both hormones and antibiotics can be released to the environment through sewage or runoff (Giger et al. 2003; Lindberg et al. 2007; Sun et al. 2014). Excess steroids can disrupt reproductive behavior, morphology, or physiology with known effects on laboratory rats and mice, as well as wild fish and reptiles (Guillette, Gunderson 2001). Antibiotics could directly disrupt the commensal microbiota of mammals or have indirect effects through changes in the environmental microbiota (Sarmah et al. 2006). Exploration of specific effects in wild mammals is an open field for investigation.

Miscellaneous Environmental Disruptors

Humans have significantly changed the acoustic and visual environments females inhabit. Noise pollution in marine systems can disrupt foraging, communication, and navigation (Fair, Becker 2000). Artificial nocturnal illumination may disrupt reproduction. For instance, light pollution suppressed melatonin levels and delayed births in wild tammar wallabies (*Macropus eugenii*) near an urban area (Robert et al. 2015). Similarly, females of the nocturnal mouse lemur (*Microcebus*) altered the initiation of their reproductive activity when exposed to light pollution (LeTallec et al. 2015).

Light and noise have been well studied because humans are keyed to visual and auditory stimuli, but olfaction is a primary sensory system in other mammals. We have no information on anthropogenic changes to the olfactory environment. Consequently, we also do not have information on any resultant changes to reproduction. Measuring the odor landscape is a significant challenge (Riffell et al. 2008). However, scientists studying insect behavior may be laying a foundation for eventual work with mammals.

In many cases, how or which environmental stresses and chemicals cause abnormalities in reproduction is unclear. For instance, rhinos have low reproductive outputs; that is, they have only a few offspring over long periods of time. Poaching for rhino horns has decimated populations with concomitant concern over a lack of genetic diversity, especially for Indian rhinos (*Rhinoceros unicornis*). Such loss is likely a concern for other rhino species because four of the five species of rhino are listed as at risk for extinction by the IUCN (Hermes et al. 2014). Although Indian rhinos can live up to 40 years in captivity, females stop reproducing at around 18 years of age due to the prevalence of vaginal and cervical tumors (leiomyomas). The causes of these reproductive tumors, often called fibroids in humans, are unknown but may be a response to some altered environmental feature (Hermes et al. 2014).

Another example in which the environmental cause is unknown is the sudden advent of a transmissible cancer in Tasmanian devils (*Sarcophilus harrisii*). Devil facial tumor disease, a fatal disease, was undocumented prior to 1996 but, as of 2009, was observed in upward of 83% of captured devils (Hawkins et al. 2009). Reproductive adults are the key infected age class, and females now only reproduce once prior to death (Jones et al. 2008). Devils were extirpated from the Australian mainland, and the Tasmanian population is seriously at risk.

Humans, their pets, and livestock transmit parasites and diseases to wild mammals. Even if these diseases had original zoonotic origins, the density of humans and human-affiliated species, as well as the mobility of humans, results in a much more rapid spread than would likely be observed in natural settings. Such anthroponotic diseases include giardia in beavers (*Castor*), *Echinococcus granulosus* in macropods (Macropodidae), and toxoplasmosis in sea otters (*Enhydra lutris*; Conrad et al. 2005; Thompson et al. 2010). Domestic goats (*Capra hircus*) are implicated in scabies infections in chamois (*Rupicapra*) and Spanish ibex (*Capra pyrenaica*; Fuchs et al. 2000). Canine distemper and parvovirus have found their way from pet dogs into wild carnivores.

A major disruption to the North American bat population may have been caused by recreational cavers bringing a European fungus to bat caves in the northeastern United States (Frick et al. 2010). The resultant disease, white nose syndrome, affects bats during their winter hibernation, when females are in a critical reproductive period, namely, mating and sperm storage (Wai-Ping, Fenton 1988). The result is massive die-offs and low recruitment of new offspring to replace the waning adults (Daszak et al. 2000; Frick et al. 2010).

Just as humans transmit diseases to wildlife, the reverse can occur. Bats have been implicated in outbreaks of emergent diseases, including Ebola, Marburg, SARS, Nipah, and others (Weller et al. 2009; Wibbelt et al. 2010). These diseases, while obviously important in terms of human health, are likely to affect the animals that carry them as well. As human populations increase, people will encroach more and more on other mammals and the exchange of zoonotic diseases will also rise. Also public perception of mammals (e.g., bats) as the carriers of diseases puts those mammals at further conservation risk, for instance, when bat roosts are deliberately destroyed for fear of disease transmission, such as rabies.

At this point, the reader can surely appreciate that the effects of human disturbance on reproduction are wide ranging. Hunting and road kills directly remove females from a population, but if the female is lactating or pregnant, her offspring also die. Overall, even small things we do can harm wildlife, but small things we do can also be helpful, including support of local, regional, or global conservation efforts. We move our discussion to these conservation efforts as they might be aimed at the reproductive female.

Reproduction and Conservation Efforts

Although humans have diverse negative impacts on the lives of mammals, they are also trying to improve those lives. Conservation biology is devoted to preserving species and habitats, as well as maintaining as much biodiversity as possible. In situ conservation is aimed at maintaining natural populations, for example, through the creation of reserves. The goals of these efforts include replacing lost habitat, conserving any remaining habitat, or linking fragmented habitats. Monitoring reproduction of populations within these habitats may be necessary. If habitat destruction is extensive or if political and economic influences prevent the formation of extensive reserves, then conservation efforts must turn their focus to live animals or their tissues in zoos, botanical gardens, and frozen-tissue collections. Captive breeding and assisted reproductive technologies are major contributors to this arm of conservation biology.

In some cases, the conservation issue is an overabundance of a given species, often an introduced species. Consequently, another aspect of reproduction in conservation biology is the regulation of invasive species, e.g., rabbits (*Oryctolagus cuniculus*) in Australia or gray tree squirrels (*Sciurus carolinensis*) in the United Kingdom and in Europe, as well as species whose abundance has skyrocketed because predators have been excluded from the area, e.g., white-tailed deer (*Odocoileus virginianus*) in the absence of wolves (*Canis lupus*) in the northeast United States. Conservation efforts in these cases may be aided by wildlife contraception.

Captive Breeding

Research in reproductive biology can inform conservation efforts by revealing areas on which we might focus our efforts. Agencies developing management strategies need data on mammalian reproduction to determine who to manage, when, and how. Currently, debate in the field of wildlife conservation is focused on two main questions related to reproduction. First, which sex is most important for conservation efforts? Second, how effective is reproductive biotechnology for conservation efforts? The answer to the first question may inform the second. Using reproductive biotechnology for conservation efforts is likely to be easier with one sex than the other. For example, preserving sperm may be similar to preserving oocytes, but obtaining sperm is much easier than obtaining oocytes. The ease of obtaining sperm from males is obvious, but so, too, is the importance of females, especially for viviparous therians. The uterine environment is critical to gestation but needs a living female. Unanswered questions related to conservation efforts include, how similar are the uterine environments of related species, and can one be a surrogate for the other?

What about monotremes? Although the platypus (*Ornithorhynchus anatinus*) and short-beaked echidnas (*Tachyglossus*) are not currently at risk, all three long-beaked echidnas (*Zaglossus*) are critically endangered with decreasing populations (IUCN Red List). Assisted reproduction for these egg-laying mammals requires knowing when and where conceptions occur and how conception is related to egg-shell deposition. Just having sperm or a genetic blueprint would not be enough. Unfortunately, the basic reproductive biology is unknown.

Captive versus Wild

How well does the biology of captive animals match that in the wild? Orcas are an example. An orca (*Orcinus orca*), given the name Granny, who was likely born in 1911, the year before the Titanic sank, was spotted in 2016. Thus, by best estimates, she was then over 105 years old: an impressive age for any mammal, our long-lived species included. Although zoos have had orcas in captivity since the early 1960s, the longest captive longevity was 34 years (Weigl 2005). Granny's ripe old age illustrates how very little we know about the life histories of many mammals, even familiar species kept in captivity, such as orcas. However, early in 2017 Granny was conspicuously absent from her social group and is presumed dead. Thus, 105 years may be close to actual longevity in this species (Close, January 3, 2017). Nevertheless, this is but a single individual and much remains a mystery about orcas. Without knowing longevity, conservation biologists cannot effectively estimate how long a population might need in order to recover from a rapid decline.

Orca populations are estimated at only 50,000 individuals. These mammals are subject to hunting, habitat destruction, disturbance (from boats), and, when in competition with the fishing industry, they risk being shot (Taylor et al. 2013). In addition, their placement at the top of their food chain means they experience bioaccumulation (i.e., they concentrate any pollutants absorbed by their prey). PCBs as mentioned above relative to bottlenosed dolphins occur in the blubber of orcas, but the amounts vary by sex and reproductive status with females having

lower levels than males (Ross et al. 2000). All of these environmental disturbances contribute to the negative impacts on orca reproduction. Pup mortality is
especially high, 50%. In captivity, orcas reproduce much earlier than in the wild,
but this has not translated to successful captive reproduction perhaps because
mothers are too young and inexperienced to properly care for their calves. Details
of reproduction in marine mammals are difficult to obtain especially those related to behavior. For terrestrial mammals, the environment provided in captivity
may be more similar to wild conditions, especially for small mammals. In these
cases, technology may allow for assisted reproduction.

One additional conservation risk is skewed sex ratios (populations with an excess of either males or females). For example, captive Indian rhinos and black
rhinos (*Diceros bicornis*) give birth to more males than females (Wildt, Wemmer
1999). This presents an obvious issue if females and their uteri are necessary for
successful production of future offspring! The uterus is a limiting resource for zoo
biologists, one that quickly becomes a problem if sex ratios are male skewed.

Assisted Reproduction

Many conservation strategies rely heavily on assisted reproduction. Electro-
ejaculation, artificial insemination, petri-dish conceptions, and embryo transfer
are among the assisted technologies used to conserve wild mammals. With these
methods, captive animals can be paired artificially rather than face-to-face. Technology avoids issues of individual mate choice and difficulties achieving copulation
in stressful habitats. Thus, transporting animals and keeping potential mating
pairs in proximity are no longer necessary.

An unexpected technological breakthrough of assisted reproduction efforts
was the perfection of techniques for preservation of genetic material in liquid
nitrogen. Cryopreservation is a bourgeoning field, not just for human fertility
endeavors but also for conservation efforts. Cryopreservation offers the chance to
freeze biodiversity. Similar to the agricultural "doomsday" seed banks set aside in
remote areas of the world, such as Svalbard in the Arctic, frozen zoos have the
potential to conserve mammalian biodiversity. Frozen zoos (genetic resource
banks), such as the one at the San Diego Zoo in California, store sperm, oocytes,
and embryos, as well as other isolated tissues. They also preserve germ-line stem-
cells, such as primordial germ cells (see chapter 6).

Because obtaining ova is more difficult (in fact requiring surgery) than obtaining sperm from induced ejaculations, conservation biologists initially focused on
cryopreservation of semen. That methodology is now fairly effective. Unfortunately, oocytes not only are more difficult to obtain but also are more difficult to
preserve. This is, in part, because oocytes are packaged in maternal cells (the
cumulus oophorus), and these maternal cells are key to oocyte maturation. This
is not to imply that females are entirely ignored. Cryopreservation of female gametes has been attempted with limited success on a few species, such as felids, and

with nominal success on others (Pope et al. 2012). One problem may be that oocyte collection circumvents the normal processes by which ova are chosen for ovulation. A technician rather than the female herself is choosing which oocyte to contribute to the next generation and any innate selective mechanisms for releasing the most viable oocytes are lost. Any sperm selection by a female is also removed with in vitro conception. This is also an issue for human-assisted reproduction.

When female gametes can be harvested, they can be mixed with sperm and resulting embryos cryopreserved. At some later point, the embryos must be thawed and placed in the uterus of a surrogate female, but continuation of development has a low and variable probability. For example, although 116 embryos were transferred to 12 caracals (*Caracal caracal*), only four females continued gestation, and they produced five kittens total. The same procedure applied to the endangered fishing cat (*Prionailurus viverrinus*) was unfortunately less successful: 146 embryos were transferred to 12 females, but only one continued gestation, and she produced only one kitten (Pope et al. 2006).

Assisted reproduction has been implemented in conservation efforts of several endangered and at-risk taxa (Pope 2000). The Spanish ibex has been successfully carried to term in the uterus of a closely related species (Fernández-Arias et al. 1999). In addition, an extinct subspecies of the Spanish ibex was cloned, although the offspring died minutes after birth (Folch et al. 2009). Such "conservation cloning" entails taking nuclear DNA from a non-reproductive cell and inserting it into an enucleated oocyte of a related species. After a period of test-tube development, cloned blastocysts are implanted into the related surrogate for the remainder of gestation. This technology has been attempted for the endangered mouflon (*Ovis orientalis*), but the lambs did not survive (Hajian et al. 2011). Given the maternal contribution to the oocyte besides DNA (chapter 6), complications due to parent-of-origin gene expression (chapter 2), and the extensive cross talk between the uterus and the embryo (chapter 7), closely related surrogates are critical for this technique to succeed.

What about species for which no closely related taxa are available? The giant panda (*Ailuropoda melanoleuca*) is the only living member of the subfamily Ailuropodinae and is the conservation poster child for the World Wildlife Fund. Conservation efforts for giant pandas have focused on reproduction. Captive breeding began in 1978 but gained notoriety for generally being unsuccessful (Zhang et al. 2009). A major problem was with males: many would mount but not copulate (Zhang et al. 2004). Eventually, advances in artificial insemination alleviated that issue and changes in infant care and captive animal management improved neonatal survival (Zhang et al. 2009). As of 2006, about 260 giant pandas were born in captivity.

Has assisted technology in aid of panda conservation been successful? The answer is mixed. Captive reproduction of giant pandas is still difficult. Females have a short, annual time frame when they will mate and conceive. Males are still reluctant to mate. Delayed implantation and a long-lived corpus luteum make detection of pregnancy difficult (Zhang et al. 2009). Finally, with artificial insemination, many females give birth to twins but can only care for one pup and often abandon the other. Luckily, zookeepers can care for this second pup. In short, captive breeding will be sufficient to maintain a captive population but whether it will allow the panda to once again roam in large numbers in the wild is less certain.

Unfortunately, while artificial insemination works well for cattle, these techniques are not as successful for pandas or cheetahs. Much of this book is devoted to the diversity of mammalian biology. Understanding that diversity, particularly that of female reproduction, is necessary for viable assisted reproductive technologies particularly for mammals with nuanced reproductive patterns. We must extend our knowledge beyond the lab mouse and domesticated cow before other mammals can benefit from these techniques. Non-invasive hormonal sampling of field-collected urine and feces may contribute to understanding the endocrine physiology of mammals in the wild. In the meantime, construction and maintenance of genome and tissue banks (frozen zoo) may conserve genetic diversity.

Wildlife Contraception and Management

Although the usual goal for conservation is to improve reproduction, curtailing population growth is occasionally necessary either in captivity or in the wild. In captivity, economic or spatial considerations may limit the number of animals that can be adequately supported, and population size may have a set limit. In the wild, several situations may necessitate population control. First, introduced species may have expanded to the detriment of endemics. European rabbits, black rats (*Rattus rattus*), red foxes (*Vulpes vulpes*), domestic cats (*Felis catus*), and gray tree squirrels all have all been introduced to island populations and either outcompeted or killed the native wildlife. Even introductions into larger areas can limit endemics, as the introduced gray tree squirrels have reduced native European red squirrel (*Sciurus vulgaris*) populations across Europe (Bertolino et al. 2008). Second, release of companion or domestic animals into the wild may produce populations that damage the ecosystem or threaten wildlife. Wild horses (*Equus caballus*) from the western United States to Assateague Island, Maryland, as well as wild pigs (*Sus scrofa*) across the United States from Florida to Arizona, are examples. Third, removal of a top predator may allow prey animals to expand without limit; white-tailed deer in the urban northeastern United States no longer have wolves or mountain lions (*Puma concolor*) to curtail population growth. Fourth, other forms of anthropogenic disturbance may release wildlife from their natural controls, as is the case for African elephants (*Loxodonta africana*), due to restricted dispersal from secure game parks, and Australian kangaroos (*Macropus*), due to increased food from crop land (Allen 2006; Herbert 2004). In such cases, particular species may out-step the carrying capacity of the environment, throwing the ecosystem out of balance and altering populations of other species.

Trapping or culling excess animals is "not always legal, wise, safe, or publicly acceptable" (Kirkpatrick et al. 2011:40). Other methods, such as the surgical sterilization of large numbers of females, are not always practical. Contraception, however, is non-lethal, reliable, and may be practical. Among the species whose populations are managed by wildlife fertility control are wild horses, urban white-tailed deer, bison (*Bison*), water buffalo (*Bubalus bubalis*), wapiti (elk, *Cervus elaphus*), and African elephants (Kirkpatrick 2007; Kirkpatrick et al. 2011). Several contraceptive avenues are available.

Early approaches to wildlife fertility control administered hormonal treatments that mimicked those in human birth control. These steroids successfully reduced fertility but ultimately failed for diverse reasons, ranging from unrealized toxicity to behavioral changes, health risks, and regulatory bureaucracy (Kirkpatrick et al. 2011). A second approach was to inhibit hypothalamic hormones that induce ovulation via the hypothalamic-pituitary axis (see chapter 5). This approach inhibits gamete production in both sexes and has been used for humans, domestic mammals, as well as species such as koala (*Phascolarctos cinereus*), kangaroos, lions (*Panthera leo*), wild dogs (*Lycaon pictus*), and Hawaiian monk seals (*Monachus schauinslandi*; Herbert 2004). A downside is that any estrogen- or testosterone- related behaviors are likely inhibited in addition to fertility (Herbert 2004). If these behaviors are key to social interactions, the social structure of the population would be altered (Gray, Cameron 2010). A third method is immunocontraception. Operating via a vaccine this strategy works by immunizing females against the zona pellucida (the non-cellular coat around the female gamete; see chapter 4). Vaccines can be delivered in low volumes without handling the animals, are safe for pregnant females, and do not influence behavior. The downside is that immunity is not long-lasting, and annual booster shots are necessary for at least 3 years (Kirkpatrick et al. 2011). Although all the approaches have side effects (Gray, Cameron 2010), as of 2011, 67 zoos were managing 76 species by immunocontraception (Kirkpatrick et al. 2011).

For long-lived species, contraception requires years to effectively reduce populations sizes, but for short-lived species, such as many rodents and rabbits, it may be more immediately effective (Fagerstone et al. 2006). Clearly, species-specific differences and habitat-specific needs will determine the most effective methods for reducing populations. However, one more hurdle may remain to disrupt conservation actions.

In the United States, regulation of contraception is controlled by the Environmental Protection Agency (EPA) using an extensive, rigorous, and costly registration process (Fagerstone et al. 2006). In addition, "the EPA registered wildlife contraceptives as 'pesticides'" (Cohn, Kirkpatrick 2015:27). At least in the United

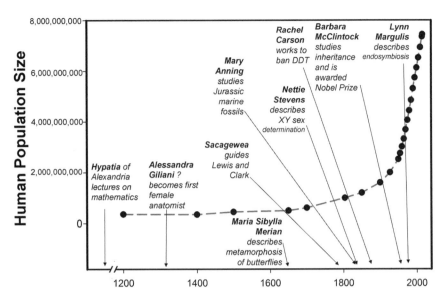

Figure 14.1. Human population growth. The number of humans begins an exponential rise in about 1700, around the time of the birth (1706) of the mathematician Émilie du Châtelet, lauded as "a great man whose only fault was being a woman" (François-Marie Arouet [aka Voltaire] in a letter to Frederick the Great; Hamel 1911:370). Figure by Teri Orr. Population size data from http://www.worldometers.info/world-population/.

States, the sociopolitical dimension of contraceptive prevents wildlife agencies from effectively reducing overabundant populations, even though they have the means to do so (Cohn, Kirkpatrick 2015). As Rutberg (2013:S38) points out "wildlife contraception has its ethical roots in animal welfare," but the conservation community and its values have been absent from the sociopolitical discussion of wildlife contraception. Perhaps the time has come to enter the dialog.

The Naked Ape: Where Do We Fit?

> The mother of the year should be a sterilized woman with two adopted children.
> —Paul Erlich, quoted in *LIFE*, April 17, 1970

While many species are experiencing lower than replacement level reproductive outputs, this is not the case for humans. Our population size is ever bigger (figure 14.1). Population booms were predicted by Paul Erlich and others, all of whom suggested that human overpopulation of the Earth would drive food shortages and wreak havoc on planetary resources. They advocated zero population growth and foretold massive population jumps to the detriment of the natural world and ourselves (Ehrlich, Holdren 1971).

Did this predicted boom occur? Yes. According to the international database of the U.S. Census Bureau, in 1970, the global population was 3.7 billion; today at 7.5 billion, it has doubled. The most populous countries by far are China (1.4 billion) and India (1.3 billion), while the United States is a distant third with 325 million. Will the numbers rise? Yes. Population growth reflects birth and death rates. If the two are equal, the population is stable. Currently, all three of the most populous countries are adding more individuals than they are losing: China: 12 births to 7 deaths; India: 20 births to 7 deaths; the United States: 13 births to 8 deaths (per 1,000 people; World Bank, data.worldbank.org). Clearly, human birth-rates must decline to maintain a stable global population. The politics and sociology of this issue are key, but so, too, is understanding the reproduction of women as mammals, the subject of our final chapter.

CHAPTER FIFTEEN

Women as Mammals

The naked ape who watches her day in and day out is odd, but she has grown accustomed to it (figure 15.1). Once it captured her and she woke up with a marking on her ear. Now, it mostly photographs her and records her with video cameras. Strangely, this beast is bipedal and seems smaller than the other naked apes—the ones that have fur on their faces. She has also seen others like this one, extremely small ones that seem helpless to move on their own, ones that need to be carried. They must be pups, but they can't run and romp on their own as her, much furrier, pups do. Is this full-sized naked ape a mammal like herself, like hyraxes, lions, and other furred beasts? Is it related to the other naked denizens of the savannah, the elephants, rhinos, and hippos? Or is this naked ape—this "woman"—something entirely different?

Figure 15.1. Kay Holekamp observing hyenas (*Crocuta crocuta*), as they observe her. Photograph from Kay Holekamp.

I was driven to understand my past. For we are not ready-made out of somebody's rib. We are composites of many different legacies, put together from leftovers in an evolutionary process that has been going on for billions of years. Even the endorphins that made my labor pains tolerable came from molecules that humans still share with earthworms.—Hrdy 1999:xv

As members of the genus *Homo* and order Primates, we too are mammals. In this chapter, we discuss women as reproducing mammals and ask, Are we very different from our mammalian counterparts? Are we "special," and if we are, how so and why? Although, we only scratch the surface of the vast biomedical, sociological, behavioral, and evolutionary literature on women as mammals, we can apply the framework of the previous chapters to our own biology. Following the general order of the proceeding chapters, we briefly outline how women fit within those topics from the inner circles to the elements of the reproductive cycle. In a bit more depth, we explore traits considered unique to women: menstruation, concealed estrus, menopause, C-sections, and birth control. These are only a few of many aspects of women as mammals. For more detail, we refer the reader to Sarah Blaffer Hrdy's three excellent, stimulating, and often provocative books: *The Woman That Never Evolved* (1991), *Mother Nature* (1999), and *Mothers and Others* (2009).

Inheritance and Evolution, Anatomy and Physiology

Women most closely resemble the other apes in our family Hominidae, including gorillas (*Gorilla*), chimps (*Pan*), and orangutans (*Pongo*). Originating as anatomically modern humans some 200,000–60,000 MYA in Africa, we are now a truly cosmopolitan species. Admittedly, we might appear to be quite novel for we do not lay eggs as the platypus does nor do we care for our young in pouches as a kangaroo mother might. Also, unlike many rodents and carnivores we do not give birth to large litters. Instead, like elephants (Elephantidae) and many bats (Chiroptera), we have singleton young, long maternal care, and an extended social system. To a large extent, our life histories are not all that different from three main lineages of mammals: our own (Primates), as well as bats (Chiroptera) and cetaceans (Cetacea). Key similarities with these taxa span a range of traits.

The genetics of human sex determination is largely the same as in other mammals and does not necessitate additional consideration here. However, now that the human genome is sequenced, interesting molecular-based questions about reproduction in mammals may be done using humans as a model system. That is, we can use humans as models for understanding other organisms. Indirect and direct genetic factors are linked to many reproductive phenomena both normal and aberrant, such as breast cancer, endometrial cancer, endometriosis, female sexual dysfunction, menarche, age at menopause, ovarian cancer, ovarian reserve, polycystic ovary syndrome, pre-eclampsia, premature ovarian failure, and uterine fibroids (Montgomery et al. 2014). Studies of entire genomes also suggest propensities for various reproductive diseases (Montgomery et al. 2014). Because we are among the "better studied mammals," we understand more of the inheritance of reproductive traits in humans than in other mammals. Similarly, the importance (e.g., medical relevance) of being able to treat women means our anatomy and reproductive pathologies are also well understood. We may now be a better model system for understanding other mammals, than they are for us.

Anatomically, women have amygdaloidal (almond-shaped) ovaries with a thick tunica (exterior cover). These characters are not unique and are found in a range of other mammals, including white-tailed deer (*Odocoileus*) and reedbuck (*Redunca*; Els 1991; Mossman, Duke 1973). A similarity between humans and some bats (e.g., *Carollia*) is a highly invasive (hemochorial) placenta (Rasweiler et al. 2011). Humans also have a simplex (single-bodied) uterus similar to some bats (e.g., many phylostomids; Rasweiler, Badwaik 2000) and a few other mammals such as sloths (Hayssen 2010), as well as other primates. In women, uterine developmental abnormalities can result in additional morphologies, the most common of which is a duplex uterus due to the incomplete fusion of the parts comprising the usual simplex morphology. In this, we revert to a possible ancestral morphology.

Sometimes our similarities to other mammals are surprising and helpful. Because the innervation is similar, the genital anatomy of female hyenas (*Crocuta*) provides a model for the clitoral enlargement in women with congenital adrenal hyperplasia. The detailed knowledge of the how the hyena clitoris is innervated helps reduce the side effects of surgical reconstruction in women with abnormally large clitorises (Baskin et al. 2006).

What about external traits? That women lack beards intrigued Darwin and led, in part, to his formulation of the concept of sexual selection. As naked apes, our secondary sexual characteristics include differences in the presence of facial hair and mammary gland development. Distributions of body fat also vary by sex (Wells 2007), as do canine size, voice, and pelvic shape and tilt. Even the brain can differ; for instance, women have more gray matter than men (Cosgrove et al. 2007). We exhibit size dimorphism with an average of ~7% difference in body size with men being (on average) larger than women (Gustafsson, Lindenfors 2004).

Becoming bipedal necessitated anatomical changes to the birth canal and the resultant constraints affected both offspring cranial size and maternal pelvic anatomy. These differences are what allow forensic scientists and paleontologists to assign sex to human remains. For each sex difference, we can find analogs in other mammalian species (McPherson, Chenoweth 2012). In one regard, however, we are unique, and this specialness is often demonstrably bad for the health of women.

A unique feature of humans is our desire to modify our anatomy. Such modifications include simple changes in hair color, more extensive changes via cosmetic surgery, or the drastic and unfortunate practices associated with genital mutilation. The cultural drive for the modifications varies but key is that these modifications are generally not beneficial to our health, especially in the case of genital mutilation, which may be carried out in clandestine locations and under questionable conditions. Infections and death are common end-products of this unnecessary practice. We hope that by the time this book finds its way into the hands of readers the numbers of girls undergoing genital mutilation has dropped to zero and that young women are allowed to mature in an unaltered and healthy state. In this regard, we humans have deviated far from the natural state of a healthy well-functioning anatomical framework within which a normal physiology may operate.

Physiologically, we are just another mammal. Our ovaries are formed under the same suite of molecular interactions and endocrine control as those of other mammals, as described in chapter 3. Estrogens and androgens organize and ultimately activate our brains and behavior as described in chapter 5. Our adult hormone cycles resemble those seen in other apes. Also, in the general ways in which our bodies work, we are (again) just average mammals.

Oogenesis to Menopause

Oogenesis in women follows the pattern described in chapter 6, but one key item is that the number of ova in humans may or may not be fixed, an area of extensive debate. Folliculogenesis in women is similar to that of other mammals. Multiple waves of follicles develop before one is chosen, and ovulation is followed within a few days by menses (Mihm et al. 2011). This full sequence often takes 26–35 days but is highly variable in duration both within individuals and across women (Arey 1939). Ovulation is triggered by luteinizing hormone from the pituitary in response to a releasing hormone from the hypothalamus. But this only happens after our brains assess our body condition, e.g., fat reserves, and determine that ovulation might lead to a successful pregnancy. This integration is part of the reason why anorexia inhibits ovulation. Stopping reproductive function is a normal and necessary response in times of food shortage or stress, not a sign of disease (Södersten et al. 2006).

Women are considered to be spontaneous ovulators and monotocous. Thus, ovulation usually results in one ovum at a time being released. After ovulation and conception, the conceptus must ensure a continued secretion of ovarian progesterone to maintain the pregnancy. This happens about 8–10 days later, around the time implantation begins (Bazer et al. 2010). Once maternal recognition of pregnancy is achieved, ovarian progesterone maintains the pregnancy for the first few weeks until placental progesterone takes over. This transition starts at 6–8 weeks (Tuckey 2006). The complexity of these initial steps is part of the reason why most embryo rejection occurs early in pregnancy (before 13 weeks; Avalos et al. 2012). The remainder of gestation is a constant endocrinological and materials cross talk between mother, placenta, and embryo, and this complex cross talk continues through birth (Myatt, Sun 2010) and into lactation.

Most females nurse their own young with their own breast milk. Many women do, but not all. Formula is something unique to humans. Even women who do breast feed have large, cross-cultural variations in milk composition, mostly due to different diets (Skibiel et al. 2013). The duration of lactation also varies culturally, but, in general, weaning occurs much later in humans than in many other mammals (chapter 9). Other primates, some bats, and cetaceans also have long lactation lengths (Kurta, Kunz 1987).

Menstruation

Age at first reproduction in most mammals is assessed by first ovulation, first mating, or first pregnancy, but in women, this is marked by menarche, the first menstrual period, when the uterine lining prepared for pregnancy is discarded. What other (if any) mammals menstruate? Menstruation (aka eumenorrhea) is generally uncommon across mammals but is present in other primates, as well as quolls (*Dasyurus*), colugos (*Cynocephalus*), hedgehogs (*Erinaceus*), elephant shrews (Macroscelididae), tree shrews (e.g., *Tupaia*), at least four families of bats (Pteropodidae, Phyllostomidae, Molossidae, Vespertilionidae), carnivorans (e.g., dogs), and at least one rodent (*Acomys*; Bellofiore et al. 2017; Emera et al. 2012; Wang et al. 2008; Zhang et al. 2007). Some classifications of this trait aim to assess whether bleeding is overt or covert, i.e., reabsorbed. Given this description one might think human menstruation is entirely overt. However, menstruating women absorb a good portion (about two-thirds) of the developed endometrium.

Why might menstruation have evolved? This question remains largely unanswered but has sparked the curiosity of numerous scientists and generated multiple hypotheses. For instance, menstruation might serve (1) as some sort of signal to conspecifics, (2) as a mechanism to combat pathogens, (3) as a way to avoid metabolic costs, (4) as a means to prevent cancer, (5) as a competency test for newly conceived zygotes, or (6) as a by-product for a highly invasive endometrium selected for by human brain evolution (Emera et al. 2012; Strassmann 1996).

Despite their differences, all of these hypotheses have received some support, as well as scrutiny. First, if menstruation is a signal, exactly what is being signaled by the menses and is that signal important? If important, why is it not present in more taxa? Second, the "cleansing hypothesis" (Profet 1993) suggests that the loss of the uterine lining helps remove potentially deleterious pathogens. However, no evidence (that we know of) has been found that women who do not menstruate are more likely to incur infections. Furthermore, menstruation could leave the uterus exposed to infection. The third hypothesis focuses on metabolic costs associated with maintaining a fully "prepared" uterus (Strassman 1996).

Indeed, extensive increases in metabolism (roughly 14% above a non-active/prepared uterus) are associated with maintaining a fully active endometrium. This translates to half a month's metabolism per year and may be significant.

Because cells associated with the endometrium are so prolific, removal of these cells may help avoid uterine cancer (hypothesis 4). However, this is inconsistent with the prevalence of both cervical and uterine cancer in women. To our knowledge, no one has compared rates of pathology in women with amenorrhea to those that menstruate. A fifth untested hypothesis suggests that females use menstruation to evaluate zygotes (Barash, Lipton 2009). Zygote assessment and subsequent embryo acceptance or rejection would be another cryptic mechanism of female choice (Thornhill 1983), which is exciting given the paucity of data for this concept (Birkhead 1998). Finally, the human-centric sixth hypothesis is that our big brains demand an invasive endometrium and that menstruation has evolved due to various by-products associated with this demanding tissue. This hypothesis not only ignores other taxa with both large brains and no menstruation, such as dolphins (an odontocete), but also ignores taxa with smaller brains that also menstruate, such as tree shrews.

Concealed Ovulation, Continuous Estrus, and Mate Choice

Both concealed ovulation and continuous estrus refer to situations in which ovulation is not obvious by a female's behavior. But is this a human-specific trait? Of course, *concealed ovulation* is clearly a weighted term and *loss of estrus* maybe more appropriate (Shaw, Darling 1985). However, while concealed connotes deception, loss of estrus is disingenuous given that women certainly want to mate, and the desire to do so may vary relative to ovulation. In addition, near ovulation, body scent may change, as well as behavior (Miller et al. 2007), and men may be able to pick up the cues generated by ovulatory women. The famous "stripper study" demonstrated that women indirectly indicate their fertility in the way they dress and behave (Miller et al. 2007). Thus, behaviorally, our estrus signaling may be as unconcealed as ever.

Do human females typically mate only when physiologically prepared for conception? The answer is clearly "no." Women mate during pregnancy and the luteal phase (Sillén-Tullberg, Møller 1993). Some argue that such "continuous receptivity" (note word choice presenting female as the inactive party) strengthens social bonds between women and their mates. In addition, if a male does not know when a female conceives, he will not know his relationship to the offspring. Paternity confusion may lead to mate guarding and the presence of males over a longer period than just during ovulation with multiple social consequences. These arguments receive mixed support (Sillén-Tullberg, Møller 1993).

Of course, noncyclic mating is not unique to women. It also occurs in captive macaques (e.g., *Macaca*) and gelada baboons (*Theropithecus gelada*), as well as wild orangutans (*Pongo*) and chimps (*Pan*; Hrdy 1999). Even pregnant langur monkeys (e.g., *Semnopithecus*) solicit matings (Hrdy 1999). Among non-primates, elephants and dolphins do it. As the limbic system is hard-wired to reward behaviors such as eating and sex, taking pleasure in mating without a direct reproductive outcome is a reasonable proximate mechanism. Evolutionary rewards probably also exist connected with female choice and sexual selection.

Did concealed ovulation and continuous estrus lead to permanently prominent breasts and fatty buttocks? Changes related to bipedalism removed the usual quadrupedal avenues for displaying estrus (Szalay, Costello 1991). For instance, some female primates have posterior visual displays, sexual skins, which change in color and size coincident with ovulation (see chapter 6). Also, the mammary glands of most primates are only obvious during nursing. Both these phenomena are temporary in other primates, whereas the similar morphology of women is relatively stable. This fact suggested to some that women exhibit permanent estrus displays, or "continuous attractivity" (Szalay, Costello 1991). In other words, women now are permanently attractive! Again, the passive female–active male gestalt is present.

The proposed scenario for the morphological change is that the ancestral vulva-focused estrus swelling is currently mimicked by fatty buttocks and hairless skin "making permanent the signal paraphernalia of the ancestral estrus state" (Szalay, Costello 1991:439). Further, pendulous breasts mimic the ancestral swollen perineal area (Szalay, Costello 1991). Of course, the development of fat stores is not unique to women, nor is a sex difference in fat stores. In addition, body fat varies by age, activity, and food availability (Dufour, Sauther 2002). Breast size is also highly variable. Soft tissues do not fossilize well. Until we have more empirical data on the morphology of our ancestors, the evolutionary history of breasts and buttocks will remain a mystery. However, if the basis for the hypothesis is concealed ovulation, we already know that women provide cues on multiple levels, either soliciting or rebutting potential mates. Finally, a persistent, stable cue is likely to cause habituation to the stimulus; if so, a permanent feature may reduce "attractivity."

Mating systems in humans are as diverse as in other mammals. To a large degree, this variation is associated with culture. For example, social monogamy is present in many western European cultures but is interspersed with cases of polygamy, for example, royal mistresses or "multiple wives." Cultural polyandry is the least common but occurs in Nepal and Tibet where women often wed brothers. Finally, partnerships with multiple participants, both male and female (polygynandry), occur in cultures throughout the world in formal or informal contexts. Perhaps, at first glance, such diversity within a single species might seem unique, but we share this flexibility (sexual behavior with both same- and opposite-sex partners) with our cousins, the bonobos (*Pan paniscus*) and with bottlenose dolphins (*Tursiops truncatus*; Bailey, Zuk 2009). Flexibility in mating systems can be associated with habitat as in lions (*Panthera leo*); in drier regions, prides associate with a single male, whereas in better-watered areas, prides associate with coalitions of two to four males (Patterson 2007). Variation in mating and in mating systems may be a feature of all mammals.

An additional aspect of human sexuality that we share with our animal kin is multiple sexualities: asexual, heterosexual, bisexual, and homosexual. Again, the huge diversity of sexuality within a single species is certainly not unique to us. Many distantly related taxa have wide-ranging sexual behaviors, including bats, dolphins, and again our close cousins, the bonobos (Bagemihl 1999).

Freeing Women from Biology?

Contraception is a feature unique to humans. It has a deep history in many cultures often in the form of herbal medicines. This is a feature we now use for wildlife population regulation in a broad range of taxa (chapter 14). In humans, contraceptives give women the freedom to separate sex from fertility. Freed from reproduction, women can develop bonds through sex without parental consequences or the risk of dying in childbirth. They are also freed from the invasive procedures, common in current obstetric practice, associated with reproduction. The use of oral contraception by more than 100 million women worldwide suggests that it is a key component of modern life (Dhont 2010). However, hormonal contraception alters natural hormone levels and may result in subsequent health problems.

Cultural mandates surrounding birth vary from culture to culture. In western societies, modern women usually give birth in a medical facility outside a hospital, whereas in much more rural societies birth may be at home. Consuming the placenta after birth rarely occurs in any country, but this may be a recent change. These are just two examples of how we now differ from our non-industrial foremothers.

The practice of Cesarean section (C-section) births has likely made substantial impacts on our biology in a variety of unexpected and largely unquantified ways. One consequence of C-sections discussed elsewhere (chapter 12) has to do with microbiomes. C-sections both prevent the vaginal microflora from colonizing the offspring and introduce new microflora into the mother (Prince et al. 2014a, 2014b). The cost-benefit analysis for C-sections is largely inconclusive (Hyde, Modi 2012). C-sections have side effects, such as an increased risk for ectopic pregnancies and placental previa, as well as ruptures due to scarring (Greene et al. 1997; Hemminki, Meriläinenb 1996)

Women in Context, Outside Influences

As humans, we like to consider ourselves impendent from our environment, able to control and, in some cases, conquer it. So what about women? Are women less tied to outside influences when compared to other mammals? Have we removed ourselves from the abiotic world: from light, temperature, substrate, trace elements, pressure, and weather? Of course not, even though our technology has ameliorated these influences. For instance, in the United States, births have both a latitudinal and a seasonal distribution (Martinez-Bakker et al. 2014). Births in northern states peak in spring-summer, whereas in the south the peak in births is in autumn. Why? One hypothesis is that complex interactions with disease vectors and parasites have seasonal cycles that subsequently alter conception rates (Martinez-Bakker et al. 2014). If so, conceptions and resultant births may be tied to both abiotic and biotic influences. Socio-economic status, age, and parity also influence seasonal birth peaks (Haandrikman, van Wissen 2008). Current work on genetically programmed cell and tissue responses to seasonal change can provide mechanistic reasons for some of these patterns (Stevenson et al. 2015).

Returning to biotic influences, viruses, bacteria, and parasites can all reduce fertility (Pellati et al. 2008). Over the course of human history, parasites and diseases, such as AIDS, bubonic plague, cholera, influenza, herpes, malaria, smallpox, syphilis, tuberculosis, and yellow fever have had significant effects on human populations (Sherman 2007). The effects of diseases include mortality, but they may also influence reproduction or are themselves fueled by reproduction, e.g., AIDS, syphilis, and herpes. Other pathogens limit their influence to one stage or age group. For instance, rubella (German measles) causes birth defects such as deafness and eye and heart abnormalities when contracted by mothers in the first trimester (Atreya et al. 2004).

Parasites can reduce fertility. Trichomonas can disrupt placental function (Secor et al. 2014). Toxoplasmosis can cause congenital abnormalities and embryo rejection. Filarial parasites can disrupt implantation and microfilarial have been found in follicular fluid (Bazi et al. 2006; Brezina et al. 2011). Clearly, we are not free from our heterospecific foes. But many of our heterospecific interactions are positive, for instance the food we eat and the microbiome we carry with us. We covered these beneficial interactions in chapter 13. What about conspecific interactions? Women are social mammals.

> The word's out: I'm a woman, and I'm going to have trouble backing off on that. I am
> what I am. I'll go out and talk to people about what's happening to their families,
> and when I do that, I'm a mother. I'm a grandmother.
> —Elizabeth Warren, *The Daily Beast*, October 24, 2011

Our care giving to kin and non-kin alike has been well studied by a slew of excellent scientists, such as Sarah Blaffer Hrdy (Hrdy 2009), Kristen Hawkes (Hawkes, Coxworth 2013), and Virpi Lummaa (Maklakoy, Lummaa 2013). Our social systems delegate reproductive roles and childcare in diverse ways. From kibbutz to boarding school, nuclear families to communes, orphanages to street gangs, human offspring are reared by diverse caretakers and changing female roles. As the primary milk producer a mother is often the main caregiver, but the advent of formula has changed this role and allowed other caregivers to support even neonates. Although formula is certainly unique to our species, in other ways we resemble other female mammals. Our fascination with the many roles for caregivers in other mammals may be due the human-centric names we give them (e.g., aunt or nanny). Caregiving by post-reproductive females often takes a central role in humans. Over one-third of the lives of many women is after menopause. In contrast, most wild mammals reproduce until death. In addition to long post-reproductive lives, women typically live longer than men.

Sex differences in longevity are not unique to humans but also occur in taxa as diverse as mice and fruit flies. Female longevity may be the norm (Maklakov, Lummaa 2013). Two main hypotheses have been put forth to explain this: "unguarded X" and "Mother's curse." The unguarded X hypothesis is based on the fact that women have two X chromosomes, whereas men only have one (chapter 3). If mortality due to aging is a result of the gradual accumulation of detrimental alleles that are not guarded by redundant information on another chromosome (the second X), then males would be at greater risk of these deleterious effects. This argument, however, is mammal specific, as in other taxa, such as birds, females have the unguarded chromosome and thus would be expected to be the shorter-lived sex.

The Mother's curse hypothesis is based on the notion that we gain our mitochondria and mitochondrial-associated DNA from our mothers and that this DNA could have detrimental mutations (Maklakov, Lummaa 2013; sidebar on page 272). But males as well as females receive this maternal DNA, so how this would cause sex differences is unclear. Recently, sexual conflict has been put forth as yet another explanation for the sex-biased differences in longevity (Maklakov, Lummaa 2013). Conflict is expected to occur if sex-specific fitness optima (that is, trade-offs between reproduction and survival) differ between females and males (Maklakov, Lummaa 2013). In practice, conflict would result in neither sex successfully reaching its fitness optimum and would increase genetic variation in the male versus female genome relative to life spans (Maklakov, Lummaa 2013). Another suggestion is that androgens (assumed here to be higher in males) result in shorter life spans (Gems 2014), but this would be a mechanism for the difference, not an adaptive explanation. Also, females may in some cases have high levels of androgens. Thus, we can see from all of these current arguments that we have no clear understanding of longevity differences between the sexes.

The earliest age at which a woman has given birth is 5 years old in a case of precocious puberty (Revel et al. 2009); more usually puberty can be as early as 10 years. Meanwhile, the oldest age at which a woman with natural conception gave birth may be 59 according to a newspaper account (*The Telegraph*, August 20, 2007, United Kingdom). However, most women cease having children around 45, with 50 being a general maximum age. Between the ages of 35 and 42, fertility plummets (Aiken 2014). However, the human female continues living for many

Of Feast, Famine and Finnish Females

Virpi Lummaa

Virpi Lummaa has changed the way we think about human reproduction. With over 110 publications, Lummaa is an expert on natural and sexual selection in modern humans. Central to her research is the female perspective in the context of life-history evolution. She is a Royal Society University Research Fellow at the University of Sheffield and also works at the University of Turku in Finland.

Using detailed data from microfiche records of weddings, births, and deaths in rural Finland (1730–1880), Lummaa compiled one of the most valuable data sets for humans to date. Her research reveals patterns in birthrates (reproduction) and survival in pre-modern humans in the context of boom or bust (famine) conditions spanning 350 years. Records for 12 generations coupled with data on ecological conditions (climate and harvests) allowed Lummaa to elucidate much about reproductive patterns in our own species during our transition to modernization. Her studies are longitudinal and have grown to include data on humans in other parts of the world, such as Mongolia, India, Africa, and Canada.

Questions of interest to Lummaa fall well within the paradigm of this book. Essentially, she asks, "What does it mean to be a female mammal, if that mammal happens to be a human?" To address this general question, her research investigates several key issues, such as sex differences in life-history traits. She (and co-authors) found that longevity in humans may relate to mating system but also to the sex of her offspring. Mothers who bear sons suffer a long-term survival cost relative to those who have daughters (Helle et al. 2002).

Two other findings deserve mention. First is that same-sex siblings, either male or female, are more likely to disperse than those of the opposite sex (Nitsch et al. 2016). Second is identifying consequences of extreme famine on reproductive performance and survival. During harsh times, mothers spontaneously abort sons more than daughters (Bruckner et al. 2015). The effects of famine are long-lasting (even across generations) and increase adult disease, thus decreasing longevity (Lummaa, Clutton-Brock 2002).

What makes Lummaa's research distinctive is that she successfully treats us (women) as just another mammal. In doing so, she questions key aspects of our biology and derives new insights regarding our evolution. Here are a few of our favorite insights. First, human births may be more seasonal than we usually think. Second, male twins reduce the fitness of sister co-twins. Third, infant crying may be adaptive, as crying may prevent infanticide or may blackmail others to care for you (Lummaa et al. 1998). Fourth, daughters cost mothers less both in terms of longevity, as well as in the chance of successful pregnancies during stressful times. Finally, oral contraceptives may alter mate choice preferences (Alvergne, Lummaa 2010).

In recent years, Lummaa has studied other long-lived mammals in particular elephants. She found a relationship between age-specific senescence and reproduction in elephants and concluded that unlike humans and orcas, elephants do not completely stop reproduction with age; that is, they may not have a post-reproductive life. Lummaa's focus has become more integrative and resulted in such broad publications as "Maternal Effects in Cooperative Breeders: From Hymenopterans to Humans" (Russell, Lummaa 2009).

Already recognized in 2006 as an outstanding young investigator (Christopher Barnard Award) by the Association for the Study of Animal Behaviour, in 2016 she received the Zoological Society of London Scientific Medal for her work on human and elephant life-history evolution.

(Photo courtesy of V. Lummaa.)

years after this reproductive senescence. This post-reproductive life span is often considered a human-specific trait, and it may be unusual for primates. That was the conclusion of a comparison of mortality and reproduction in seven other species of primates (Alberts et al. 2013). Orcas (*Orcinus orca*) and elephants may also have an extensive post-reproductive life, but among our much closer relatives (other primates), we are unusual.

What is the adaptive value of life after reproduction? One hypothesis is that selection actually operates to increase longevity (or lower death rate), and the reproductive process cannot keep up, thus reproductive senescence is more apparent. So, the question becomes, Why has the reproductive life span in humans not done a better job of keeping up with the evolution of the somatic life span (Alberts et al. 2013)? Reasons why the somatic life span of female humans exceeds that of the reproductive life span can be grouped into ultimate (evolutionary) and proximate (mechanistic) hypotheses. We cover each in turn.

The major evolutionary suggestion for post-reproductive life is the "Grandmother hypothesis." This hypothesis suggests that caring for grandchildren will increase the grandmother's inclusive fitness. In this case, the long inter-birth intervals and long dependency periods in humans select for caregiving assistance both from siblings and from older kin (Hawkes 2004; Hawkes, Coxworth 2013).

The major proximate hypothesis is that the life span of oocytes is set, and thus, with age the number of ova becomes limiting. Some support comes from long-lived orca, which also experience a decay in the reproductive life span. Contrary evidence comes from African elephants (*Loxodonta africana*) that have viable ova well into their 50s or even 60s. (Alberts et al. 2013). For humans, the reproductive tracts of 50- and 60-year-old women can sustain pregnancies with healthy results if donor oocytes are used (Aiken 2014). Their aged oocytes, however, appear to lose their developmental abilities and are more likely to experience chromosomal nondisjunction (a genetic abnormality during cell division), while resulting embryos are more likely to undergo programmed cell death (apoptosis; Aiken 2014). For many modern women, freezing ova before reaching senescence is a good strategy if they want children later in life. Indeed, many women in western cultures are delaying reproduction and increasing their earnings potential when able to do so (Miller 2011).

Reproductive Biology with a Cause: Conservation

The future of humans like all other mammals relies on how we are able to navigate the many changes we have imposed on our environment: changes in habitat, air, water quality, and climate. Our population size has boomed, and we inhabit every corner of the Earth. Birthrates vary substantially by which corner one finds oneself in and are often higher in developing countries relative to more economically stable ones. Meanwhile, the availability of assisted reproduction, the use of C-sections, advances in care of pre-term babies, increased longevity, and later ages of active reproduction push the envelope of our reproductive lives. These processes also have clear global distribution. In economically stable countries, women can choose to have families later in life, while elsewhere, the age at first reproduction remains low and longevity is short. Where will all of this lead? Of course, we do not know. We only hope that the intelligence and resilience that led us to this point can be focused on maintaining a sustainable planet for all reproducing females.

Glossary

Abiotic, non-living; e.g., water, temperature, and volcanoes are non-living features of the Earth; chapter 11

Abortifacient, something (usually a chemical substance) that is able to cause an abortion; chapter 12

Albumin, a protein that can coagulate with heat, such as ovalbumin in the egg white of bird eggs; chapter 4

Allantois, an extraembryonic membrane often modified for the collection of waste during gestation; chapters 2, 4

Allele, a genetic variant; chapter 3

Alloparental care, care of offspring other than one's own; chapter 13

Amnion, an extraembryonic membrane often modified for protection of the embryo; chapters 2, 4

Amniote tetrapods, vertebrates that originally evolved shelled eggs with complex extraembryonic tissues (external to the embryo but within the egg); chapter 2

Androgens, steroid hormones, such as testosterone, usually associated with males but also important in female physiology; chapter 5

Apoptosis, apoptotic, programmed cell death, cell death associated with a natural process; chapters 6, 15

Atresia, atretic, the degeneration of an ovarian follicle; chapters 4, 6

Biome, a major region of the earth distinguished by a characteristic climate and a typical set of living organisms. Examples include deciduous forests, hot deserts, and tundra; chapter 11

Biotic, living; all organisms that reproduce using genetic material contribute to the biotic component of Earth; chapters 12, 13

Blastocyst, one of the very early stages of conceptus development; chapters 4, 6, 7

Blubber, a thick layer of fat below the skin (subcutaneous), used primarily by marine mammals for insulation; chapter 11

Brown adipose tissue, BAT, a type of fatty tissue prominent in newborn and hibernating mammals for the generation of heat in lieu of shivering; chapter 8

Capacitation, a set of chemical reactions by which females render sperm fit for conception; chapter 6

Chorion, an extraembryonic membrane often modified for gas exchange or protection of the embryo, chapters 2, 4

Chorionic gonadotropin, hCG (human), CG (general), a protein hormone, similar to luteinizing hormone, produced by early blastocysts near implantation that aids in pregnancy recognition; chapter 5

Chromosome, a visible structure (microscopically) within a cell made up of DNA and thus genes. Most chromosomes called autosomes are the same in both sexes. Sex chromosomes differ. In most mammals, females have two identical sex chromosomes (usually called X-chromosomes); males have un-identical chromosomes (an X and a Y); chapter 3

Ciliated, ciliary, having cilia, projections from cells that microscopically look like eyelashes. Cells with cilia line the inside of the oviduct and their movement creates currents in the fluid within the oviduct; chapters 4, 6, 7

Clade, a monophyletic lineage, an ancestor and all its descendants; box 2.3

Cleidoic egg, shelled egg, an adaptation of tetrapods to prevent desiccation of embryos on land; chapter 2

Cloaca, a combined opening for the digestive, urinary, and reproductive tracts; chapter 4

Codon, a sequence of three nucleotides in DNA or RNA that form a unit of genetic code; chapter 3

Colostrum, an antibody rich milk produced late in gestation and early in lactation; chapter 12

Commensal, commensalism, a physically close relationship between two species that is either neutral to both or neutral to one and beneficial to the other; chapter 12

Conception, the union of an ovum and a sperm; chapter 6

Conceptus, the whole product of conception including all the cells that form the embryo and the placenta especially very early in development

Conspecific, member of the same species; chapters 10, 13

Contralateral, anatomically, on opposite sides; chapter 7

Convergent evolution, a situation in which unrelated organisms evolve similar traits as with wings in insects, birds, and bats; chapter 2

Coprophagy, ingesting fecal matter; chapters 11, 12

Corpus luteum, corpora lutea, CL, an ovarian gland that produces progesterone; often this ball of cells originally contained a female gamete before ovulation; chapter 6

Cytoplasm, all the material inside a cell but outside the nucleus; chapters 4, 6

Deme, localized population of conspecific organisms; chapter 3

Diploid cell, a cell containing pairs of chromosomes, all the non-gametic cells in the body; chapter 3

Dispersal, natal dispersal, juveniles moving away from the territory of their birth; chapter 10

Distal, an anatomical term referring to something away from a particular reference point; chapter 4

Dystocia, a difficult or obstructed birth, as occurs when a neonate's appendage or body part gets stuck in the birth canal; chapter 13

Ectoparasite, a parasite living on the skin of its host; chapter 12

Endemic, native to an area; chapter 14

Endometrium, the cellular lining of the uterus; chapters 1, 4

Endoparasite, a parasite living within a host; chapter 12

Estrus (noun), Estrous (adj), sometimes referred to as "heat," estrus is the period when females actively search for mating opportunities (oestrus, oestrous are the British spellings)

Epigenetics, can refer to modifications of proteins that are not a direct result of genetic information as well as to the inheritance of such modifications; chapter 3

Epipubic bones, an ancestral pair of bones attached to the pelvis that supports the contents of the lower abdomen, formerly thought to be related to the presence of a marsupial pouch; chapter 2

Estrogens, steroid hormones usually associated with female reproduction but also important in male physiology; chapter 5

Evolution, the process by which the inherited part of a lineage changes over time; natural selection is one of the mechanisms by which evolutionary change can occur, mutation is a second, chance events are a third, migration is a fourth

Extant, living, as opposed to extinct; chapter 2

Extra-embryonic membranes/tissues, the amnion, chorion, allantois, and yolk sac. Temporary tissues that coordinate interactions between the embryo and its environment (uterus or shell). They become the placenta in many mammals; chapter 2, 4

Follicle, a ball of ovarian cells which contains a female gamete until ovulation and then involutes to form a corpus luteum or a non-functional remnant; chapters 4, 6

Follicle stimulating hormone, FSH, a gonadotropin usually produced in the pituitary that stimulates aspects of folliculogenesis; chapter 6

Follicular phase, a stage of the reproductive cycle in which the ovary produces waves of maturing female gametes; chapter 6

Folliculogenesis, the changes in ovarian cells nearest an oocyte as the oocyte matures; chapter 6

Genital ridge, genital tubercle, the part of the early embryo that will eventually become the gonad (ovary or testis); chapters 1, 3, 4, 6

Genome, the entire genetic complement of a species or individual; chapter 3

Genotype, the genetic information that results in a particular trait; chapter 3

Gestation, in therian mammals, the process between conception and birth; chapter 7

Gonadotropin, gonadotrophin, a group of large protein hormones produced in the pituitary or placenta that stimulate the production of ovarian hormones; chapter 6

Gonadotropin releasing hormone, GnRH, a hypothalamic protein hormone that triggers release of gonadotropins from the pituitary; chapter 6

Grade, a group of taxa lumped together due to a similar trait but not similar ancestry, e.g., pinnipeds, the fin-footed, aquatic carnivores; box 2.3

Haploid cell, a cell containing only one set of chromosomes; only gametes are haploid; chapters 3, 4

Heterospecific, member of a different species; chapter 12

Homeothermy, physiological maintenance of a consistent body temperature; chapter 5

Homologous characters have a similar evolutionary origin but may not have a similar function, such as a bat's wing, a fur seal's flipper, and a gorilla's arm; chapter 3

Humoral immunity, immunity related to antibodies (and other immune-related proteins) transported through the circulation; chapter 12

Implantation, the first attachment of the embryo to the uterus; chapter 7

Ipsilateral, anatomically, on the same side; chapter 7

Isolating mechanisms, *see* reproductive isolation

Karyotype, the number and shape of the chromosomes in a cell; chapter 3

Kin selection, a subset of natural selection whereby increasing the reproductive success of kin increases your own reproductive success; chapters 10, 13

K-selection (vs. r-selection), a hypothetical mechanism for adaptations such as low litter size, long times to sexual maturity, that may be related to species living at their carrying capacity; box 10.1

Lactation, the part of the mammalian reproductive cycle in which milk is produced; chapter 9

Life history, the constellation of reproductive traits specific to a species, including the major reproductive stages and events such as litters per year, length of gestation and lactation, or litter size; box 5.1

Lordosis, a posture used by some female rodents and cats to facilitate mating; chapter 6

Luteal phase, a stage of the reproductive cycle in which a corpus luteum is actively producing progesterone; chapters 1, 5, 6

Luteinizing hormone, LH, a gonadotropin usually produced in the pituitary that often triggers ovulation and the subsequent transition of the ovulatory follicle to a corpus luteum; chapter 6

Luteolytic, an agent that precipitates lysis (breakdown, degradation) of the corpus luteum; chapter 5

Luteotrophic, luteotropic, an agent that promotes maintenance or growth of the corpus luteum; chapter 5

Mammalia, the taxonomic class of vertebrates to which mammals belong

Meconium, the earliest fecal matter a neonate produces, the final waste products from material ingested in utero; chapter 12

Meiosis, cell division of cells destined to become gametes; box 4.1 and associated figure

Microbe, microbial, bacteria, protists, fungi, and other organisms too small to be seen without a microscope; chapter 12

Microbiome, microbiota, the constellation of microbes that are closely associated with an individual or species; chapter 12

Microcystins, a set of peptide (~7 amino acids) liver toxins produced by cyanobacteria; chapter 14

Mitosis, cell division of cells not destined to become gametes; box 4.1

Monogamy, generally suggests one mate at a time; chapters 1, 6

Monotocous, producing a single offspring per reproductive attempt; chapter 7

Mutualism, mutualistic, interactions between species or individuals that are mutually beneficial; chapter 13

Mycoestrogen, an estrogenic steroid produced by fungi; chapter 12

Natural selection, a mechanism of evolution that involves differential reproduction. In other words, different females leave different numbers of offspring, and those females that leave the most offspring contribute more of their heritable material to the next generation

Neo-oogenesis, the formation of oocytes from primordial germ cells in adult females; chapter 6

Niche, a species role in and interactions with its abiotic and biotic environment; the range of environmental variables, such as temperature, a species can live within; chapters 11, 12

Oogenesis, the formation of female gametes; chapter 6

Oogonia, oogonium, oocyte, ova, ovum, egg, names given to female gametes; chapter 4, box 4.1

Oviparity, egg laying, the release of embryos within shells after a period of uterine development, as happens in monotremes; chapter 2

Ovulation, the release of a female gamete from ovarian tissue; chapter 6

Parous, Parity, having given birth; multiparous, having given birth multiple times; nulliparous, never having given birth; primiparous, first time giving birth.

Parthenogenesis, a type of asexual reproduction; chapter 3

PCBs, polychlorinated biphenyls, a category of toxic and persistent pollutants; chapter 14

Pelage, fur, the fur coat shielding most mammals from the elements and helping in insulation; chapters 8, 12

Phenotype, the often visible result of genetic information; chapter 3

Pheromone, a chemical signal produced to influence the behavior of a conspecific; chapter 6

Philopatry, philopatric, living in the same area, the opposite of dispersal; chapter 13

Phylogeny, the inferred evolutionary relationships among organisms; chapter 2

Placenta, a temporary organ of embryonic origin that is the interface between the embryo and the mother during gestation; chapter 4

Polar body, polar bodies, cells resulting from oogenesis that contain only small amounts of maternal cytoplasm; box 4.1 and associated figure

Polyandry, mating with more than one male; chapters 1, 6

Polyovular, either (a) a follicle with more than one female gamete, (b) an ovary that ovulates more than one gamete, or (c) a species that ovulates more than one gamete (could be from different ovaries); chapters 4, 6

Polytocous, producing more than one offspring per reproductive attempt; chapter 7

Primordial germ cells, embryonic cells that are the progenitors of all female gametes; chapters 4, 6

Progestogens, steroid hormones, such as progesterone usually, associated with gestation but have diverse physiological functions and are precursors to other steroid hormones

Pronucleus, the nucleus of a gamete before conception

Proteomics, the comparison of the entire complement of proteins across different tissues or at different times in the same tissue

Proto-lacteal secretions, the precursors to milk; chapter 7

Reproductive isolation, when lineages become distinct because individuals from different populations can no longer mate and produce viable progeny

Reproductive success, aka fitness, leaving more offspring than other females in a population

Reproductive value, a female's current and expected future contribution to the next generation

RNA, ribonucleic acid, chemically similar to DNA and used to transmit information from DNA to cellular organelles outside the nucleus; chapters 3, 5

r-selection (vs. K-selection), a hypothetical mechanism for adaptations for fast population growth, such as larger litter sizes and short times to sexual maturity; box 10.1

Secondary hard palate, tissue on the roof of the mouth that helps separate the oral and nasal cavities and thus facilitates suckling; chapter 3

Selection, usually shorthand for natural selection

Selection pressure, an aspect of an organism's biology that might influence reproductive success

Sexual selection, a subcategory of natural selection that explores the consequences of intersexual processes (e.g., female mate choice) and intra-sexual competition (e.g., male-male competition); chapter 1

Singleton, having one offspring per reproductive cycle

Speciation, the evolutionary process of forming reproductively isolated lineages

Species, usually a reproductively isolated lineage of organisms although other definitions are used in the literature

Symbiotic, symbiosis, a physically close, long-term relationship between two species that is beneficial to at least one of them; chapter 12

Syncytium, syncytia, a cell with more than one nucleus; chapter 6

Systemic, system-wide, in physiology, often referring to hormones or other chemicals present throughout the body usually via the circulatory system

Thermoregulation, control of body temperature usually through physiological or behavioral processes

Transcription and translation, molecular steps in the conversion of genetic information into protein structure; chapter 3

Trophoblast, the cells of the early conceptus that become the extra-embryonic membranes; chapters 4, 7

Urogenital sinus/opening, a combined opening for both urine and reproductive products; chapters 3, 4

Vaginal plug, congealed secretions remaining in the reproductive tract after mating; chapter 6

Viviparity, the release (birth) of embryos without shells after a period of embryonic growth in a female's uterus; the opposite of egg laying (oviparity); chapter 2

Zona pellucida, an external, non-cellular coating surrounding the oocyte and early conceptus; chapter 4

Zygote, in most regards, a synonym for conceptus with a more technical restriction in the field of embryology; chapter 7

Literature Cited

Aagaard, K., et al. 2014. The placenta harbors a unique microbiome. Science Translational Medicine 6:237ra65.

Adamczewski, J.Z., et al. 1997. Seasonal patterns in body composition and reproduction of female muskoxen (*Ovibos moschatus*). Journal of Zoology, London 241:245–269.

Adamczewski, J.Z., et al. 1998. The influence of fatness on the likelihood of early-winter pregnancy in muskoxen (*Ovibos moschatus*). Theriogenology 50:605–614.

Adams, G.P., et al. 1990. Effects of lactational and reproductive status on ovarian follicular waves in llamas (*Lama glama*). Journal of Reproduction and Fertility 90:535–545.

Adams, G.P., M.H. Ratto. 2013. Ovulation-inducing factor in seminal plasma: a review. Animal Reproduction Science 136:148–156.

Adkins-Regan, E. 2005. Hormones and animal social behavior. Princeton University Press, Princeton, NJ.

Adolph, E.F., F.W. Heggeness. 1971. Age changes in body water and fat in fetal and infant mammals. Growth 35:55–63.

Aiken, R.J. 2014. Age, the environment and our reproductive future: bonking baby boomers and the future of sex. Reproduction 147:S1–S11.

Aitken, R.J., et al. 2015. Are sperm capacitation and apoptosis the opposite ends of a continuum driven by oxidative stress? Asian Journal of Andrology 17:633–639.

Akinloye, A.K., B.O. Oke. 2010. Characterization of the uterus and mammary glands of the female African giant rats (*Cricetomys gambianus*, Waterhouse) in Nigeria. International Journal of Morphology 28:93–96.

Albertini, D.F. 2015. The mammalian oocyte. Pp. 59–97 in Knobil and Neill's physiology of reproduction (T.M. Plant, et al., eds.). Academic Press, New York, NY.

Alberts, S.C., et al. 2013. Reproductive aging patterns in primates reveal that humans are distinct. Proceedings of the National Academy of Sciences 110:13440–13445.

Alcorn, G.T., E.S. Robinson. 1983. Germ cell development in female pouch young of the tammar wallaby (*Macropus eugenii*). Journal of Reproduction and Fertility 67:319–325.

Allen, W.R. 2006. Ovulation, pregnancy, placentation and husbandry in the African elephant (*Loxodonta africana*). Philosophical Transactions of the Royal Society B 361:821–834.

Allen, W.R., et al. 2005. Placentation in the African elephant, *Loxodonta africana*. IV. Growth and function of the fetal gonads. Reproduction 130:713–720.

Allen-Blevins, C.R., et al. 2015. Milk bioactives may manipulate microbes to mediate parent-offspring conflict. Evolution, Medicine, and Public Health, 2015.1:106–121.

Alligood, C.A., et al. 2008. Pup development and maternal behavior in captive Key Largo woodrats (*Neotoma floridana smalli*). Zoo Biology 27:394–405.

Alvergne, A., V. Lummaa. 2010. Does the contraceptive pill alter mate choice in humans? Trends in Ecology and Evolution 25:171–179.

Amoroso, E.C., et al. 1951. Reproductive organs of near-term and newborn seals. Nature 168:771–772.

Amoroso, E.C., et al. 1965. Reproductive and endocrine organs of foetal, newborn and adult seals. Journal of Zoology, London 147:430–486.

Amstislavsky, A., Y. Ternovskaya. 2000. Reproduction in mustelids. Animal Reproduction Science 60–61:571–581.

Andersen, I.L., et al. 2011. Maternal investment, sibling competition, and offspring survival with increasing litter size and parity in pigs (*Sus scrofa*). Behavioral Ecology and Sociobiology 65:1159–1167.

Anderson, M.J., et al. 2006. Mammalian sperm and oviducts are sexually selected: evidence for co-evolution. Journal of Zoology, London 270:682–686.

Andres, D., et al. 2013. Sex differences in the consequences of maternal loss in a long-lived mammal, the red deer (*Cervus elaphus*). Behavioral Ecology and Sociobiology 67:1249–1258.

Anthony, L.L., D.T. Blumstein. 2000. Integrating behaviour into wildlife conservation: the multiple ways that behaviour can reduce N_e. Biological Conservation 95:303–315.

Archer, M., et al. 1985. First Mesozoic mammal from Australia—an early Cretaceous monotreme. Nature 318:363–366.

Archie, E.A., et al. 2006. The ties that bind: genetic relatedness predicts the fission and fusion of social groups in wild African elephants. Proceedings of the Royal Society B 273:513–522.

Arey, L.B. 1939. The degree of normal menstrual irregularity: an analysis of 20,000 calendar records from 1,500 individuals. American Journal of Obstetrics and Gynecology 37:12–29.

Arman, P., et al. 1974. The composition and yield of milk from captive red deer (*Cervus elaphus* L.). Journal of Reproduction and Fertility 37:67–84.

Arnould, J.P.Y., et al. 2003. The comparative energetics and growth strategies of sympatric Antarctic and subantarctic fur seal pups at Îles Crozet. Journal of Experimental Biology 206:4497–4506.

Asdell, S.A. 1946. Patterns of mammalian reproduction. Cornell University Press, Ithaca, NY.

Asher, G.W. 2011. Reproductive cycles of deer. Animal Reproduction Science 124:170–175.

Asher, M., et al. 2013. Large males dominate: ecology, social organization, and mating system of wild cavies, the ancestors of the guinea pig. Behavioral Ecology and Sociobiology 62:1509–1521.

Ashwell, K.W.S. 2013. Neurobiology of monotremes. CSIRO, Collingwood, Victoria, Australia.

Atanda, S.A., et al. 2012. Mycotoxin management in agriculture: a review. Journal of Animal Science Advances 2(Supplement 3.1):250–260.

Atkinson, S. 1997. Reproductive biology of seals. Reviews of Reproduction 2:175–194.

Atkinson, S., et al. 1994. Reproductive morphology and status of female Hawaiian monk seals (*Monachus schauinslandi*) fatally injured by adult male seals. Journal of Reproduction and Fertility 100:225–230.

Atreya, C.D., et al. 2004. Rubella virus and birth defects: molecular insights into the viral teratogenesis at the cellular level. Birth Defects Research, Part A 70:431–437.

Atwood, T.C., H.P. Weeks. 2003. Sex-specific patterns of mineral lick preference in white-tailed deer. Northeastern Naturalist 10:409–414.

Aurich, C. 2011. Reproductive cycles of horses. Animal Reproduction Science 124: 220–228.

Austin, C.R., R.V. Short (eds.). 1982. Reproduction in mammals. Cambridge University Press, Cambridge, England.

Avalos, L.A., et al. 2012. A systematic review to calculate background miscarriage rates using life table analysis. Birth Defects Research, Part A 94:417–423.

Avise, J.C. 2013. Evolutionary perspectives on pregnancy. Columbia University Press, New York, NY.

Baerwald, A.R., et al. 2005. Form and function of the corpus luteum during the human menstrual cycle. Ultrasound in Obstetrics and Gynecology 25:498–507.

Baerwald, A.R., R.A. Pierson. 2004. Ovarian follicular development during the use of oral contraception: a review. Journal of Obstetrics and Gynecology Canada 26:19–24.

Bagemihl, B. 1999. Biological exuberance: animal homosexuality and natural diversity. Macmillan, London, England.

Bailey, N.W., M. Zuk. 2009. Same-sex sexual behavior and evolution. Trends in Ecology and Evolution 24:439–446.

Bainbridge, D.R.J., R.N. Jabbour. 1999. Source and site of action of anti-luteolytic interferon in red deer (*Cervus elaphus*): possible involvement of extra-ovarian oxytocin secretion in maternal recognition of pregnancy. Journal of Reproduction and Fertility 116:305–313.

Baird, D.D., E.C. Birney. 1985. Bilateral distribution of implantation sites in small mammals of 22 North American species. Journal of Reproduction and Fertility 75:381–392.

Baker, B.E., et al. 1970. Muskox (*Ovibos moschatus*). I. Gross composition, fatty acid, and mineral constitution. Canadian Journal of Zoology 48:1345–1347.

Baker, R.R., M.A. Bellis. 1993. Human sperm competition: ejaculate manipulation by females and a function for female orgasm. Animal Behaviour 46:997–909.

Baker, T.G. 1982. Oogenesis and ovulation. Pp. 17–45 in Reproduction in mammals. 1. Germ cells and fertilization (C.R. Austin, R.V. Short, eds.). Cambridge University Press, London, England.

Bakker, J., M.J. Baum. 2000. Neuroendocrine regulation of GnRH release in induced ovulators. Frontiers in Neuroendocrinology 21:220–262.

Bakloushinskaya, I., et al. 2013. A new form of the mole vole *Ellobius tancrei* Blasius, 1884 (Mammalia, Rodentia) with the lowest chromosome number. Comparative Cytogenetics 7:163–169.

Balke, J.M.E., et al. 1988a. Reproductive anatomy of three nulliparous female Asian elephants: the development of artificial breeding techniques. Zoo Biology 7:99–113.

Balke, J.M.E., et al. 1988b. Anatomy of the reproductive tract of the female African elephant (*Loxodonta africana*) with reference to development of techniques for artificial insemination. Journal of Reproduction and Fertility 84:485–492.

Banci, V., A. Harestad. 1988. Reproduction and natality of wolverine (*Gulo gulo*) in Yukon. Annales Zoologici Fennici 25:265–270.

Banerjee, A., et al. 2009. Melatonin regulates delayed embryonic development in the short-nosed fruit bat, *Cynopterus sphinx*. Reproduction 138:935–944.

Barclay, R.M.R. 1994. Constraints on reproduction by flying vertebrates: energy and calcium. American Naturalist 144:1021–1031.

Barclay, R.M.R., et al. 2000. Foraging behaviour of the large-footed myotis, *Myotis moluccarum* (Chiroptera: Vespertilionidae) in south-eastern Queensland. Australian Journal of Zoology 48:385–392.

Barash, D.P., J.E. Lipton. 2009. How women got their curves and other just-so stories: evolutionary enigmas. Columbia University Press, New York, NY.

Barrett, J., et al. 1999. Extension of reproductive suppression by pheromonal cues in subordinate female marmoset monkeys, *Callithrix jacchus*. Journal of Reproduction and Fertility 90:411–418.

Barry, R.E., P.J. Mundy. 2002. Seasonal variation in the degree of heterospecific association of two syntopic hyraxes (*Heterohyrax brucei* and *Procavia capensis*) exhibiting synchronous parturition. Behavioral Ecology and Sociobiology 52:177–181.

Bartholomew G.A. 1970. A model for the evolution of pinniped polygyny. Evolution 24:546–559.

Bartholomew, G.A. 1972. Body temperature and energy metabolism. Pp. 290–368 in Animal physiology: principles and adaptations, 2nd edition (M.S. Gorden, ed.). Macmillan, New York, NY.

Bartholomew, G.A., P.G. Hoel. 1953. Reproductive behavior of the Alaska fur seal, *Callorhinus ursinus*. Journal of Mammalogy 34:417–436.

Bartol, F.F., et al. 2013. Lactocrine signaling and developmental programming. Journal of Animal Science 91:696–705.

Baskin, L.S., et al. 2006. A neuroanatomical comparison of humans and spotted hyena, a natural animal model for common urogenital sinus: clinical reflections on feminizing genitoplasty. Journal of Urology 175:276–283.

Batzli, G.O., et al. 1974. Growth and survival of suckling brown lemmings, *Lemmus trimucronatus*. Journal of Mammalogy 55:828–831.

Batzli, G.O., et al. 1980. Nutritional ecology of microtine rodents: resource utilization near Atkasook, Alaska. Arctic and Alpine Research 12:483–499.

Bautista, A., et al. 2015. Intrauterine position as a predictor of postnatal growth and survival in the rabbit. Physiology and Behavior 138:101–106.

Baverstock, P., B. Green. 1975. Water recycling in lactation. Science 187:657–658.

Bazer, F.W. 2013. Pregnancy recognition signaling mechanisms in ruminants and pigs. Journal of Animal Science and Biotechnology 4:23.

Bazer, F.W., et al. 2009. Comparative aspects of implantation. Reproduction 118:195–209.

Bazer, F.W., et al. 2010. Novel pathways for implantation and establishment and maintenance of pregnancy in mammals. Molecular Human Reproduction 16:135–152.

Bazi, T., et al. 2006. Filariasis infection is a probable cause of implantation failure in *in vitro* fertilization cycles. Fertility and Sterility 85:1822:E13–E15.

Beard, L.A., G.C. Grigg. 2000. Reproduction in the short-beaked echidna, *Tachyglossus aculeatus*: field observations at an elevated site in south-east Queensland. Proceedings of the Linnean Society of New South Wales 122:89–99.

Beauplet, G., et al. 2005. Interannual variation in the post-weaning and juvenile survival of subantarctic fur seals: influence of pup sex, growth rate and oceanographic conditions. Journal of Animal Ecology 74:1160–1172.

Bedford, J.M., et al. 1997a. Unusual ampullary sperm crypts, and behavior and role of the cumulus oophorus, in the oviduct of the least shrew, *Cryptotis parva*. Biology of Reproduction 56:1255–1267.

Bedford, J.M., et al. 1997b. Ovulation induction and gamete transport in the female tract of the musk shrew, *Suncus murinus*. Journal of Reproduction and Fertility 110:115–125.

Bedford, J.M., et al. 1997c. Novel sperm crypts and behavior of gametes in the fallopian tube of the white-toothed shrew, *Crocidura russula* Monacha. Journal of Experimental Zoology 277:262–273.

Bedford, J.M., et al. 1999. Reproductive features of the eastern mole (*Scalopus aquaticus*) and star-nosed mole (*Condylura cristata*). Journal of Reproduction and Fertility 117:345–353.

Bedford, J.M., et al. 2004. Novelties of conception in insectivorous mammals (Lipotyphla), particularly shrews. Biological Reviews 79:891–909.

Beery, A.K., I. Zucker. 2011. Sex bias in neuroscience and biomedical research. Neuroscience and Biobehavioral Reviews 35:565–572.

Bellofiore, N., et al. 2017. First evidence of a menstruating rodent: the spiny mouse (*Acomys cahirinus*). American Journal of Obstetrics and Gynecology 216:p40.e1-40.e11.

Bensley, B.A. 1910. Practical anatomy of the rabbit. University of Toronto Press, Blakiston's Son & Co., Philadelphia.

Benson, B.N., et al. 1992. Vocalizations of infant and developing tree shrews (*Tupaia belangeri*). Journal of Mammalogy 75:106–119.

Benson-Amram, S., K.E. Holekamp. 2016. Innovative problem solving by wild spotted hyenas. Proceedings of the Royal Society B 279:4087–4095.

Benson-Amram, S., et al. 2016. Brain size predicts problem-solving ability in mammalian carnivores. Proceedings of the National Academy of Sciences 113:2532–2537.

Bercovitch, F.B., P.S.M. Berry. 2012. Herd composition, kinship and fission-fusion social dynamics among wild giraffe. African Journal of Ecology 51:206–216.

Bermejo-Alvarez, P., et al. 2010. Sex determines the expression level of one third of the actively expressed genes in bovine blastocysts. Proceedings of the National Academy of Science 107:3394–3399.

Berta, A., et al. 2015. Eye, nose, hair, and throat: external anatomy of the head of a neonate gray whale (Cetacea, Mysticeti, Eschrichtiidae). Anatomical Record 298:648–659.

Bertolino, S., et al. 2008. Predicting the spread of the American grey squirrel (*Sciurus carolinensis*) in Europe: a call for a co-ordinated European approach. Biological Conservation 141:2564–2574.

Betteridge, K.J., et al. 1982. Development of horse embryos up to twenty two days after ovulation: observations on fresh specimens. Journal of Anatomy 135:191–209.

Beukeboom, L.W., N. Perrin. 2014. The evolution of sex determination. Oxford University Press, Oxford, England.

Bhattacharya, K. 2013. Ovulation and rate of implantation following unilateral ovariectomy in mice. Journal of Human Reproductive Sciences 6:45–48.

Bielby, J., et al. 2007. The fast-slow continuum in mammalian life history: an empirical reevaluation. American Naturalist 169:748–757.

Birdsall, D.A., D. Nash. 1973. Occurrence of successful multiple insemination of females in natural populations of deer mice (*Peromyscus maniculatus*). Evolution 27:106–110.

Birkhead, T.R. 1998. Cryptic female choice: criteria for establishing female sperm choice. Evolution 52:1212–1218.

Birney, E.C., D.D. Baird. 1985 Why do some mammals polyovulate to produce a litter of two? American Naturalist 126:136–140.

Blackburn, D.G. 2015. Evolution of vertebrate viviparity and specializations for fetal nutrition: a quantitative and qualitative analysis. Journal of Morphology 276:961–990.

Blake, B.H. 1992. Estrous calls in captive Asian chipmunks, *Tamias sibiricus*. Journal of Mammalogy 73:597–603.

Blake, B.H. 2012. Ultrasonic calling in 2 species of voles, *Microtus pinetorum* and *M. pennsylvanicus*, with different social systems. Journal of Mammalogy 93:1051–1060.

Blix, A.S., J.W. Lentfer. 1979. Modes of thermal protection in polar bear cubs—at birth and on emergence from the den. American Journal of Physiology 236:R67–R74.

Blumstein, D.T. 1997. Infanticide among golden marmots (*Marmota caudate aurea*). Ethology, Ecology and Evolution 9:169–173.

Blumstein, D.T. 2000. The evolution of infanticide in rodents: a comparative analysis. Pp. 178–197 in Infanticide by males and its implications (C.P. van Schaik, C.H. Janson, eds.). Cambridge University Press, Cambridge, England.

Boellstorff, D.E., et al. 1994. Reproductive behaviour and multiple paternity of California ground squirrels. Animal Behaviour 47:1057–1064.

Bonner, W.N. 1984. Lactation strategies in pinnipeds: problems for a marine mammalian group. Symposia of the Zoological Society of London 51:253–272.

Boness, D.J., W.D. Bowen. 1996. The evolution of maternal care in pinnipeds. BioScience 46:645–654.

Boness, D.J., et al. 2002. Life history and reproductive strategies. Pp. 278–324 in Marine mammal biology: an evolutionary approach (A.R. Hoelzel, ed.). Blackwell Science, Oxford, England.

Boone, W.R., et al. 2004. Evidence that bears are induced ovulators. Theriogenology 61:1163–1169.

Boulva, J. 1971. Observations on a colony of whelping harbor seals, *Phoca vitulina concolor*, on Sable Island, Nova Scotia. Journal of the Fisheries Research Board, Canada 28:755–759.

Boutin, S., et al. 2000. Anticipatory parental care: acquiring resources for offspring prior to conception. Proceedings of the Royal Society B 267:2081–2085.

Bowen, W.D., et al. 1987. Mass transfer from mother to pup and subsequent mass loss by the weaned pup in the hooded seal, *Cystophora cristata*. Canadian Journal of Zoology 65:1–8.

Bowler, M., et al. 2014. Refining reproductive parameters for modelling sustainability and extinction in hunted primate populations in the Amazon. PLOS ONE 9:e93625.

Bowles, J., P. Koopman. 2010. Sex determination in mammalian germ cells: extrinsic versus intrinsic factors. Reproduction 139:943–958.

Boyd, I.L. 1991. Environmental and physiological factors controlling the reproductive cycles in pinnipeds. Canadian Journal of Zoology 69:1135–1148.

Boyd, I.L. 1996. Individual variation in the duration of pregnancy and birth date in Antarctic fur seals: the role of environment, age, and sex of fetus. Journal of Mammalogy 77:124–133.

Boydston, E.E., et al. 2001. Sex differences in territorial behavior exhibited by the spotted hyena (Hyaenidae, *Crocuta crocuta*). Ethology 107:369–385.

Bradshaw, F.J., D. Bradshaw. 2011. Progesterone and reproduction in marsupials: a review. General and Comparative Endocrinology 170:18–40.

Bradshaw, W.E., C.M. Holzapfel. 2006. Evolutionary response to rapid climate change. Science 312:1477–1478.

Braude, S. 2000. Dispersal and new colony formation in wild naked mole-rats: evidence against inbreeding as the system of mating. Behavioral Ecology 11:7–12.

Brezina, P.R., et al. 2011. Description of the parasite *Wucheria bancrofti* microfilariae identified in follicular fluid following transvaginal oocyte retrieval. Journal of Assisted Reproduction and Genetics 28:433–436.

Broekhuizen, S., F. Maaskamp. 1980. Behaviour of does and leverets of the European hare (*Lepus europaeus*) whilst nursing. Journal of Zoology, London 191:487–501.

Broekhuizen, S., et al. 1986. Variation in timing of nursing in the brown hare (*Lepus europaeus*) and the European rabbit (*Oryctolagus cuniculus*). Mammal Review 16:139–144.

Bronson, F.H. 2009. Climate change and seasonal reproduction in mammals. Philosophical Transactions of the Royal Society B 364:3331–3340.

Brookshier, J.S., W.S. Fairbanks. 2003. The nature and consequences of mother daughter associations in naturally and forcibly weaned bison. Canadian Journal of Zoology 81:414–423.

Broussard, D.R., et al. 2005. The effects of capital on an income breeder: evidence from female Columbian ground squirrels. Canadian Journal of Zoology 83:546–552.

Brown, B.W. 2000. A review on reproduction in South American camelids. Animal Reproduction Science 58:159–195.

Browne, P., et al. 2006. Endocrine differentiation of fetal ovaries and testes of the spotted hyena (*Crocuta crocuta*): timing of androgen-independent versus androgen-driven genital development. Reproduction 132:649–659.

Browning, J.Y., et al. 1980. Comparison of serum progesterone, 10α-dihydroprogesterone, and estradiol-17β in pregnant and pseudopregnant rabbits: evidence for postimplantation recognition of pregnancy. Biology of Reproduction 23:1014–1019.

Bruce, H.M. 1959. An exteroceptive block to pregnancy in the mouse. Nature 184:105.

Bruce, N.W., J.R. Wellstead. 1992. Spacing of fetuses and local competition in strains of mice with large, medium and small litters. Journal of Reproduction and Fertility 95:783–789.

Bruckner, T.A., et al. 2015. Culled males, infant mortality and reproductive success in a pre-industrial Finnish population. Proceedings of the Royal Society B 282:20140835.

Brunton, P.J. 2013. Effects of maternal exposure to social stress during pregnancy: consequences for mother and offspring. Reproduction 146:R175–R189.

Bryden, M.M. 1969. Growth of the southern elephant seal, *Mirounga leonine* (Linn.). Growth 33:69–82.

Buchanan, G.D., et al. 1956. Implantation in armadillos ovariectomized during the period of delayed implantation. Journal of Endocrinology 14:121–128.

Buffenstein, R. 2008. Negligible senescence in the longest living rodent, the naked mole-rat: insights from a successfully aging species. Journal of Comparative Physiology B 178:439–445.

Bull, J.J., M.G. Bulmer. 1981. The evolution of XY females in mammals. Heredity 47:347–365.

Burton, F.D., et al. 1995. Preliminary report on *Presbytis francoisi leucocephalus*. International Journal of Primatology 16:311–317.

Buzzio, O.L., A. Castro-Vázquez. 2002. Reproductive biology of the corn mouse, *Calomys musculinus*, a Neotropical sigmodontine. Mastozoología Neotropical 9:135–158.

Callahan, J.R. 1981. Vocal solicitation and parental investment in female *Eutamias*. American Naturalist 118:872–875.

Cameron, E.Z. 2004. Facultative adjustment of mammalian sex ratios in support of the Trivers-Willard hypothesis: evidence for a mechanism. Proceedings of the Royal Society B 271:1723–1710.

Cann, R.L., et al. 1987. Mitochondrial DNA and human evolution. Nature 325:31–36.

Cannon, B., J. Nedergaard. 2004. Brown adipose tissue: function and physiological significance. Physiological Reviews 84:277–359.

Capellini, C. 2012. The evolutionary significance of placental interdigitation in mammalian reproduction: contributions from comparative studies. Placenta 33:763–768.

Carling, M.D., et al. 2003. Microsatellite analysis reveals multiple paternity in a population of wild pronghorn antelopes (*Antilocapra americana*). Journal of Mammalogy 84:1237–1243.

Carlini, A.R., et al. 2000. Energy gain and loss during lactation and postweaning in southern elephant seal pups (*Mirounga leonine*) at King George Island. Polar Biology 23:437–440.

Caro, T. 2005. Antipredator defenses in birds and mammals. University of Chicago Press, Chicago, IL.

Caro, T., et al. 2012. Pelage coloration in pinnipeds: functional considerations. Behavioral Ecology 23:765–774.

Carter, A.M. 2012. Evolution of placental function in mammals: the molecular basis of gas and nutrient transfer, hormone secretion, and immune responses. Physiological Reviews 92:1543–1576.

Carter, A.M., et al. 2013. A new form of rodent placentation in the relict species, *Laonastes aenigmamus* (Rodentia: Diatomyidae). Placenta 34:548–558.

Case, T.J. 1978. On the evolution and significance of postnatal growth rates in the terrestrial vertebrates. Quarterly Review of Biology 52:243–282.

Casida, L.E. 1968. Studies on the postpartum cow. Wisconsin Agricultural Experiment Station Research Bulletin 270:1–54.

CDC (Centers for Disease Control and Prevention). 2000. Birth to 36 months: girls. http://www.cdc.gov/growthcharts/data/set1clinical/cj41l018.pdf.

Cerqueira, R., M. Lara. 1991. Rainfall and reproduction of cricetid rodents in northeastern Brazil. Pp. 545–549 in Global trends in wildlife management (K. Bobek, et al., eds.). Swiat Press, Kraʹkow, Poland.

Cesario, M.D., S.M.M. Matheus. 2008. Structural and ultrastructural aspects of folliculogenesis in *Didelphis albiventris*, the South-American opossum. International Journal of Morphology 26:113–120.

Cetica, P.D., et al. 2005. Morphology of female genital tracts in Dasypodidae (Xenarthra, Mammalia): a comparative survey. Zoomorphology 124:57–65.

Champagne, F.A., J.P. Curley. 2009. Epigenetic mechanisms mediating the long-term effects of maternal care on development. Neuroscience and Biobehavioral Reviews 33:593–600.

Chapais, B. 1995. Alliances as a means of competition in primates: evolutionary, developmental, and cognitive aspects. Yearbook of Physical Anthropology 38:115–136.

Chaplin, R.K., M. Follenbensbee. 1993. Milk composition and production from hand-milked muskoxen. Rangifer 12:61–63.

Chapman, C.A. 1991. Reproductive biology of captive capybaras. Journal of Mammalogy 72:206–208.

Chapman, H.C. 1881. Observations upon the hippopotamus. Proceedings of the Academy of Natural Sciences of Philadelphia 33:126–148.

Charlesworth, B., N.D. Dempsey. 2001. A model of the evolution of the unusual sex chromosome system of *Microtus oregoni*. Heredity 86:387–394.

Charnov, E.L. 1991. Evolution of life history variation among female mammals. Proceedings of the National Academy of Sciences 88:1134–1137.

Charnov, E.L. 1993. Life history invariants. Oxford Series in Ecology and Evolution, Oxford University, Oxford, England.

Chen, D.D. 2014. PAH-induced activation of aryl hydrocarbon receptor signaling and its effects on neural crest development in zebrafish. BA thesis, Smith College, Northampton, MA.

Chevalier-Clément, F. 1989. Pregnancy loss in the mare. Animal Reproduction Science 20:231–244.

Christiansen, E., et al. 1978. Morphological variations in the preputial gland of wild bank voles, *Clethrionomys glareolus*. Holarctic Ecology 1:321–325.

Clancy, K.B.H., et al. 2009. Endometrial thickness is not independent of luteal phase day in a rural Polish population. Anthropological Science 117:157–153.

Clark, M.M., et al. 1994. Differences in the sex ratios of offspring originating in the right and left ovaries of Mongolian gerbils (*Meriones unguiculatus*). Journal of Reproduction and Fertility 101:393–396.

Clarke, F.M., C.G. Faulkes. 2001. Intracolony aggression in the eusocial naked mole-rat, *Heterocephalus glaber*. Animal Behaviour 61:311–324.

Clauss, M., et al. 2014. Low scaling of a life history variable: analyzing eutherian gestation periods with and without phylogeny-informed statistics. Mammalian Biology 79:9–16.

Clemente, M., et al. 2009. Progesterone and conceptus elongation in cattle: a direct effect on the embryo or an indirect effect via the endometrium. Reproduction 138:507–517.

Clements, M.N., et al. 2011. Gestation length variation in a wild ungulate. Functional Ecology 25:691–703.

Close, K. 2017. The world's oldest killer whale, Granny, is believed dead. Time Magazine January 3, 2017. http://time.com/4620481/oldest-killer-whale-granny-dead/?xid=time _socialflow_facebook.

Cloutier, D., D.W. Thomas. 1992. *Carollia perspicillata*. Mammalian Species 417:1–9.

Clutton-Brock, T.H., E. Huchard. 2013. Social competition and its consequences in female mammals. Journal of Zoology, London 289:151–171.

Clutton-Brock, T.H., et al. 2006. Intrasexual competition and sexual selection in cooperative mammals. Nature 444:1065–1068.

Cockcroft, V.G., W. Sauer. 1990. Observed and inferred epimeletic (nurturant) behavior in bottlenose dolphins. Aquatic Mammals 16:31–32.

Cohen, A.A. 2004. Female post-reproductive lifespan: a general mammalian trait. Biological Reviews 79:733–750.

Cohn, P., J.F. Kirkpatrick. 2015. History of the science of wildlife fertility control: reflection of a 25-year international conference series. Applied Ecology and Environmental Sciences 3:22–29.

Compagna, C., et al. 1988. Pup abduction and infanticide in southern sea lions. Behaviour 107:44–60.

Conaway, C.H. 1971. Ecological adaptation and mammalian reproduction. Biology of Reproduction 4:239–247.

Conley, A.J., et al. 2007. Placental expression and molecular characterization of aromatase cytochrome P450 in the spotted hyena (*Crocuta crocuta*). Placenta 28:668–675.

Conrad, P.A., et al. 2005. Transmission of toxoplasma: clues from the study of sea otters as sentinels of *Toxoplasma gondii* flow into the marine environment. International Journal of Parasitology 35:1155–1168.

Coopersmith, C.B., E.M. Banks. 1983. Effects of olfactory cues on sexual behavior in the brown lemming, *Lemmus trimucronatus*. Journal of Comparative Psychology 97:120–126.

Corona, R., F. Lévy. 2015. Chemical olfactory signals and parenthood in mammals. Hormones and Behavior 68:77–90.

Cosgrove, K.P., et al. 2007. Evolving knowledge of sex differences in brain structure, function and chemistry. Biological Psychiatry 62:847–855.

Coureaud, G., et al. 2010. A pheromone to behavior, a pheromone to learn: the rabbit mammary pheromone. Journal of Comparative Physiology A 196:779–790.

Cowie, A.T. 1974. Overview of the mammary gland. Journal of Investigative Dermatology 63:2–9.

Coy, P., et al. 2012. Roles of the oviduct in mammalian fertilization. Reproduction 144:649–660.

Craig, S.F., et al. 1997. The "paradox" of polyembryony: a review of the cases and a hypothesis for its evolution. Evolutionary Ecology 11:127–143.

Crawford, J.C., et al. 2008. Microsatellite analysis of mating and kinship in beavers (*Castor canadensis*). Journal of Mammalogy 89:575–581.

Creel, S., N.M. Creel. 2002. The African wild dog: behavior, ecology and conservation. Princeton University Press, Princeton, NJ.

Creel, S., et al. 2009. Glucocorticoid stress hormones and the effect of predation risk on elk reproduction. Proceedings of the National Academy of Sciences 106:12388–12393.

Cretegny, C., M. Genoud. 2006. Rate of metabolism during lactation in small terrestrial mammals (*Crocidura russula*, *Mus domesticus*, and *Microtus arvalis*). Comparative Biochemistry and Physiology, Part A 144:125–134.

Crichton, E.G., P.H. Krutzsch. 1987. Reproductive biology of the female little mastiff bat, *Mormopterus planiceps* (Chiroptera: Molossidae) in southeast Australia. American Journal of Anatomy 178:369–386.

Cross, B.A. 1977. Comparative physiology of milk removal. Symposia of the Zoological Society of London 41:193–210.

Crowell-Davis, S.L. 2007. Sexual behavior of mares. Hormones and Behavior 52:12–17.

Cunha, G.R., et al. 2003. Urogenital system of the spotted hyena (*Crocuta crocuta* Erxleben): a functional histological study. Journal of Morphology 256:205–218.

Cunha, G.R., et al. 2005. The ontogeny of the urogenital system of the spotted hyena (*Crocuta crocuta* Erxleben). Biology of Reproduction 73:554–564.

Dailey, M.D. 1985. Diseases of Mammalia: Cetacea. Pp. 805–847 in Diseases of marine mammals. Volume 4. Part 2. Reptilia, Aves, Mammalia (O. Kinne, ed.). Biologische Anstalt Helgoland, Hamburg, Germany.

Dalton, A.J.M., et al. 2014. Broad thermal capacity facilitates the primarily pelagic existence of northern fur seals (*Callorhinus ursinus*). Marine Mammal Science 30:994–1013.

Daszak, P., et al. 2000. Emerging infectious diseases of wildlife—threats to biodiversity and human health. Science 287:443–449.

Davies-Morel, M.C.G. 2008. Equine reproductive physiology, breeding and stud management, 3rd edition. CAB International, Wallingford, England.

Davis, D.D., H.E. Story. 1949. The female external genitalia of the spotted hyena. Fieldiana Zoology 31:287–283.

Deanesly, R., A.S. Parkes. 1933. The reproductive processes of certain mammals. Part 4: The oestrous cycle of the grey squirrel (*Sciurus carolinensis*). Philosophical Transactions of the Royal Society B 22:47–96.

Debier, C., et al. 2012. Differential changes of fat-soluble vitamins and pollutants during lactation in northern elephant seal mother-pup pairs. Comparative Biochemistry and Physiology A 162:323–330.

De Felici, M., F. Barrios. 2013. Seeking the origin of female germline stem cells in the mammalian ovary. Reproduction 146:R125–R130.

Degen, A.A. 1997. Ecophysiology of small desert mammals. Springer, Berlin, Germany.

Degen, A.A., et al. 2002. Energy requirements during reproduction in female common spiny mice (*Acomys cahirinus*). Journal of Mammalogy 83:645–651.

De la Iglesia, H.O., W.J. Schwartz. 2006. Timely ovulation: circadian regulation of the female hypothalamo-pituitary-gonadal axis. Endocrinology 147:1148–1153.

Delgado, R., et al. 2007. Paternity assessment in free-ranging wild boar (*Sus scrofa*): are littermates full-sibs. Mammalian Biology 73:169–176.

Demmers, K.J., et al. 2000. Production of interferon by red deer (*Cervus elaphus*) conceptuses and the effects of roIFN-τ on the timing of luteolysis and the success of asynchronous embryo transfer. Journal of Reproduction and Fertility 118:387–395.

Denker, H.-W. 2000. Structural dynamics and function of early embryonic coats. Cells Tissues Organs 166:180–207.

Derrickson, E.M. 1992. Comparative reproductive strategies of altricial and precocial eutherian mammals. Functional Ecology 6:57–65.

Derrickson, E.M., et al. 1996. Milk composition of two precocial, arid-dwelling rodents, *Kerodon rupestris* and *Acomys cahirinus*. Physiological Zoology 69:1402–1418.

Derocher, A.E., et al. 1993. Aspects of milk composition and lactation in polar bears. Canadian Journal of Zoology 71:561–567.

Deschner, T., et al. 2003. Timing and probability of ovulation in relation to sex skin swelling in the wild West African chimpanzees, *Pan troglodytes verus*. Animal Behaviour 66:551–560.

De Villena, F.P., C. Sapienza. 2001. Female meiosis drives karyotypic evolution in mammals. Genetics 159:1179–1189.

DeYoung, R.W., et al. 2002. Multiple paternity in white-tailed deer (*Odocoileus virginianus*) revealed by DNA microsatellites. Journal of Mammalogy 83:884–892.

Dhont, M. 2010. History of oral contraception. European Journal of Contraception and Reproductive Health Care 15:S12–S18.

Dickins, T.E., Q. Rahman. 2012. The extended evolutionary synthesis and the role of soft inheritance in evolution. Proceedings of the Royal Society B 279:2913–2921.

Diedrich, V., et al. 2014. Djungarian hamsters (*Phodopus sungorus*) are not susceptible to stimulating effects of 6-methoxy-2-benzoxazolinone on reproductive organs. Naturwissenschaften 101:115–121.

Dixson, A.F. 2013. Male infanticide and primate monogamy. Proceedings of the National Academy of Sciences 110:E4937.

Dixson, A.F., M.J. Anderson. 2002. Sexual selection, seminal coagulation and copulatory plug formation in primates. Folia Primatologica 73:63–69.

Dloniak, S.M., et al. 2006. Rank-related maternal effects of androgens on behavior in wild spotted hyaenas. Nature 440:1190–1193.

Dobzhansky, T. 1973. Nothing in biology makes sense except in the light of evolution. The American Biology Teacher 35:125–129.

Drake, A., et al. 2008. Parent-offspring resource allocation in domestic pigs. Behavioral Ecology and Sociobiology 62:309–319.

Drake, S.E., et al. 2015. Sensory hairs in the bowhead whale, *Balaena mysticetes* (Cetacea, Mammalia). Anatomical Record 298:1327–1335.

Drea, C.M., et al. 1996. Aggression decreases as play emerges in infant spotted hyaenas: preparation for joining the clan. Animal Behaviour 51:1323–1336.

Drea, C.M., et al. 1998. Androgens and masculinization of genitalia in the spotted hyaena (*Crocuta crocuta*). 2. Effects of prenatal anti-androgens. Journal of Reproduction and Fertility 113:117–127.

Drews, B., et al. 2013. Free blastocyst and implantation stages in the European brown hare: correlation between ultrasound and histological data. Reproduction, Fertility and Development 25:866–878.

Druart, X. 2012. Sperm interaction with the female reproductive tract. Reproduction in Domestic Animals 47(Supplement 4):348–352.

Duchesne, D., et al. 2011. Habitat selection, reproduction and predation of wintering lemmings in the Arctic. Oecologia 167:967–980.

Dufour, C.M.S., et al. 2015. Ventro-ventral copulation in a rodent: a female initiative? Journal of Mammalogy 96:1017–1023.

Dufour, D.L., M.L. Sauther. 2002. Comparative and evolutionary dimensions of the energetics of human pregnancy and lactation. American Journal of Human Biology 14:584–602.

Duke, K.L. 1951. The external genitalia of the pika, *Ochotona princeps*. Journal of Mammalogy 32:169–173.

Dunbar, R.I.M. 1980. Determinants and evolutionary consequences of dominance among female gelada baboons. Behavioral Ecology and Sociobiology 7:253–265.

Dwyer, P.D. 1963. Seasonal changes in pelage of *Miniopterus schreibersi blepotis* (Chiroptera) in north-eastern New South Wales. Australian Journal of Zoology 11:290–300.

Eadie, W.R. 1948. Corpora amylacea in the prostatic secretion and experiments on the formation of a copulatory plug in some insectivores. Anatomical Record 102:259–267.

East, M.L., H. Hofer. 1991a. Loud calling in a female dominated society. I. Structure and composition of whooping bouts of spotted hyenas, *Crocuta crocuta*. Animal Behaviour 42:637–649.

East, M.L., H. Hofer. 1991b. Loud calling in a female dominated society. II. Contexts and functions of whooping of spotted hyenas, *Crocuta crocuta*. Animal Behaviour 42:651–669.

East, M.L., H. Hofer. 2001. Male spotted hyenas (*Crocuta crocuta*) queue for status in social groups dominated by females. Behavioral Ecology 12:558–568.

East, M.[L.], et al. 1989. Functions of birth dens in spotted hyaenas (*Crocuta crocuta*). Journal of Zoology, London 219:690–697.

East, M.L., et al. 1993. The erect "penis" is a flag of submission in a female-dominated society: greetings in Serengeti spotted hyenas. Behavioral Ecology and Sociobiology 33:355–370.

East, M.L., et al. 2003. Sexual conflicts in spotted hyenas: male and female mating tactics and their reproductive outcome with respect to age, social status and tenure. Proceedings of the Royal Society B 270:1247–1254.

East, M.L., et al. 2009. Maternal effects on offspring social status in spotted hyenas. Behavioral Ecology 20:478–483.

Eberhard, W.G. 1996. Female control: sexual selection by cryptic female choice. Princeton University Press, Princeton, NJ.

Ecke, D.H., A. R. Kinney. 1956. Aging meadow mice, *Microtus californicus*, by observations of molt progression. Journal of Mammalogy 37:249–254.

Ecroyd, H., et al. 2009. Testicular descent, sperm maturation and capacitation: lessons from our most distant relatives, the monotremes. Reproduction, Fertility and Development 21:992–1001.

Edson, M.A., et al. 2009. The mammalian ovary from genesis to revelation. Endocrine Reviews 30:624–712.

Ehrlich, P.R., J.P. Holdren. 1971. Impact of population growth. Science 171:1212–1217.

Eichel, E.W., R.J. Ablin. 2013. The female prostate: correcting the G-spot. Journal of Sex Medicine 10:611–619.

Eisenberg, J.F., et al. 1971. Reproductive behavior of the Asiatic elephant (*Elephas maximus maximus* L.). Behaviour 38:193–225.

Elchlepp, J.G. 1952. The urogenital organs of the cottontail rabbit (*Sylvilagus floridanus*). Journal of Mammalogy 91:169–198.

Ellis, H. 2011. Anatomy of the uterus. Anaesthesia and Intensive Care Medicine 12:99–101.

Els, D.A. 1991. Aspects of reproduction in mountain reedbuck from Rolfontein Nature Reserve. South African Journal of Wildlife Research 21:43–46.

ElWishy, A.B. 1987. Reproduction in the female dromedary (*Camelus dromedaries*): a review. Animal Reproduction Science 15:273–287.

Emera, D., et al. 2012. The evolution of menstruation: a new model for genetic assimilation. BioEssays 34:26–35.

Emmons, L.H., A. Biun. 1991. Malaysian tree shrews. Maternal behavior of a wild tree shrew, *Tupaia tana*, in Sabah. National Geographic Research and Exploration 7:70–81.

Enders, A.C., et al. 1958. Histological and histochemical observations on the armadillo uterus during the delayed and post-implantation periods. Anatomical Record 130:639–657.

Enders, A.C., et al. 2005. Structure of the ovaries of the Nimba otter shrew, *Micropotamogale lamottei*, and the Madagascar hedgehog, *Echinops telfairi*. Cells Tissues Organs 179:179–191.

Engelhardt, H., et al. 2002. Conceptus influences the distribution of uterine leukocytes during early porcine pregnancy. Biology of Reproduction 66:1875–1880.

Engh, A.L., et al. 2000. Mechanisms of maternal rank "inheritance" in the spotted hyaena, *Crocuta crocuta*. Animal Behaviour 40:323–332.

Engh, A.L., et al. 2002. Reproductive skew among males in a female-dominated mammalian society. Behavioral Ecology 13:193–200.

Engh, A.L., et al. 2003. Coprologic survey of parasites of spotted hyenas (*Crocuta crocuta*) in the Masai Mara National Reserve, Kenya. Journal of Wildlife Diseases 39:224–227.

Enjapoori, A.K., et al. 2014. Monotreme lactation protein is highly expressed in monotreme milk and provides antimicrobial protection. Genome Biology and Evolution 6:2754–2773.

Eppig, J.J., et al. 2002. The mammalian oocyte orchestrates ovarian follicular development. Proceedings of the National Academy of Sciences 99:2890–2894.

Erskine, M.S. 1989. Solicitation behavior in the estrous female rat: a review. Hormones and Behavior 23:473–502.

Espinosa, M.B., et al. 2011. The ovary of *Lagostomus maximus* (Mammalia, Rodentia): an analysis by confocal microscopy. Biocell 35:37–42.

Estienne, M.J., et al. 2008. Dietary supplementation with a source of omega-3 fatty acids increases sperm number and the duration of ejaculation in boars. Theriogenology 70:70–76.

Ewer, R.F. 1973. The carnivores. Cornell University Press, Ithaca, NY.

Fadem, B.H., et al. 1982. Care and breeding of the gray, short-tailed opossum (*Monodelphis domestica*). Laboratory Animal Science 32:405–409.

Fagerstone, K.A., et al. 2006. When, where and for what wildlife species will contraception be a useful management approach? Proceedings of the Vertebrate Pest Conference 22:45–54.

Fair, P.A., P.R. Becker. 2000. Review of stress in marine mammals. Journal of Aquatic Ecosystem Stress and Recovery 7:335–354.

Fairbanks, L.A. 2000. Maternal investment throughout the life span in Old World monkeys. Pp. 341–367 in Old world monkeys (P.F. Whitehead, C.J. Jolly, eds.). Cambridge University Press, Cambridge, England.

Farah, Z. 1993. Composition and characteristics of camel milk. Journal of Dairy Research 60:603–626.

Farley, S.D., C.T. Robbins. 1995. Lactation, hibernation, and mass dynamics of American black bears and grizzly bears. Canadian Journal of Zoology 73:2216–2222.

Faulkes, C.G., N.C. Bennett. 2001. Family values: group dynamics and social control of reproduction in African mole-rats. Trends in Ecology and Evolution 16:184–190.

Faurie, A.S., et al. 2004. Peripartum body temperatures in free-ranging ewes (*Ovis aries*) and their lambs. Journal of Thermal Biology 29:115–122.

Favoretto, S.M., et al. 2015. Reproductive system of brown-throated sloth (*Bradypus variegatus*, Schinz 1825, Pilosa, Xenarthra): anatomy and histology. Anatomia Histologia Embryologia, doi:10.1111/ahe.12193.

Fay, F.H. 1985. *Odobenus rosmarus*. Mammalian Species 238:1–7.

Fenelon, J.C., et al. 2014. Embryonic diapause: development on hold. International Journal of Developmental Biology 58:163–174.

Ferguson-Smith, M.A., W. Rens. 2010. The unique sex chromosome system in platypus and echidna. Russian Journal of Genetics 46:1160–1164.

Fernandes, R.A., et al. 2012. Placental tissues as sources of stem cells—review. Open Journal of Animal Sciences 2:166–173.

Fernandez, R., et al. 2002. Mapping the SRY gene in *Microtus cabrerae*: a vole species with multiple SRY copies in males and females. Genome 45:600–603.

Fernández-Arias, A., et al. 1999. Interspecies pregnancy of Spanish ibex (*Capra pyrenaica*) fetus in a domestic goat (*Capra hircus*) recipients induces abnormally high plasmatic levels of pregnancy-associated glycoprotein. Theriogenology 51:1419–1430.

Ferner, K., et al. 2014. The placentation of Eulipotyphla—reconstructing a morphotype of the mammalian placenta. Journal of Morphology 275:1122–1144.

Ferretti, M.P. 2007. Evolution of bone-cracking adaptations in hyaenids (Mammalia, Carnivora). Swiss Journal of Geosciences 100:41–52.

Finkenwirth, C., et al. 2016. Oxytocin is associated with infant-care behavior and motivation in cooperatively breeding marmoset monkeys. Hormones and Behavior 80:10–18.

Fisher D.O., et al. 2002. Convergent maternal care strategies in ungulates and macropods. Evolution 56: 167–176.

Flamini, M.A., et al. 2002. Morphological characterization of the female prostate (Skene's gland or paraurethral gland) of *Lagostomus maximus maximus*. Annals of Anatomy 184:341–345.

Fleming, T.H. 1971. *Artibeus jamaicensis*: delayed embryonic development in a Neotropical bat. Science 171:402–404.

Fletcher, Q.E., et al. 2012. Oxidative damage increases with reproductive energy expenditure and is reduced by food-supplementation. Evolution 67:1527–1536.

Flint, A.P.F., et al. 1990. The maternal recognition of pregnancy in mammals. Journal of Zoology, London 221:327–341.

Flint, A.P.F., et al. 1997. Blastocyst development and conceptus sex selection in red deer *Cervus elaphus*: studies of a free-living population on the Isle of Rum. General and Comparative Endocrinology 106:374–383.

Flowerdew, J.R. 1987. Mammals: their reproductive biology and population ecology. Edward Arnold, London, England.

Folch J., et al. 2009. First birth of an animal from an extinct subspecies (*Capra pyrenaica pyrenaica*) by cloning. Theriogenology 71:1026–1034.

Fouda, M.M., et al. 1990. Maternal-infant relationships in captive sika deer (*Cervus nippon*). Small Ruminant Research 3:199–209.

Fourvel, J.-B., et al. 2015. Large mammals of Fouvent-Saoint-Andoche (Haut-Saône, France): a glimpse into a late Pleistocene hyena den. Geodiversitas 37:237–266.

Fox, C.A., B. Fox. 1971. A comparative study of coital physiology, with special reference to the sexual climax. Journal of Reproduction and Fertility 24:319–336.

Francis, C.M., et al. 1994. Lactation in male fruit bats. Nature 367:681–692.

Frank, L.G. 1997. Evolution of genital masculinization: why do female hyaenas have such a large "penis"? Trends in Ecology and Evolution 12:58–62.

Frank, L.G., S.E. Glickman. 1994. Giving birth through a penile clitoris: parturition and dystocia in the spotted hyaena (*Crocuta crocuta*). Journal of Zoology, London 234:659–665.

Frank, L.G., et al. 1985. Testicular origin of circulating androgen in the spotted hyaena, *Crocuta crocuta*. Journal of Zoology, London 207a:613–615.

Frank, L.G., et al. 1989. Ontogeny of female dominance in the spotted hyaena: perspectives from nature and captivity. Symposia of the Zoological Society of London 61:127–146.

Frank, L.G., et al. 1990. Sexual dimorphism in the spotted hyaena (*Crocuta crocuta*). Journal of Zoology, London 221:308–313.

Frank, L.G., et al. 1991. Fatal sibling aggression, precocial development, and androgens in neonatal spotted hyenas. Science 252:702–704.

Frank, L.G., et al. 1995. Masculinization costs in hyaenas. Nature 377:584–585.

Frankenberg, S., L. Selwood. 2001. Ultrastructure of oogenesis in the brushtail possum. Molecular Reproduction and Development 56:297–306.

Frazer, J.F.D., A.S.G. Huggett. 1974. Species variations in the foetal growth rates of eutherian mammals. Journal of Zoology, London 174:481–509.

Fredga, K. 1994. Bizarre mammalian sex-determining mechanism. Pp. 419–431 in The differences between the sexes (R.V. Short, E. Balaban, eds.), Cambridge University Press, Cambridge, England.

Freeland, C.A. 1987. Aristotle on bodies, matter, and potentiality. Pp. 392–407 in Philosophical issues in Aristotle's biology (J.G. Lennox, A. Gotthelf, eds.). Cambridge University Press, Cambridge, England.

Freeman, M.E., et al. 2000. Prolactin: structure, function and regulation of secretion. Physiological Reviews 80:1523–1631.

Frick, W.F., et al. 2010. An emerging disease causes regional population collapse of a common North American bat species. Science 329:679–682.

Friebe, A., et al. 2014. Factors affecting date of implantation, parturition and den entry estimated from activity and body temperature in free-ranging brown bears. PLOS ONE 9(7):e101410.

Fuchs, E., S. Corbach-Söhle. 2010. Tree shrews. Pp. 263–275 in The UFAW handbook on the care and management of laboratory and other research animals, 8th ed. (R. Hubrecht, J. Kirkwood, eds.). Wiley-Blackwell, Oxford, England.

Fuchs, K., et al. 2000. Detection of space-time clusters and epidemiological examinations of scabies in chamois. Veterinary Parasitology 92:63–73.

Funkhouser, L.J., S.R. Bordenstein. 2013. Mom knows best: the universality of maternal microbial transmission. PLOS Biology 11(8):e10016131.

Galbreath, G.J. 1985. The evolution of monozygotic polyembryony in *Dasypus*. Pp. 243–246 in The evolution and ecology of armadillos, sloths, and vermilinguas (G.G. Montgomery, ed.). Smithsonian Institution Press, Washington, DC.

Garratt, M., et al. 2014. Female promiscuity and maternally dependent offspring growth rates in mammals. Evolution 68:1207–1215.

Garshelis, D.L. 2004. Variation in ursid life histories. Pp. 53–73 in Giant pandas: biology and conservation (D. Lindburg, K. Baragona, eds.). University of California Press, Berkeley, CA.

Gélin, U., et al. 2013. Offspring sex, current and previous reproduction affect feeding behaviour in wild eastern grey kangaroos. Animal Behaviour 86:885–891.

Gemmell, R.T., et al. 2002. Birth in marsupials. Comparative Biochemistry and Physiology 131B:621–630.

Gems, D. 2014. Evolution of sexually dimorphic longevity in humans. Aging 6:84–91.

Genoud, M., P. Vogel. 1990. Energy requirements during reproduction and reproductive effort in shrews (Soricidae). Journal of Zoology, London 220:41–60.

Gentry, R.L., G.L. Kooyman (eds.). 1986. Fur seals: maternal strategies on land and at sea. Princeton University Press, Princeton, NJ.

George, J.C., et al. 1999. Age and growth estimates of bowhead whales (*Balaena mysticetes*) via aspartic acid racemization. Canadian Journal of Zoology 77:571–580.

Georges, J.-Y., et al. 2001. Milking strategy in subantarctic fur seals *Arctocephalus tropicalis* breeding on Amsterdam Island: evidence from changes in milk composition. Physiological and Biochemical Zoology 74:548–559.

Gérard, P. 1932 Etudes sur l'ovogenèse et l'ontogenèses chez les lémuriens du genre *Galago*. Archives de Biologie, Liège 43:93–151.

Gero, S., H. Whitehead. 2007. Suckling behavior in sperm whale calves: observations and hypotheses. Marine Mammal Science 23:398–413.

Gervasi, M.G., et al. 2009. The endocannabinoid system in bull sperm and bovine oviductal epithelium: role of anandamide in sperm-oviduct interaction. Reproduction 137:403–414.

Gidley-Baird, A.A. 1981. Endocrine control of implantation and delayed implantation in rats and mice. Journal of Reproduction and Fertility 29(Supplement):97–109.

Giger, W., et al. 2003. Occurrence and fate of antibiotics as trace contaminants in wastewaters, sewage sludges, and surface waters. Chimia 57:485–491.

Gilbert, A.N. 1986. Mammary number and litter size in Rodentia: the "one-half rule." Proceedings of the National Academy of Science 83:4828–4830.

Gilbert, A.N. 1995. Tenacious nipple attachment in rodents: the sibling competition hypothesis. Animal Behaviour 50:881–891.

Gilbert, J.A., et al. 2014. The Earth microbiome project: successes and aspirations. BMC Biology 12.1:69.

Gilbert, S.F. 2014. Developmental biology, 10th edition. Sinauer, Sunderland, MA.

Gilg, O. 2002. The summer decline of the collared lemming, *Dicrostonyx groenlandicus*, in high arctic Greenland. Oikos 99:499–510.

Gill, J. 2012. Happy Ada Lovelace Day! Honoring Dr. Evelyn Chrystalla Pielou [blog]. https://contemplativemammoth.com/2012/10/16/happy-ada-lovelace-day-honoring-dr-evelyn-chrystalla-pielou/.

Gitschier, J. 2010. The gift of observation: an interview with Mary Lyon. PLOS Genetics 6:e1000813.

Gittleman, J.L. 1988. Behavioral energetics of lactation in a herbivorous carnivore, the red panda (*Ailurus fulgens*). Ethology 79:13–24.

Givens, M.D., M.D.S. Marley. 2008. Infectious causes of embryonic and fetal mortality. Theriogenology 70:270–285.

Gjøstein, H., et al. 2004. Milk production and composition in reindeer (*Rangifer tarandus*): effect of lactational stage. Comparative Biochemistry and Physiology A 137:649–656.

Glickman, S.E., et al. 1987. Androstenedione may organize or activate sex-reversed traits in female spotted hyenas. Proceedings of the National Academy of Sciences 84:3444–3447.

Glickman, S.E., et al. 1998. Androgens and masculinization of genitalia in the spotted hyaena (*Crocuta crocuta*). 3. Effects of juvenile gonadectomy. Journal of Reproduction and Fertility 113:129–135.

Glickman, S.E., et al. 2005. Sexual differentiation in three unconventional mammals: spotted hyenas, elephants and tammar wallabies. Hormones and Behavior 48:403–417.

Glickman, S.E., et al. 2006. Mammalian sexual differentiation: lessons from the spotted hyena. Trends in Endocrinology and Metabolism 17:349–356.

Golightly, E., et al. 2011. Endocrine immune interactions in human parturition. Molecular and Cellular Endocrinology 335:52–59.

Golla, W., et al. 1999. Within-litter sibling aggression in spotted hyaenas: effect of maternal nursing, sex and age. Animal Behaviour 58:715–726.

Gombe, S. 1985. Short term fluctuations in progesterone, oestradiol and testosterone in pregnant and non-pregnant hyaena, *Crocuta crocuta* (Erxleben). African Journal of Ecology 23:269–271.

Gomez-Lopez, N., et al. 2013. Evidence for a role for the adaptive immune response in human term parturition. American Journal of Reproductive Immunology 69:212–230.

Gosling, L.M., et al. 1984. Differential investment by female coypus (*Myocastor coypus*) during lactation. Symposia of the Zoological Society, London 51:273–300.

Gottschang, J.L. 1956. Juvenile molt in *Peromyscus leucopus noveboracensis*. Journal of Mammalogy 37:516–520.

Gowans, S., et al. 2001. Social organization in northern bottlenose whales, *Hyperoodon ampullatus*: not driven by deep-water foraging? Animal Behaviour 62:369–377.

Goymann, W., et al. 2001. Androgens and the role of female "hyperaggressiveness" in spotted hyenas (*Crocuta crocuta*). Hormones and Behavior 39:83–92.

Grant, J., A. Hawley. 1991. Some observations on the mating behaviour of captive American pine martens *Martes americana*. Acta Theriologica 41:439–442.

Graves, J.A.M. 1996. Mammals that break the rules: genetics of marsupials and monotremes. Annual Review of Genetics 30:233–260.

Graves, J.A.M., M.B. Renfree. 2013. Marsupials in the age of genomics. Annual Review of Genomics and Human Genetics 14:393–420.

Gray, C.A., et al. 2001. Developmental biology of uterine glands. Biology of Reproduction 65:1311–1323.

Gray, M.E., E.Z. Cameron. 2010. Does contraceptive treatment in wildlife result in side effects? a review of quantitative and anecdotal evidence. Reproduction 139:45–55.

Greene, R., et al. 1997. Long-term implications of cesarean section. American Journal of Obstetrics and Gynecology 176:254–255.

Greenwald, G.S., R.D. Peppler. 1968. Prepubertal and pubertal changes in the hamster ovary. Anatomical Record 161:447–457.

Griffin, J., et al. 2006. Comparative analysis of follicle morphology and oocyte diameter in four mammalian species (mouse, hamster, pig, and human). Journal of Experimental and Clinical Assisted Reproduction, doi:10.1186/1743-1050-3-2.

Griffin, P.C., et al. 2005. Mortality by moonlight: predation risk and the snowshoe hare. Behavioral Ecology 16:938–944.

Griffiths, M., et al. 1973. Observations of the comparative anatomy and ultrastructure of mammary glands and on the fatty acids of the triglycerides in platypus and echidna milk fats. Journal of Zoology, London 169:255–279.

Grosser, O. 1909. Vergleichende Anatomie und Entwicklungsgeschichte der Eihäute und der Placenta. W. Braumüller, Vienna, Austria.

Grosser, O. 1927. Frühentwicklung, Eihautbildung und Placentation des Menschen und der Säugetiere. J.F. Bergmann, Munich, Germany.

Grützner, F., et al. 2006. How did the platypus get its sex chromosome chain? A comparison of meiotic multiples and sex chromosomes in plants and animals. Chromosoma 115:75–88.

Guillette, J. J., Jr., M.P. Gunderson. 2001. Alterations in development of reproductive and endocrine systems of wildlife populations exposed to endocrine-disrupting contaminants. Reproduction 122:857–864.

Gustafsson, A., P. Lindenfors. 2004. Human size evolution: no evolutionary allometric relationship between male and female stature. Journal of Human Evolution 47:253–266.

Gyllensten, U., et al. 1991. Paternal inheritance of mitochondrial DNA in mice. Nature 352:255–257.

Haandrikman, K., L.J.G. van Wissen. 2008. Effects of the fertility transition on birth seasonality in the Netherlands. Journal of Biosocial Science 40:655–672.

Hafez, B., E.S.E. Hafez. 2000. Reproduction in farm animals, 7th edition. Lippincott Williams and Wilkins, Philadelphia, PA.

Haig, D. 1996. Placental hormones, genomic imprinting, and maternal-fetal communication. Journal of Evolutionary Biology 9:357–380.

Haig, D. 1999. What is a marmoset? American Journal of Primatology 49:285–296.

Hajian, M., et al. 2011. "Conservation cloning" of vulnerable Esfahan mouflon (*Ovis orientalis isphahanica*): in vitro and in vivo studies. European Journal of Wildlife Research 57:959–969.

Hamel, F. 1911. An eighteenth-century marquise: a study of Emilie du Châtelet and her times. James Pott & Co. New York, NY.

Hamilton, P.K., et al. 1998. Age structure and longevity in North Atlantic right whales *Eubalaena glacialis* and their relation to reproduction. Marine Ecology and Progress Series 171:285–292.

Hamilton, W.J., et al. 1986. Sexual monomorphism in spotted hyenas, *Crocuta crocuta*. Ethology 71:63–73.

Hammond, K.A., J. Diamond. 1992. An experimental test for a ceiling on sustained metabolic rate in lactating mice. Physiological Zoology 65:952–977.

Handley, L.J.L., H. Perrin. 2007. Advances in our understanding of mammalian sex-biased dispersal. Molecular Ecology 16:1559–1578.

Hansen, R.M. 1957. Development of young varying lemmings (*Dicrostonyx*). Arctic 10:105–117.

Harrison Matthews, L. 1935. The oestrous cycle and intersexuality in the female mole (*Talpa europaea* Linn.). Proceedings of the Zoological Society of London 105:347–383.

Harrison Matthews, L. 1939. Reproduction in the spotted hyaena, *Crocuta crocuta* (Erxleben). Philosophical Transactions of the Royal Society B 230:1–78.

Harrison Matthews, L. 1954. [No title.] Proceedings of the Zoological Society of London 124:198.

Hartman, C.G. 1924. Observations on the motility of the opossum genital tract and the vaginal plug. Anatomical Record 27:293–303.

Hartman, C.G. 1957. How do sperms get into the uterus? Fertility and Sterility 8:403–427.

Hartung, T.G., D.A. Dewsbury. 1978. A comparative analysis of copulatory plugs in muroid rodents and their relationship to copulatory behavior. Journal of Mammalogy 59:717–723.

Haselton, M.G., K. Gildersleeve. 2016. Human ovulation cues. Current Opinion in Psychology 7:120–125.

Hasler, J.F., E.M. Banks. 1975. The influence of mature males on sexual maturation in female collared lemmings (*Dicrostonyx groenlandicus*). Journal of Reproduction and Fertility 42:583–586..

Hasler, J.F., E.M. Banks. 1985. Reproductive performance and growth in captive collared lemmings (*Dicrostonyx groenlandicus*). Canadian Journal of Zoology 53:777–787.

Hasler, J.F., et al. 1974. Ovulation and related phenomena in the collared lemmings (*Dicrostonyx groenlandicus*). Journal of Reproduction and Fertility 38:21–28.

Hasler, J.F., et al. 1976. The influence of photoperiod on growth and sexual function in male and female collared lemmings (*Dicrostonyx groenlandicus*). Journal of Reproduction and Fertility 46:323–329.

Hastings, K.K., J.W. Testa. 1998. Maternal and birth colony effects on survival of Weddell seal offspring from McMurdo Sound, Antarctica. Journal of Animal Ecology 67:722–740.

Havera, S.P. 1979. Energy and nutrient cost of lactation in fox squirrels. Journal of Wildlife Management 43:958–965.

Hawkes, K. 2004. The grandmother effect. Nature 428:128–129.

Hawkes, K., J.E. Coxworth. 2013. Grandmothers and the evolution of human longevity: a review of findings and future directions. Evolutionary Anthropology 22:294–302.

Hawkins, C.E., P.A. Racey. 2009. A novel mating system in a solitary carnivore: the fossa. Journal of Zoology, London 277:196–204.

Hawkins, C.E., et al. 2002. Transient masculinization in the fossa, *Cryptoprocta ferox* (Carnivora, Viverridae). Biology of Reproduction 66:610–615.

Hawkins, C.E., et al. 2006. Emerging disease and population decline of an island endemic, the Tasmanian devil *Sarcophilus harrisii*. Biological Conservation 131:307–324.

Hay, M.F., W.R. Allen. 1975. An ultrastructural and histochemical study of the interstitial cells in the gonads of the fetal horse. Journal of Reproduction and Fertility, Supplement 23:557–561.

Haynie, M.L., et al. 2003. Parentage, multiple paternity, and breeding success in Gunnison's and Utah prairie dogs. Journal of Mammalogy 84:1244–1253.

Hayssen, V. 1984. Mammalian reproduction: constraints on the evolution of infanticide. Pp. 105–123 in Infanticide: comparative and evolutionary perspectives (G. Hausfater, S.B. Hrdy, eds.). Aldine, New York, NY.

Hayssen, V. 1985. A comparison of the reproductive biology of metatherian (marsupial) and eutherian (placental) mammals with special emphasis on sex differences in the behavior of the opossum *Didelphis virginiana*. PhD dissertation, Cornell University, Ithaca, NY.

Hayssen, V. 1993. Empirical and theoretical constraints on the evolution of lactation. Journal of Dairy Science 75:3213–3233.

Hayssen, V. 2008a. Patterns of body and tail length and body mass in Sciuridae. Journal of Mammalogy 89:852–873.

Hayssen, V. 2008b. Reproductive effort in squirrels: ecological, phylogenetic, allometric, and latitudinal patterns. Journal of Mammalogy 89:582–606.

Hayssen, V. 2008c. Reproduction within marmotine ground-squirrels (Sciuridae, Xerinae, Marmotini): patterns among genera. Journal of Mammalogy 89:607–616.

Hayssen, V. 2009. *Bradypus tridactylus* (Pilosa: Bradypodidae). Mammalian Species 839:1–9.

Hayssen, V. 2010. *Bradypus variegatus* (Pilosa: Bradypodidae). Mammalian Species 42(850):19–32.

Hayssen, V. 2011. *Choloepus hoffmanni* (Pilosa: Megalonychidae). Mammalian Species 43(873):37–55.

Hayssen, V. 2016. Reproduction in grey squirrels: from anatomy to conservation. Pp. 115–130 in The grey squirrel: ecology and management of an invasive species in Europe (C. Shuttleworth, et al., eds.). European Squirrel Initiative, W.O. Jones, Llangefni, Wales.

Hayssen, V., D.G. Blackburn. 1985. α-Lactalbumin and the origins of lactation. Evolution 39:1147–1149.

Hayssen, V., T.H. Kunz. 1996. Allometry of litter mass in bats: comparisons with maternal size, wing morphology, and phylogeny. Journal of Mammalogy 77:476–490.

Hayssen, V., et al. 1985. Metatherian reproduction: transitional or transcending? American Naturalist 126:617–632.

Hayssen, V., et al. 1993. Asdell's patterns of mammalian reproduction: a compendium of species-specific data. Cornell University Press, Ithaca, NY.

Hearn, J.P. 1974. The pituitary gland and implantation in the tammar wallaby, *Macropus eugenii*. Journal of Reproduction and Fertility 39:235–241.

Heideman, P.D. 1989. Delayed development in Fischer's pygmy fruit bat, *Haplonycteris fischeri*, in the Philippines. Journal of Reproduction and Fertility 85:363–382.

Heideman, P.D., F.H. Bronson. 1992. A pseudo seasonal reproductive strategy in a tropical rodent, *Peromyscus nudipes*. Journal of Reproduction and Fertility 95:57–67.

Helle, S., et al. 2002. Sons reduced maternal longevity in preindustrial humans. Science 296:1085.

Hemminki, E., J. Meriläinenb. 1996. Long-term effects of cesarean sections: ectopic pregnancies and placental problems. American Journal of Obstetrics and Gynecology 174:1569–1574.

Henry, O. 1997. The influence of sex and reproductive state on diet preference in four terrestrial mammals of the French Guianan rain forest. Canadian Journal of Zoology 75:929–935.

Henschel, J.R., J.D. Skinner. 1991. Territorial behavior by a clan of spotted hyaenas *Crocuta crocuta*. Ethology 88:223–235.

Herbert, C.A. 2004. Long-acting contraceptives: a new tool to manage overabundant kangaroo populations in nature reserves and urban areas. Australian Mammalogy 26:67–74.

Herbst, M., et al. 2004. A field assessment of reproductive seasonality in the threatened wild Namaqua dune mole-rat (*Bathyergus janetta*). Journal of Zoology, London 263:259–268.

Hermes, R., et al. 2014. Reproductive tract tumours: the scourge of woman reproduction ails Indian rhinoceroses. PLOS ONE 9:e92595.

Herzing, D.L., C.M. Johnson. 1997. Interspecific interactions between Atlantic spotted dolphins (*Stenella frontalis*) and bottlenose dolphins (*Tursiops truncatus*) in the Bahamas, 1985–1995. Aquatic Mammals 23:85–99.

Heske, E.J., P.M. Jensen. 1993. Social structure in *Lemmus lemmus* during the breeding season. Pp. 387–395 in The biology of lemmings (N.C. Stenseth, R.A. Ims, eds.). Academic Press, New York, NY.

Hess (Baerwald), A.R., et al. 2000. Vascular characteristics of the human corpus luteum in the first trimester of pregnancy. Ultrasound International 6:2–10.

Hewson, R. 1976. A population study of mountain hares (*Lepus timidus*) in north-east Scotland from 1956–1969. Journal of Animal Ecology 45:395–414.

Hildebrandt, T.B., et al. 2011. Reproductive cycle of the elephant. Animal Reproduction Science 124:176–183.

Hill, A. 1980. Hyaena provisioning of juvenile offspring at the den. Mammalia 44:594–595.

Hinde, K., et al. 2009. Rhesus macaque milk: magnitude, sources, and consequences of individual variation over lactation. American Journal of Physical Anthropology 138:148–157.

Hinde, K., et al. 2014. Cortisol in mother's milk across lactation reflects maternal life history and predicts infant temperament. Behavioral Ecology 26:269–281.

Hobson, B.M., I.L. Boyd. 1984. Gonadotrophin and progesterone concentrations in placentae of grey seals (*Halichoerus grypus*). Journal of Reproduction and Fertility 72:521–528.

Hoeck, H.N. 1989. Demography and competition in hyrax: a 17 years [sic] study. Oecologia 79:353–360.

Hoelzel, A.R., et al. 1999. Alpha-male paternity in elephant seals. Behavioral Ecology and Sociobiology 46:298–306.

Hofer, H., M.L. East. 1993. The commuting system of Serengeti spotted hyaenas: how a predator copes with migratory prey. III. Attendance and maternal care. Animal Behaviour 46:575–589.

Hofer, H., M.L. East. 1997. Skewed offspring sex ratios and sex composition of twin litters in Serengeti spotted hyaenas (*Crocuta crocuta*) are a consequence of siblicide. Applied Animal Behaviour Science 51:307–316.

Hofer, H., M.L. East. 2003. Behavioral processes and costs of co-existence in female spotted hyenas: a life history perspective. Evolutionary Ecology 17:315–331.

Holekamp, K.E., S.M. Dloniak. 2011. Intraspecific variation in the behavioral ecology of a tropical carnivore, the spotted hyena. Advances in the Study of Behavior 42:189–229.

Holekamp, K.E., L. Smale. 1990. Provisioning and food sharing by lactating spotted hyenas, *Crocuta crocuta* (Mammalia: Hyaenidae). Ethology 86:191–202.

Holekamp, K.E., L. Smale. 1991. Dominance acquisition during mammalian social development: the "inheritance" of maternal rank. American Zoologist 31:306–317.

Holekamp, K.E., L. Smale. 1993. Ontogeny of dominance in free-living spotted hyaenas: juvenile rank relations with other immature individuals. Animal Behaviour 46:451–466.

Holekamp, K.E., L. Smale. 1995. Rapid change in offspring sex ratios after clan fission in the spotted hyena. American Naturalist 145:261–268.

Holekamp, K.E., L. Smale. 1998. Behavioral development in the spotted hyena. BioScience 48:997–1005.

Holekamp, K.E., et al. 1993. Fission of a spotted hyena clan: consequences of prolonged female absenteeism and causes of female emigration. Ethology 93:285–299.

Holekamp, K.E., et al. 1996a. Rank and reproduction in the female spotted hyaena. Journal of Reproduction and Fertility 108:229–237.

Holekamp, K.E., et al. 1996b. Patterns of association among female spotted hyenas (*Crocuta crocuta*). Journal of Mammalogy 78:55–64.

Holekamp, K.E., et al. 1999a. Association of seasonal reproductive patterns with changing food availability in an equatorial carnivore, the spotted hyaena (*Crocuta crocuta*). Journal of Reproduction and Fertility 116:87–93.

Holekamp, K.E., et al. 1999b. Vocal recognition in the spotted hyaena and its possible implications regarding the evolution of intelligence. Animal Behaviour 58:383–395.

Holekamp, K.E., et al. 2007. Social intelligence in the spotted hyena (*Crocuta crocuta*). Philosophical Transactions of the Royal Society B 363:523–538.

Holekamp, K.E., et al. 2012. Society, demography and genetic structure in the spotted hyena. Molecular Ecology 21:613–632.

Holland, N., S.M. Jackson. 2002. Reproductive behaviour and food consumption associated with the captive breeding of platypus (*Ornithorhynchus anatinus*). Journal of Zoology, London 256:279–288.

Holt, W.V., A. Fazeli. 2016. Sperm selection in the female mammalian reproductive tract. Focus on the oviduct: hypotheses, mechanisms, and new opportunities. Theriogenology 85:105–112.

Holt, W.V., et al. 2003. Reproductive science and integrative conservation, volume 8. Cambridge University Press, Cambridge, England.

Hood, C.S. 1989. Comparative morphology and evolution of the female reproductive tract in macroglossine bats (Mammalia, Chiroptera). Journal of Mammalogy 199:207–221.

Hood, W.R., et al. 2014. Milk composition and lactation strategy of a eusocial mammal, the naked mole-rat. Journal of Zoology, London 293:108–118.

Hoogland, J.L. 1985. Infanticide in prairie dogs: lactating females kill offspring of close kin. Science 230:1037–1040.

Hooper, E.T. 1972. A synopsis of the rodent genus *Scotinomys*. Occasional Papers of the Museum of Zoology University of Michigan 665:1–32.

Hooper, E.T., M.D. Carleton. 1976. Reproduction, growth and development in two contiguously allopatric rodent species, genus *Scotinomys*. Miscellaneous Publications, Museum of Zoology, University of Michigan 151:1–52.

Hooper, L.V. 2004. Bacterial contributions to mammalian gut development. Trends in Microbiology 12:129–134.

Höner, O.P., et al. 2007. Female mate-choice drives the evolution of male-biased dispersal in a social mammal. Nature 448:798–801.

Hosken, D., T.H. Kunz. 2009. But is it male lactation or not? Trends in Ecology and Evolution 24:355.

Houston, A.I., et al. 2007. Capital or income breeding? A theoretical model of female reproductive strategies. Behavioral Ecology 18:241–250.

Hradecky, P. 1982. Uterine morphology in some African antelopes. Journal of Zoo Animal Medicine 13:132–136.

Hrdy, S.B. 1991 (1999). The woman that never evolved: with a new preface and bibliographical updates, revised edition. Harvard University Press, Cambridge, MA.

Hrdy, S.B. 1999. Mother nature: maternal instincts and how they shape the human species. Ballantine Books, New York, NY.

Hrdy, S.B. 2000. The optimal number of fathers. Evolution, demography, and history in the shaping of female mate preferences. Annals of the New York Academy of Sciences 907:75–96.

Hrdy, S.B. 2009. Mothers and others: the evolutionary origins of mutual understanding. Belknap Press of Harvard University Press, Cambridge, MA.

Hudson, R., H. Distel. 2013. Fighting by kittens and piglets during suckling: what does it mean? Ethology 119:353–359.

Huggett, A.S.G., W.F. Widdas. The relationship between mammalian foetal weight and conception age. Journal of Physiology 114:306–317.

Hughes, R.L. 1993. Monotreme development with particular reference to the extraembryonic membranes. Journal of Experimental Zoology 266:480–494.

Hughes, R.L., L.S. Hall. 1998. Early development and embryology of the platypus. Philosophical Transactions: Biological Sciences 353:1101–1114.

Humphrey, L.T. 2010. Weaning behavior in human evolution. Seminars in Cell and Developmental Biology 21:453–461.

Husson, A.M. 1978. Mammals of Suriname. E.J. Brill, Leiden, Netherlands.

Hyde, M.J., N. Modi. 2012. The long-term effects of birth by caesarean section: the case for a randomised controlled trial. Early Human Development 88:943–949.

Ickowicz, D., et al. 2012. Mechanism of sperm capacitation and the acrosome reaction: role of protein kinases. Asian Journal of Andrology 14:816–821.

Jabbour, H., et al. 1997. Conservation of deer: contributions from molecular biology, evolutionary ecology, and reproductive physiology. Journal of Zoology, London 243:461–484.

Jainudeen, M.R., E.S.E Hafez. 1980. Reproductive failure in females. Pp. 449–470 in Reproduction in farm animals (E.S.E. Hafez, ed.). Lea and Febiger, Philadelphia.

Jamieson, D.J., et al. 2006. Emerging infections and pregnancy. Emerging Infectious Diseases 12:1638–1643.

Jarvis, J.U.M., P.W. Sherman. 2002. *Heterocephalus glaber.* Mammalian Species 706:1–9.

Jenkins, F.A., Jr. 1990. Monotremes and the biology of Mesozoic mammals. Netherlands Journal of Zoology 40:5–31.

Jenks, S.M., et al. 1995. Acquisition of matrilineal rank in captive spotted hyaenas: emergence of a natural social system in peer-reared animals and their offspring. Animal Behaviour 50:893–904.

Jenness, R. 1979. The composition of human milk. Seminars in Perinatology 3:225–239.

Jenness, R. 1984. Lactational performance of various mammalian species. Journal of Dairy Science 69:869–885.

Jenness, R., et al. 1981. Composition of milk of the sea otter (*Enhydra lutris*). Comparative Biochemistry and Physiology 70A:275–379.

Jensen, P.M., T.O. Gustafsson. 1984. Evidence for pregnancy failure in young *Lemmus lemmus* and *Microtus oeconomus.* Canadian Journal of Zoology 62:2568–2570.

Jöchle, W. 1975. Current research in coitus-induced ovulation: a review. Journal of Reproduction and Fertility, Supplement 22:165–207.

Johnson, G., et al. 2010. Evidence that sperm whale (*Physeter macrocephalus*) calves suckle through their mouth. Marine Mammal Science 26:990–996.

Joly, D.O., F. Messier. 2005. The effect of bovine tuberculosis and brucellosis on reproduction and survival of wood bison in Wood Buffalo National Park. Journal of Animal Ecology 74:543–551.

Jones, K.T., S.I.R. Lane. 2013. Molecular causes of aneuploidy in mammalian eggs. Development 140:3719–3730.

Jones, M.E., et al. 2008. Life-history change in disease-ravaged Tasmanian devil populations. Proceedings of the National Academy of Science 105:10023–10027.

Jones, W.T. 1987. Dispersal patterns in kangaroo rats (*Dipodomys spectabilis*). Pp. 119–127 in Mammalian dispersal patterns (B.D. Chepko-Sade, Z.T. Halpin, eds.). University of Chicago Press, Chicago, IL.

Jönsson, K.I. 1997. Capital and income breeding as alternative tactics of resource use in reproduction. Oikos 78:57–66.

Joshi, C.K., et al. 1978. Studies on oestrus cycle in Bikaneri she-camel (*Camelus dromedaries*). Indian Journal of Animal Science 48:141–145.

Just, W., et al. 2007. *Ellobius lutescens*: sex determination and sex chromosome. Sexual Development 1:211–221.

Kaneko, T., et al. 2003. Mating-induced cumulus-oocyte maturation in the shrew, *Suncus murinus*. Reproduction 125:817–826.

Kanitz, W., et al. 2001. Comparative aspects of follicular development, follicular and oocyte maturation and ovulation in cattle and pigs. Archive Tierzucht Dummerstorf 44(Special Issue):9–23.

Kaňková, Š., et al. 2007. Influence of latent toxoplasmosis on the secondary sex ratio in mice. Parasitology 134:1709–1717.

Kappeler, P. 2015. Lemur behavior informs the evolution of social monogamy. Trends in Ecology and Evolution 29:591–593.

Kasuya, T., H. Marsh. 1984. Life history and reproductive biology of the short-finned pilot whale, *Globicephala macrorhynchus*, off the Pacific coast of Japan. Reports of the International Whaling Commission 6(Special Issue):259–310.

Kayanja, F.I.B., L.H. Blankenship. 1973. The ovary of the giraffe, *Giraffa camelopardalis*. Journal of Reproduction and Fertility 34:305–313.

Kellas, L.M., et al. 1958. Ovaries of some fœtal and prepubertal giraffes (*Giraffa camelopardalis* (Linnaeus)). Nature 181:487–488.

Kelly, B.P., et al. 2010. Seasonal home ranges and fidelity to breeding sites among ringed seals. Polar Biology 33:1095–1109.

Kerth, G., et al. 2011. Bats are able to maintain long-term social relationships despite the high fission-fusion dynamics of their groups. Proceedings of the Royal Society B 278:2761-2757.

Keverne, E.B. 2015. Genomic imprinting, action, and interaction of maternal and fetal genomes. Proceedings of the National Academy of Sciences 112:6834–6840.

Keverne, E.B., J.P. Curley. 2008. Epigenetics, brain evolution and behavior. Frontiers in Neuroendocrinology 29:398–412.

Kevles, B. 1986. Females of the species: sex and survival in the animal kingdom. Harvard University Press, Cambridge, MA.

Kidder, D.L, T.R. Worsley. 2004. Causes and consequences of extreme Permo-Triassic warming to globally equable climate and relation to the Permo-Triassic extinction and recovery. Palaeogeography, Palaeoclimatology, Palaeoecology 203: 207–237.

Kimura, J., et al. 2005. Three-dimensional reconstruction of the equine ovary. Anatomia, Histologia, Embryologia 34:48–51.

King, C.M. 1983. *Mustela erminea*. Mammalian Species 195:1–8.

Kinsley, C.H., et al. 2014. The mother as hunter: significant reduction in foraging costs through enhancements of predation in maternal rats. Hormones and Behavior 66:649–654.

Kirkpatrick, J.F. 2007. Measuring the effects of wildlife contraception: the argument for comparing apples with oranges. Reproduction, Fertility and Development 19:548–552.

Kirkpatrick, J.F., et al. 2011. Contraceptive vaccines for wildlife: a review. American Journal of Reproductive Immunology 66:40–50.

Kirsch, J.A.W. 1977. The six-percent solution: second thoughts on the adaptedness of the Marsupialia. American Scientist 65"276–288.

Klein, C., M.H.T. Troedsson. 2011. Maternal recognition of pregnancy in the horse: a mystery still to be solved. Reproduction, Fertility and Development 23:952–963.

Knight, C.H., et al. 1998. Local control of mammary development and function. Reviews of Reproduction 3:104–112.

Kobayashi, T., et al. 2007. Exceptional minute sex-specific region in the XO mammal, Ryukyu spiny rat. Chromosome Research 15:175–187.

Koester, H. 1970. Ovum transport. Pp. 189–228 in Mammalian reproduction (H. Gibian, E.J. Plotz, eds.). Springer-Verlag, NY, NY.

Köhncke, M., K. Leonhardt. 1986. *Cryptoprocta ferox*. Mammalian Species 254:1–5.

Kokko, H., M.D. Jennions. 2008. Parental investment, sexual selection and sex ratios. Journal of Evolutionary Biology 21:919–948.

Kolowski, J.M., K.E. Holekamp. 2009. Ecological and anthropogenic influences on space use by spotted hyaenas. Journal of Zoology, London 277:23–36.

Koprowski, J.L. 1992. Removal of copulatory plugs by female tree squirrels. Journal of Mammalogy 73:572–576.

Koprowski, J.L. 1996. Natal philopatry, communal nesting, and kinship in fox squirrels and gray squirrels. Journal of Mammalogy 77:1006–1016.

Koprowski, J.L. 1998. Conflict between the sexes: a review of the social and mating systems of the tree squirrels. Special Publication, Virginia Museum of Natural History 6:33–41.

Korine, C., et al. 2004. Reproductive energetics of captive and free-ranging Egyptian fruit bats (*Rousettus aegyptiacus*). Ecology 85:220–230.

Koskela, E., H. Ylönen. 1995. Suppressed breeding in the field vole (*Microtus agrestis*): an adaptation to cyclically fluctuating predation risk. Behavioral Ecology 6:311–315.

Kraaijeveld-Smit, F.J.L., et al. 2002. Multiple paternity in a field population of a small carnivorous marsupial, the agile antechinus, *Antechinus agilis*. Behavioral Ecology and Sociobiology 52:84–91.

Kraaijeveld-Smit, F.J.L., et al. 2003. Paternity success and the direction of sexual selection in a field population of a semelparous marsupial, *Antechinus agilis*. Molecular Ecology 12:475–484.

Kress, A., et al. 2001. Oogenesis in the marsupial stripe-faced dunnart, *Sminthopsis macroura*. Cells Tissues Organs 168:188–202.

Krisher, R.L. 2004. The effect of oocyte quality on development. Journal of Animal Science 32(Supplement):E14-E23.

Kristal, M.B., et al. 2012. Placentophagia in humans and nonhuman mammals: causes and consequences. Ecology of Food and Nutrition, 51:177–197.

Król, E., et al. 2012. Strong pituitary and hypothalamic responses to photoperiod but not to 6-methoxy-2-benzoxazolinone in female common voles (*Microtus arvalis*). General and Comparative Endocrinology 179:289–295.

Kuhnlein, H.V., et al. 2006. Vitamins A, D, and E in Canadian Arctic traditional food and adult diets. Journal of Food Composition and Analysis 19:495–506.

Künkele, J. 2000. Energetics of gestation relative to lactation in a precocial rodent, the guinea pig (*Cavia porcellus*). Journal of Zoology, London 250:533–539.

Künkele, J., F. Trillmich. 1997. Are precocial young cheaper? Lactation energetics in the guinea pig. Physiological Zoology 70:589–596.

Kunz, T.H., D.J. Hosken. 2009. Male lactation: why, why not and is it care? Trends in Ecology and Evolution 24:80–85.

Kunz, T.H., K.S. Orrell. 2004. Reproduction, energy costs of. Encyclopedia of Energy 5:423–442.

Kunz, T.H., et al. 1994. Allomaternal care: helper-assisted birth in the Rodrigues fruit bat, *Pteropus rodricensis* (Chiroptera: Pteropodidae). Journal of Zoology, London 232:691–700.

Kunz, T.H., et al. 1996. Assessment of sex, age, and reproductive condition in mammals. Pp. 279–290 in Measuring and monitoring biological diversity: standard methods for mammals. (Wilson, D.E., et al., eds.). Smithsonian Institution Press, Washington, DC.

Kuroiwa, A., et al. 2011. Additional copies of CBX2 in the genomes of males of mammals lacking SRY, the Amami spiny rat (*Tokudaia osimensis*) and the Tokunoshima spiny rat (*Tokudaia tokunoshimensis*). Chromosome Research 19:635–644.

Kuruppath, S., et al. 2012. Monotremes and marsupials: comparative models to better understand the function of milk. Journal of Bioscience 37:581–588.

Kurta, A., T.H. Kunz. 1987. Size of bats at birth and maternal investment during pregnancy. Symposia of the Zoological Society of London 57:79–106.

Kurta, A., et al. 1989. Energetics of pregnancy and lactation in free ranging little brown bats (*Myotis lucifugus*). Physiological Zoology 62:804–818.

Kusinski, L.C., et al. 2014. Contribution of placental genomic imprinting and identification of imprinted genes. Pp 275–284 in The guide to investigation of mouse pregnancy (B.A. Croy, et al., eds.). Elsevier, New York, NY.

Kuyper, M.A. 1985. The ecology of the golden mole, *Amblysomus hottentotus*. Mammal Review 15:3–11.

Kvadsheim, P.H., J.J. Aarseth. 2002. Thermal function of phocid seal fur. Marine Mammal Science 18:952–962.

Kwiecinski, G.G., et al. 1987. Annual skeletal changes in the little brown bat, *Myotis lucifugus lucifugus*, with particular reference to pregnancy and lactation. American Journal of Anatomy 178:410–420.

Laakkonen J., et al. 1998. Dynamics of intestinal coccidia in peak density *Microtus agrestis*, *Microtus oeconomus* and *Clethrionomys glareolus* populations in Finland. Ecography 21:135–139.

Lacey, E.A. 2004. Sociality reduces individual direct fitness in a communally breeding rodent, the colonial tuco-tuco (*Ctenomys sociabilis*). Behavioral Ecology and Sociobiology 56:449–457.

Lacey, E.A., P.W. Sherman. 1997. Cooperative breeding in naked mole-rats: implications for vertebrate and invertebrate sociality. Pp. 267–301 in Cooperative breeding in mammals (N.G. Solomon, J.A. French, eds.). Cambridge University Press, New York, NY.

Lacey, E.A., et al. 2000. Life underground. University of Chicago Press, Chicago, IL.

Lambert, R.T. 2005. A pregnancy-associated glycoprotein (PAG) unique to the roe deer (*Capreolus capreolus*) and its role in the termination of embryonic diapause and maternal recognition of pregnancy. Israel Journal of Zoology 51:1–11.

Lamming, G.E. (ed.). 1994. Marshall's physiology of reproduction. Springer, New York, NY.

Langer, P. 2002. The digestive tract and life history of small mammals. Mammal Review 32:107–131.

Langer, P. 2008. The phases of maternal investment in eutherian mammals. Zoology 111:148–162.

Langer, P. 2009. Differences in the composition of colostrum and milk in eutherians reflect differences in immunoglobulin transfer. Journal of Mammalogy 90:332–339.

Larson, M.A., et al. 2001. Sexual dimorphism among bovine embryos in their ability to make the transition to expanded blastocyst and in the expression of the signaling molecule IFN-τ. Proceedings of the National Academy of Science 98:9677–9682.

Laurance, W.F., et al. 2014. Agricultural expansion and its impacts on tropical nature. Trends in Ecology and Evolution 29:107–116.

Laws, R.M. 1959. The foetal growth rates of whales with special reference to the fin whale, *Balaenoptera physalus* Linn. Discovery Reports 29:281–308.

Lebl, K., et al. 2011. Local environmental factors affect reproductive investment in female edible dormice. Journal of Mammalogy 92:926–933.

Le Boeuf, B.J. 1972. Sexual behavior in the northern elephant seal *Mirounga angustirostris*. Behaviour 41:1–26.

Le Boeuf, B.J., S.L. Mesnick. 1990. Sexual behavior of male northern elephant seals: I. Lethal injuries to adult females. Behaviour 116:143–162.

Lee, K.Y., F.J. DeMayo. 2004. Animal models of implantation. Reproduction 128:679–695.

Lee, P.C. 1996. The meanings of weaning: growth, lactation, and life history. Evolutionary Anthropology: Issues, News, and Reviews 5:87–98.

Lee, P.C., et al. 2016. The reproductive advantages of a long life: longevity and senescence in wild female African elephants. Behavioral Ecology and Sociobiology 70:337–345.

Lee, S. van der, L.M. Boot. 1956. Spontaneous pseudopregnancy in mice II. Acta Physiologica et Pharmacologica Neerlandica 5:213–214.

Lefèvre, C.M., et al. 2010. Evolution of lactation: ancient origin and extreme adaptations of the lactation system. Annual Review of Genomics and Human Genetics 11:219–238.

Lehrman, S.R., et al. 1988. Primary structure of pituitary prolactin. International Journal of Peptide and Protein Research 21:544–554.

Leiser, R., P. Kaufmann. 1994. Placental structure: in a comparative aspect. Experimental and Clinical Endocrinology 102:122–134.

Lesse, H.J. 1988. The formation and function of oviduct fluid. Journal of Reproduction and Fertility 82:843–856.

Lesse, H.J. 2012. Metabolism of the preimplantation embryo: 40 years on. Reproduction 143:417–427.

Lesse, H.J., et al. 2001. Formation of Fallopian tubal fluid: role of a neglected epithelium. Reproduction 121:339–346.

LeTallec, T., et al. 2015. Effects of light pollution on seasonal estrus and daily rhythms in a nocturnal primate. Journal of Mammalogy 96:438–445.

Levy, N., G. Bernadsky. 1991. Creche behavior of the Nubian ibex *Capra ibex nubiana* in the Negev desert highlands, Israel. Israel Journal of Zoology 37:125–137.

Lewis, R.J., P.M. Kappeler. 2005. Are Kirindy sifaka capital or income breeders? It depends. American Journal of Primatology 67:365–369.

Licht, P., et al. 1992. Hormonal correlates of 'masculinization' in female spotted hyaenas (*Crocuta crocuta*). 2. Maternal and fetal steroids. Journal of Reproduction and Fertility 95:463–474.

Licht, P., et al. 1998. Androgens and masculinization of genitalia in the spotted hyaena (*Crocuta crocuta*). 1. Urogenital morphology and placental androgen production during fetal life. Journal of Reproduction and Fertility 113:105–116.

Lidicker, W.Z. 1973. Regulation of numbers in an island population of the California vole, a problem in community dynamics. Ecological Monographs 43:271–302.

Liggins, G.C., et al. 1973. The mechanism of initiation of parturition in the ewe. Recent Progress in Hormone Research 29:111–159.

Lilia, K., et al. 2010. Gross anatomy and ultrasonographic images of the reproductive system of the Malayan tapir (*Tapirus indicus*). Anatomia, Histologia, Embryologia 39:569–575.

Lillegraven, J.A. 1975. Biological considerations of the marsupial-placental dichotomy. Evolution 29:707–722.

Lillegraven, J.A. 1979. Reproduction in Mesozoic mammals. Pp 259–276 in Mesozoic mammals (J.A. Lillegraven, et al., eds.). University of California Press, Berkeley, CA.

Lindberg, R.H., et al. 2007. Environmental risk assessment of antibiotics in the Swedish environment with emphasis on sewage treatment plants. Water Research 41:613–619.

Linnaeus, C. 1758. Systema naturae per regna tria naturae, secundum classes, ordines, genera, species, cum characteribus, differentiis, synonymis, locis. Editio decima, reformata. Volume 1. Laurentii Salvii, Stockholm, Sweden.

Lindenfors, P., et al. 2003. The monophyletic origin of delayed implantation in carnivores and its implications. Evolution 57:1952–1956.

Lindeque, M., J.D. Skinner. 1982. A seasonal breeding in the spotted hyaena (*Crocuta crocuta*, Erxleben), in southern Africa. Africa Journal of Ecology 20:271–278.

Lindeque, M., et al. 1986. Adrenal and gonadal contribution to circulating androgens in spotted hyaenas (*Crocuta crocuta*) as revealed by LHRH, hCG and ACTH stimulation. Journal of Reproduction and Fertility 78:211–217.

Linzey, D.W., A.V. Linzey. 1967a. Growth and development of the golden mouse *Ochrotomys nuttalli nuttalli*. Journal of Mammalogy 48:445–448.

Linzey, D.W., A.V. Linzey. 1967b. Maturational and seasonal molts in the golden mouse, *Ochrotomys nuttalli*. Journal of Mammalogy 48:236–241.

Lisenjohann T., et al. 2015. State-dependent foraging: lactating voles adjust their foraging behavior according to the presence of a potential nest predator and season. Behavioral Ecology and Sociobiology 69:747–754.

Liu, H., et al. 2003. Energy requirements during reproduction in female Brandt's voles (*Microtus brandtii*). Journal of Mammalogy 84:1410–1416.

Lloyd, S., et al. 1999. Reproductive strategies of a warm temperate vespertilionid, the large-footed myotis, *Myotis moluccarum* (Microchiroptera: Vespertilionidae). Australian Journal of Zoology 47:261–274.

Lochmiller, R.L., et al. 1962. Energetic cost of lactation in *Microtus pinetorum*. Journal of Mammalogy 63:475–481.

Lombardi, J. 1994. Embryo retention and evolution of the amniote condition. Journal of Morphology 220:368.

Lombardi, J. 1998. Comparative vertebrate reproduction. Kluwer Academic Publishers, Boston, MA.

Loskutoff, N.M., et al. 1990. Reproductive anatomy, manipulation of ovarian activity and non-surgical embryo recovery in suni (*Nesotragus moschatus zuluensis*). Journal of Reproduction and Fertility 88:521–532.

Lowther, A.D, S.D. Goldsworthy. 2016. When were the weaners weaned? Identifying the onset of Australian sea lion nutritional independence. Journal of Mammalogy 97:1304–1311.

Lukas, D., T. Clutton-Brock. 2012. Cooperative breeding and monogamy in mammalian societies. Proceedings of the Royal Society B 279:2151–2156.

Lukas, D., E. Huchard. 2014. The evolution of infanticide by males in mammalian societies. Science 346:841–843.

Lummaa, V., T. Clutton-Brock. 2002. Early development, survival and reproduction in humans. Trends in Ecology and Evolution 17:141–147.

Lummaa, V., et al. 1998. Why cry? Adaptive significance of intensive crying in human infants. Evolution and Human Behavior 19:193–202.

Luo, S.-M., et al. 2013. Sperm mitochondria in reproduction: good or bad and where do they go? Journal of Genetics and Genomics 40:549–556.

Luo, Z.-X., et al. 2004. Evolution of dental replacement in mammals. Bulletin of the Carnegie Museum of Natural History 36:159–175.

Luo, Z.-X., et al. 2011. A Jurassic eutherian mammal and divergence of marsupials and placentals. Nature 476:442–445.

Lyamin, O., et al. 2005. Continuous activity in cetaceans after birth. Nature 435:1177.

Ma, W., et al. 1998. Role of the adrenal gland and adrenal-mediated chemosignals in suppression of estrus in the house mouse: the Lee-Boot effect revisited. Biology of Reproduction 59:1317–1320.

MacDonald, P.C., et al. 1978. Initiation of parturition in the human female. Seminars in Perinatology 2:273–296.

MacLean, S.F. Jr., et al. 1974. Population cycles in arctic lemmings: winter reproduction and predation by weasels. Arctic and Alpine Research 6:1–12.

MacLeod, K.J., T. H. Clutton-Brock. 2015. Low costs of allonursing in meerkats: mitigation by behavioral change? Behavioral Ecology 26:697–705.

MacLeod, K.J., D. Lukas. 2014. Revisiting non-offspring nursing: allonursing evolves when the costs are low. Biology Letters 10:20140378.

Maier, W., et al. 1996. New therapsid specimens and the origin of the secondary hard and soft palate of mammals. Journal of Zoological Systematics and Evolutionary Research 34:9–19.

Maingon, L. 2016. Comox Valley loses a tiny giant of an environmentalist. http://tidechange.ca/2016/07/20/comox-valley-loses-tiny-giant-environmentalist/.

Maklakov A.A., V. Lummaa. 2013. Evolution of sex differences in lifespan and aging: causes and constraints. BioEssays 35:717–724.

Mandalaywala, T.M., et al. 2014. Physiological and behaviour responses to weaning conflict in free-ranging primate infants. Animal Behaviour 97:241–247.

Manning, T.H. 1954. Remarks on the reproduction, sex ratio, and life expectancy of the varying lemming, *Dicrostonyx groenlandicus*, in nature and captivity. Arctic 7:36–48.

Markham, A.C., et al. 2014. Rank effects on social stress in lactating chimpanzees. Animal Behaviour 87:195–202.

Marmontel, M. 1988. The reproductive anatomy of the female manatee *Trichechus manatus latirostris* (Linnaeus 1758) based on gross and histologic observations. MS thesis, University of Miami, Coral Gables, FL.

Marsh, H., T. Kasuya. 1986. Evidence for reproductive senescence in female cetaceans. Reports of the International Whaling Commission Special Issue 8:57–74.

Marshall, C.D., J.F. Eisenberg. 1996. *Hemicentetes semispinosus*. Mammalian Species 541:1–4.

Martin, R.D. 1966. Tree shrew: unique reproductive mechanism of systematic importance. Science 152:1402–1404.

Martin, R.D. 1968. Reproduction and ontogeny in tree-shrews (*Tupaia belangeri*), with reference to their general behavior and taxonomic relationships. Zeitschrift für Tierpsychologie 25:409–495.

Martin, R.D., A.M. MacLaron. 1985. Gestation period, neonatal size and maternal investment in placental mammals. Nature 313:220–223.

Martinez-Bakker, M., et al. 2014. Human birth seasonality: latitudinal gradient and interplay with childhood disease dynamics. Proceedings of the Royal Society B 281:20132438.

Maurus, M., et al. 1965. Cerebral representation of the clitoris in ovariectomized squirrel monkeys. Experimental Neurology 13:283–288.

Maxwell, C.S., M.L. Morton. 1975. Comparative thermoregulatory capabilities of neonatal ground squirrels. Journal of Mammalogy 56:821–828.

Mayor, P., et al. 2011. Functional anatomy of the female genital organs of the wild black agouti (*Dasyprocta fuliginosa*) female in the Peruvian Amazon. Animal Reproduction Science 121:240–257.

Mayor, P., et al. 2013. Functional morphology of the genital organs in the wild paca (*Cuniculus paca*) female. Animal Reproduction Science 140:206–215.

McAllan, B., et al. 2006. Photoperiod as a reproductive cue in the marsupial genus *Antechinus*: ecological and evolutionary consequences. Biological Journal of the Linnean Society 87:365–379.

McClintock, M.K. 1981. Social control of the ovarian cycle and the function of estrous synchrony. American Zoologist 21:243–256.

McCracken, G.F., M.K. Gustin. 1991. Nursing behavior in Mexican free-tailed bat maternity colonies. Ethology 89:305–321.

McCue, P.M. 1998. Review of ovarian abnormalities in the mare. American Association of Equine Practitioners Proceedings 44:125–133.

McEntee, K. 1990. Reproductive pathology of domestic mammals. Academic Press, New York, NY.

McGuire, B., S. Sullivan. 2001. Suckling behavior of pine voles (*Microtus pinetorum*). Journal of Mammalogy 82:690–699.

McLaren, A. 2003. Primordial germ cells in the mouse. Developmental Biology 262:1–15.

McLay, D.W., H.J. Clarke. 2003. Remodelling the paternal chromatin at fertilization in mammals. Reproduction 125:625–633.

McNab, B.K. 1986. The influence of food habits on the energetics of eutherian mammals. Ecological Monographs 56:1–19.

McPherson, F.J., P.J. Chenoweth. 2012. Mammalian sexual dimorphism. Animal Reproduction Science 131:109–122.

Mehrer, C.F. 1976. Gestation period in the wolverine, *Gulo gulo*. Journal of Mammalogy 57:570.

Menkhorst, E., L. Selwood. 2008. Vertebrate extracellular preovulatory and postovulatory egg coats. Biology of Reproduction 79:790–797.

Menkhorst, E., et al. 2009. Evolution of the shell coat and yolk in amniotes: a marsupial perspective. Journal of Experimental Zoology 312B:625–638.

Mihm, M., et al. 2011. The normal menstrual cycle in women. Animal Reproduction Science 124:229–236.

Millar, J.A. 1978. Energetics of reproduction in *Peromyscus leucopus*: the cost of lactation. Ecology 59:1055–1061.

Millar, J.S. 2001. Reproduction in lemmings. Ecoscience 8:145–150.

Miller, A.M. 2011. The effect of motherhood timing on career path. Journal of Population Economics 24:1071–1100.

Miller, D.L., et al. 2007. Placental structure and comments on gestational ultrasonographic examination. Pp. 331–348 in Reproductive biology and phylogeny of Cetacea (D.L. Miller, ed.). Science Publishers, Enfield, NH.

Miller, G., et al. 2007. Ovulatory cycle effects on tip earnings by lap dancers: economic evidence for human estrus? Evolution and Human Behavior 28:375–381.

Miller-Ben-Shaul, D. 1963. Short-tailed shrews (*Blarina brevicauda*) in captivity. International Zoo Yearbook 4:121–123.

Mills, M.G.L. 1982. *Hyaena brunnea*. Mammalian Species 194:1–5.

Mills, M.G.L. 1983. Mating and denning behavior of the brown hyaena *Hyaena brunnea* and comparisons with other Hyaenidae. Zeitschrift für Tierpsychologie 63:331–342.

Milner, J., et al. 1990. *Nyctinomops macrotis*. Mammalian Species 351:1–4.

Mira, A. 1998. Why is meiosis arrested? Journal of Theoretical Biology 194:275–287.

Moehlman, P.D., Hofer, H. 1997. Cooperative breeding, reproductive suppression, and body mass in canids. Pp. 76–127 in Cooperative breeding in mammals (N.G. Solomon, J.A. French, eds.). Cambridge University Press, Cambridge, England.

Montgomery, G.W., et al. 2014. The future for genetic studies in reproduction. Molecular Human Reproduction 20:1–14.

Mor, G., I. Cardenas. 2010. The immune system in pregnancy: a unique complexity. American Journal of Immunology 63:425–433.

Moran, S., et al. 2009. Multiple paternity in the European hedgehog. Journal of Zoology, London 278:349–353.

Morandini, V., M. Ferrer. 2015. Sibling aggression and brood reduction: a review. Ethology Ecology and Evolution 27:2–16.

Morbeck, M.E., et al. (eds.). 1997. The evolving female: a life history perspective. Princeton University Press, Princeton, NJ.

Morrison, D.W. 1978. Foraging ecology and energetics of the frugivorous bat *Artibeus jamaicensis*. Ecology 59:716–723.

Morrow, G., et al. 2009. Reproductive strategies of the short-beaked echidna—a review with new data from a long-term study on the Tasmanian subspecies (*Tachyglossus aculeatus setosus*). Australian Journal of Zoology 57:275–282.

Mossman, H.W., K.L. Duke. 1973. Comparative morphology of the mammalian ovary. University of Wisconsin Press, Madison, WI.

Mossman, H.W., I. Judas. 1949. Accessory corpora lutea, lutein cell origin, and the ovarian cycle in the Canadian porcupine. American Journal of Anatomy 85:1–39.

Mtango, N.R., et al. 2008. Oocyte quality and maternal control of development. International Review of Cell and Molecular Biology 258:223–290.

Mueller, N. 2015. The infant microbiome development: mom matters. Trends in Molecular Medicine 21:109–117.

Mullen, D.A. 1968. Reproduction in brown lemmings (*Lemmus trimucronatus*) and its relevance to their cycle of abundance. University of California Publications in Zoology 85:1–24.

Musser, G.C., M.D. Carleton. 2005. Superfamily Muroidea. Pp. 894–1532 in Mammal species of the world, 3rd edition (D.E. Wilson, D.M. Reeder, eds.). Johns Hopkins University Press, Baltimore, MD.

Myatt, L., K. Sun. 2010. Role of fetal membranes in signaling of fetal maturation and parturition. International Journal of Developmental Biology 54:545–553.

Mysorekar, I.U., B. Cao. 2014. Microbiome in parturition and preterm birth. Seminars in Reproductive Medicine 32:50–55.

Naaktgeboren, C. 1979. Behavior aspects of parturition. Animal Reproduction Science 2:155–166.

Nagy, T.R., et al. 1995. Endocrine correlates of seasonal body mass dynamics in the collared lemming (*Dicrostonyx groenlandicus*). American Zoologist 35:246–258.

Nathanielsz, P.W. 1978. Parturition in rodents. Seminars in Perinatology 2:223–234.

Neaves, W.B., et al. 1980. Sexual dimorphism of the phallus in spotted hyaena (*Crocuta crocuta*). Journal of Reproduction and Fertility 59:509–513.

Negus, N.C., P.J. Berger. 1977. Experimental triggering of reproduction in a natural population of *Microtus montanus*. Science 196:1230–1231.

Negus, N.C., P.J. Berger. 1998. Reproductive strategies of *Dicrostonyx groenlandicus* and *Lemmus sibiricus* in high-arctic tundra. Canadian Journal of Zoology 76:390–399.

Neill, J.D. (ed.). 2006. Knobil and Neill's physiology of reproduction, 3rd edition. Elsevier, New York, NY.

Nelson, R. J. 1987. Photoperiod-nonresponsive morphs: a possible variable in microtine population-density fluctuations. American Naturalist 130:350–369.

Nelson, R.J. 2011. An introduction to behavioral neuroendocrinology. Sinauer, Sunderland, MA.

Neuhaus P. 2003. Parasite removal and its impact on litter size and body condition in Columbian ground squirrels (*Spermophilus columbianus*). Biology Letters 270:S213–S215.

Nicolás, L., et al. 2011. Littermate presence enhances motor development, weight gain and competitive ability in newborn and juvenile domestic rabbits. Developmental Psychobiology 53:37–46.

Nicoll, M.E., P.A. Racey. 1985. Follicular development, ovulation, fertilization and fetal development in tenrecs (*Tenrec ecaudatus*). Journal of Reproduction and Fertility 74:47–55.

Nishiwaki, M., H. Marsh. 1985. Dugong. Handbook of Marine Mammals 3:1–31.

Nitsch, A., et al. 2016. Sibship effects on dispersal behaviour in a pre-industrial human population. Journal of Evolutionary Biology 29:1986–1998.

Nixon, B., et al. 2011. Understanding the evolutionary significance of epididymal sperm maturation. Journal of Andrology 32:665–671.

Nordstrom, C.A., et al. 2013. Foraging habitats of lactating northern fur seals are structured by thermocline depths and submesoscale fronts in the eastern Bering Sea. Deep-Sea Research II 88–89:78–96.

Nowak, R.A., J.M. Bahr. 1983. Maternal recognition of pregnancy in the rabbit. Journal of Reproduction and Fertility 69:623–627.

Numan, M., T.R. Insel. 2003. The neurobiology of parental behavior. Springer, New York, NY.

O'Donoghue, P.N. 1963. Reproduction in the female hyrax (*Dendrohyrax arborea ruwenzorii*). Proceedings of the Zoological Society of London 141:207–237.

Oftedal, O.T. 1997. Lactation in whales and dolphins: evidence of divergence between baleen- and toothed-species. Journal of Mammary Gland Biology and Neoplasia 2:205–230.

Oftedal, O.T. 2000. Use of maternal reserves as a lactation strategy in large mammals. Proceedings of the Nutrition Society 59:99–106.

Oftedal, O.T. 2013. Origin and evolution of the major constituents of milk. Pp. 1–42 in Advanced dairy chemistry. Volume 1A: proteins: basic aspects, 4th edition (P.L.H. McSweeney, P.F. Fox, eds.). Springer Science, New York, NY.

Oftedal, O.T., et al. 1987. The behavior, physiology, and anatomy of lactation in the Pinnipedia. Pp. 175–245 in Current mammalogy, volume 1 (H.H. Genoways, ed.). Plenum, New York, NY.

Oftedal, O.T., et al. 2014. Can an ancestral condition for milk oligosaccharides be determined? Evidence from the Tasmanian echidna (*Tachyglossus aculeatus setosus*). Glycobiology 24:826–839.

O'Gara, B.W. 1969. Unique aspects of reproduction in the female pronghorn (*Antilocapra americana* Ord). American Journal of Anatomy 125:217–231.

Olcese, J. 2012. Circadian aspects of mammalian parturition: a review. Molecular and Cellular Endocrinology 349:62–67.

Oliveira, S.F., et al. 2000. Advanced oviductal development, transport to the preferred implantation site, and attachment of the blastocyst in captive-bred, short-tailed fruit bats, *Carollia perspicillata*. Anatomy and Embryology 201:357–381.

Olsson, K. 1986. Pregnancy—a challenge to water balance. Physiology 1:131–134.

Opie, C., et al. 2013. Male infanticide leads to social monogamy in primates. Proceedings of the National Academy of Sciences 110:13328–13332, plus supplemental data.

Orr, T.J., P.L. Brennan. 2015. Sperm storage: distinguishing selective processes and evaluating criteria. Trends in Ecology and Evolution 30:261–272.

Orr, T.J., T. Garland, Jr. 2017. Complex reproductive traits and whole-organism performance. Journal of Integrative and Comparative Biology, in press.

Orr, T.J., M. Zuk. 2014. Reproductive delays in mammals: an unexplored avenue for postcopulatory sexual selection. Biological Reviews 89:889–912.

Orr, T. J., et al. 2016. Diet choice in frugivorous bats: gourmets or operational pragmatists? Journal of Mammalogy 97:1578–1588.

Osada, T., et al. Puromycin-sensitive aminopeptidase is essential for the maternal recognition of pregnancy in mice. Molecular Endocrinology 15:882–893.

Pachkowski, M., et al. 2013. Spring-loaded reproduction: effects of body condition and population size on fertility in migratory caribou (*Rangifer tarandus*). Canadian Journal of Zoology 91:473–479.

Packer, C., et al. 1992. A comparative analysis of non-offspring nursing. Animal Behaviour 43:265–281.

Padilla, M., et al. 2010. *Tapirus pinchaque* (Perissodactyla: Tapiridae). Mammalian Species 42:166–182.

Padula, A.M. 2005. The freemartin syndrome: an update. Animal Reproduction Science 87:93–109.

Padykula, H.A., J.M. Taylor. 1982. Marsupial placentation and its evolutionary significance. Journal of Reproduction and Fertility 31(Supplement):95–104.

Pan, H., et al. 2005. Transcript profiling during mouse oocyte development and the effect of gonadotropin priming and development in vitro. Developmental Biology 286:493–506.

Pangas, S.A., A. Rajkovic. 2015. Follicular development: mouse, sheep, and human model. Pp. 947–995 in Knobil and Neill's physiology of reproduction (T.M. Plant, et al., eds.). Academic Press New York, NY.

Pangle, W.M., K.E. Holekamp. 2010. Lethal and nonlethal anthropogenic effects on spotted hyenas in the Masai Mara National Reserve. Journal of Mammalogy 91:154–164.

Papaioannou, G.I., et al. 2010. Normal ranges of embryonic length, embryonic heart rate, gestational sac diameter and yolk sac diameter at 6–10 weeks. Fetal Diagnosis and Therapy 28:207–219.

Parga, J.A. 2003. Copulatory plug displacement evidences sperm competition in *Lemur catta*. International Journal of Primatology 24:889–899.

Parker, G.A., T.R. Birkhead. 2013. Polyandry: the history of a revolution. Philosophical Transactions of the Royal Society B 368:20120335.

Parker, K.L., et al. 1990. Comparison of energy metabolism in relation to daily activity and milk consumption by caribou and muskox neonates. Canadian Journal of Zoology 68:106–114.

Parkes, A.S. 1977. H.M. Bruce. Journal of Reproduction and Fertility 49:1–4.

Patterson, B.D. 2007. On the nature and significance of variability in lions (*Panthera leo*). Evolutionary Biology 34:55–60.

Pellati, D., et al. 2008. Genital tract infections and infertility. European Journal of Obstetrics and Gynecology and Reproductive Biology 140:3–11.

Pennisi, E. 2004. The birth of the nucleus. Science 305:766–768.

Perry, J.S. 1964. The structure and development of the reproductive organs of the female African elephant. Philosophical Transactions of the Royal Society of London, B 248:35–51.

Perry, J.S., I.W. Rowlands. 1962. Early pregnancy in the pig. Journal of Reproduction and Fertility 4:175–188.

Perryman, W.L., et al. 2002. Gray whale calf production 1994–2000: are observed fluctuations related to changes in seasonal ice cover? Marine Mammal Science 18:121–144.

Perven, H.A., et al. 2014. A postmortem study on the weight of the human ovary. Medicine Today 26:12–14.

Petraglia, F., et al. 2010. Neuroendocrine mechanisms in pregnancy and parturition. Endocrine Reviews 31:783–816.

Petter-Rousseaux, A., F. Bourlière, 1965. Persistence des phénomènes d'ovogénèse chez l'adulte de *Daubentonia madagascariensis* (Prosimii, Lemuriformes). Folia Primatologia 3:241–244.

Philips, S.S. 2000. Population trends and the koala conservation debate. Conservation Biology 14:650–659.

Phoenix, C.H., et al. 1959. Organizing action of prenatally administered testosterone propionate on the tissues mediating mating behavior in the female guinea pig. Endocrinology 65:369–382.

Pielmeier, K.G., et al. submitted. Reproductive investment in canids and leporids: influences of ecology and neonatal development.

Pierson, R.A., et al. 2003. Ortho EVRA™/ EVRA™ versus oral contraceptives: follicular development and ovulation in normal cycles and after an intentional dosing error. Fertility and Sterility 80:34–42.

Pioz, M., et al. 2008. Diseases and reproductive success in a wild mammal: example in the alpine chamois. Oecologia 155:691–704.

Place, N.J., et al. 2011. The anti-androgen combination, flutamide plus finasteride, paradoxically suppressed LH and androgen concentrations in pregnant spotted hyenas, but not in males. General and Comparative Endocrinology 170:455–459.

Plant, T.M., A.J. Zeleznik (eds.). 2015. Knobil and Neill's physiology of reproduction, 4th edition. Academic Press, Waltham, MA.

Plard, F., et. al. 2013. Parturition date for a given female is highly repeatable with five roe deer populations. Biology Letters 9:20120841.

Plön, S., R.T.F. Bernard. 2007. Anatomy with particular reference to the female. Pp. 147–169 in Reproductive biology and phylogeny of Cetacea (D.L. Miller, ed.). Science Publishers, Enfield, NH.

Pocock, R.I. 1924. Some external characters of *Orycteropus afer*. Proceedings of the Zoological Society of London 94:697–706.

Poiani, A. 2006. Complexity of seminal fluid: a review. Behavioral Ecology and Sociobiology 60:289–310.

Pond, C.M. 1977. The significance of lactation in the evolution of mammals. Evolution 31:177–199.

Pond, C.M. 2012. The evolution of mammalian adipose tissue. Pp. 227–269 in Adipose tissue biology (M.E. Symonds, ed.). Springer, New York, NY.

Pope, C.E. 2000. Embryo technology in conservation efforts for endangered felids. Theriogenology 53:163–174.

Pope, C.E., et al. 2006. In vitro embryo production and embryo transfer in domestic and non-domestic cats. Theriogenology 66:1518–1524.

Pope, C.E., et al. 2012. In vivo survival of domestic cat oocytes after vitrification, intracytoplasmic sperm injection and embryo transfer. Theriogenology 77: 531–538.

Poppitt, S.D., et al. 1994. Energetics of reproduction in the lesser hedgehog tenrec, *Echinops telfarir* (Martin). Physiological Zoology 67:976–994.

Pournelle, G.H. 1965. Observations on birth and early development of the spotted hyena. Journal of Mammalogy 46:503.

Power, M.L., J. Schulkin. 2012. The evolution of the human placenta. Johns Hopkins University Press, Baltimore, MD..

Pratt, D.M., V.H. Anderson. 1979. Giraffe cow-calf relationships and social development of the calf in the Serengeti. Zeitschrift für Tierpsychologie 51:233–251.

Prince, A.L., et al. 2014a. The microbiome, parturition, and timing of birth: more questions than answers. Journal of Reproductive Immunology 104–105:12–19.

Prince, A.L., et al. 2014b. The microbiome and development: a mother's perspective. Seminars in Reproductive Medicine 32:14–22.

Prince, A.L., et al. 2015. The perinatal microbiome and pregnancy: moving beyond the vaginal microbiome. Cold Spring Harbor Perspectives in Medicine 16:1–14.

Profet, M. 1993. Menstruation as a defense against pathogens transported by sperm. Quarterly Review of Biology 68:335–386.

Promislow, D.E.L., P.H. Harvey. 1990. Living fast and dying young: a comparative analysis of life-history variation among mammals. Journal of Zoology, London 220:417–437.

Prugh, L.R., C.D. Golden. 2014. Does moonlight increase predation risk? meta-analysis reveals divergent responses of nocturnal mammals to lunar cycles. Journal of Animal Ecology 83:504–514.

Puente, A.E., D.A. Dewsbury. 1976. Courtship and copulatory behavior of bottlenosed dolphins (*Tursiops truncatus*). Cetology 21:1–9.

Puget, A., C. Gouarderes. 1974. Weight gain of the Afghan pika (*Ochotona rufescens rufescens*) from birth to 19 weeks of age, and during gestation. Growth 38:117–129.

Purvis, A., P.H. Harvey. 1995. Mammal life-history evolution: a comparative test of Charnov's model. Journal of Zoology, London 237:259–283.

Racey, D.N., et al. 2009. Galactorrhoea is not lactation. Trends in Ecology and Evolution 24:354–355.

Racey, P.A., A.C. Entwistle. 2000. Life-history and reproductive strategies of bats. Pp. 363–414 in Reproductive biology of bats (E.G. Crichton, P.H. Krutzsch, eds.). Academic Press, London, England.

Racey, P.A., J.D. Skinner. 1979. Endocrine aspects of sexual mimicry in spotted hyaenas *Crocuta crocuta*. Journal of Zoology, London 187:315–326.

Rachlow, J.L., et al. 2005. Natal burrows and nests of free-ranging pygmy rabbits (*Brachylagus idahoensis*). Western North American Naturalist 65:136–139.

Racicot, K., et al. 2014. Understanding the complexity of the immune system during pregnancy. American Journal of Reproductive Immunology 72:107–116.

Ralls, K., et al. 1986. Mother-young relationships in captive ungulates: variability and clustering. Animal Behaviour 34:134–145.

Ramsay, M.A., R.L. Dunbrack. 1986. Physiological constraints on life history phenomena: the example of small bear cubs at birth. American Naturalist 127:735–743.

Rasmussen, J.L., R.L. Tilson. 1984. Food provisioning by adult maned wolves (*Chrysocyon brachyurus*). Zeitschrift für Tierpsychologie 65:346–352.

Rasmussen, L.E., B.A. Schulte. 1998. Chemical signals in the reproduction of Asian (*Elephas maximus*) and African (*Loxodonta africana*) elephants. Animal Reproduction Science 53:19–34.

Rasweiler, J.J., IV, et al. 2000. Anatomy and physiology of the female reproductive tract. Pp. 157–219 in Reproductive biology of bats (E.G. Crichton, P.H. Krutzsch, eds.). Academic Press, New York, NY.

Rasweiler, J.J., IV, et al. 2011. Ovulation, fertilization and early embryonic development in the menstruating fruit bat, *Carollia perspicillata*. Anatomical Record 294:506–519.

Ratto, M.H., et al. 2005. Local versus systemic effect of ovulation-inducing factor in the seminal plasma of alpacas. Reproductive Biology and Endocrinology 3:29 (5 pages, not paginated).

Réale, D., et al. 1996. Female-based mortality induced by male sexual harassment in a feral sheep population. Canadian Journal of Zoology 74:1812–1818.

Reddy, M.L., et al. 2001. Opportunities for using Navy marine mammals to explore associations between organochlorine contaminants and unfavorable effects on reproduction. Science of the Total Environment 274:171–182.

Reidman, M.L. 1982. The evolution of alloparental care and adoption in mammals and birds. Quarterly Review of Biology 57:405–435.

Reijnders, P.J.H., et al. 2010. Earlier pupping in harbor seals, *Phoca vitulina*. Biology Letters 6:854–857.

Rekwot, P.I., et al. 2001. The role of pheromones and biostimulation in animal reproduction. Animal Reproduction Science 65:157–170.

Rendell, L., H. Whitehead. 2001. Culture in whales and dolphins. Behavioral and Brain Sciences 24:309–382.

Renfree, M.B. 2006. Life in the pouch: womb with a view. Reproduction, Fertility and Development 18:721–734.

Renfree, M.B. 2010. Marsupials: placental mammals with a difference. Placenta 31, Supplement A, Trophoblast Research 24:S21–S26.

Renfree, M.B., G. Shaw. 2000. Diapause. Annual Reviews of Physiology 62:353–375.

Renfree, M.B., et al. 2013. The origin and evolution of genomic imprinting and viviparity in mammals. Philosophical Transactions of the Royal Society B 368:20120151.

Revel, A., et al. 2009. At what age can human oocytes be obtained? Fertility and Sterility 92:458–463.

Reynolds, J.E., et al. 2005. Marine mammal research: conservation beyond crisis. Johns Hopkins University Press, Baltimore, MD.

Richards, J.S., et al. 2015. Ovulation. Pp. 997–1021 in Knobil and Neill's physiology of reproduction (T.M. Plant, et al., eds.). Academic Press, New York, NY.

Richardson, B.E., R. Lehmann. 2010. Mechanisms guiding primordial germ cell migration: strategies from different organisms. Nature Reviews 11:37–49.

Riedelsheimer, B., et al. 2007. Histological study of the cloacal region and associated structures in the hedgehog tenrec *Echinops telfairi*. Zeitschrift für Säugetierkunde 72:330–341.

Riffell, J.A., et al. 2008. Physical processes and real-time chemical measurement of the insect olfactory environment. Journal of Chemical Ecology 34:837–853.

Rinkenberger, J.L., et al. 1997. Molecular genetics of implantation in the mouse. Developmental Genetics 21:6–20.

Rismiller, P.D., M.W. McKelvey. 2009. Activity and behaviour of lactating echidnas (*Tachyglossus aculeatus multiaculeatus*) from hatching of egg to weaning of young. Australian Journal of Zoology 57:265–273.

Robert, K.A., et al. 2015. Artificial light at night desynchronizes strictly seasonal reproduction in a wild mammal. Proceedings of the Royal Society B 282(1816):20151745.

Robbins, C.T., et al. 2012. Maternal condition determines birth date and growth of newborn bear cubs. Journal of Mammalogy 93:540–546.

Robbins, J.R., A.I. Bakardjiev. 2012. Pathogens and the placental fortress. Current Opinions in Microbiology 15:36–43.

Robeck, T.R., et al. 2004. Reproductive physiology and development of artificial insemination technology in killer whales (*Orcinus orca*). Biology of Reproduction 71:650–660.

Roff, D.A. 2002. Life history evolution. Sinauer, Amherst, MA.

Romero, T., F. Aureli. 2008. Reciprocity of support in coatis (*Nasua nasua*). Comparative Psychology 122:19–25.

Ross, P.S., et al. 2000. High PCB concentrations in free ranging pacific killer whales, *Orcinus orca*: effects of age, sex and dietary preference. Marine Pollution Bulletin 40:504–515.

Rossi, L.F., et al. 2011. Female reproductive tract of the lesser anteater (*Tamandua tetradactyla*, Myrmecophagidae, Xenarthra): anatomy and histology. Journal of Morphology 272:1307–1313.

Rothchild, I. 2002. The yolkless egg and the evolution of eutherian viviparity. Biology of Reproduction 68:337–357.

Rubio-Casillas, A., E.A. Jannini. 2011. New insights from one case of female ejaculation. Journal of Sex Medicine 8:3500–3504.

Russell, A.F., V. Lummaa. 2009. Maternal effects in cooperative breeders: from hymenopterans to humans. Philosophical Transactions of the Royal Society B 364:143–1167.

Rutberg, A.T. 1987. Adaptive hypotheses of birth synchrony in ruminants: an interspecific test. American Naturalist 130:692–710.

Rutland, A.T. 2013. Managing wildlife with contraception: why is it taking so long? Journal of Zoo and Wildlife Medicine 44:S38–S46.

Sadleir, R.M.F.S. 1969. The ecology of reproduction in wild and domestic mammals. Methuen, London, England.

Sadlier, R.M.F.S. 1982. Energy consumption and subsequent partitioning in lactating black-tailed deer. Canadian Journal of Zoology 60:382–286.

Saito, C. 1998. Cost of lactation in the Malagasy primate *Propithecus verreauxi*: estimates of energy intake in the wild. Folio Primatologica 69(Supplement 1):414.

Sale, M.G., et al. 2013. Multiple paternity in the swamp Antechinus (*Antechinus minimus*). Australian Mammalogy 35:227–230.

Sánchez-Villagra, M.R. 2010. Developmental palaeontology in synapsids: the fossil record of ontogeny in mammals and their closest relatives. Proceedings of the Royal Society B 277:1139–1147.

Santos, F.C.A., et al. 2003. Structure, histochemistry, and ultrastructure of the epithelium and stroma in the gerbil (*Meriones unguiculatus*) female prostate. Tissue and Cell 35:447–457.

Santos, F.C.[A.], et al. 2006. Testosterone stimulates growth and secretory activity of the female prostate in the adult gerbil (*Meriones unguiculatus*). Biology of Reproduction 75:370–379.

Sarmah, A.K., et al. 2006. A global perspective on the use, sales, exposure pathways, occurrence, fate and effects of veterinary antibiotics (VAs) in the environment. Chemosphere 65:725–759.

Scaramuzzi, R.J., et al. 2011. Regulation of folliculogenesis and the determination of ovulation rate in ruminants. Reproduction, Fertility and Development 23:444–467.

Schareff, C.M. 2007. Anatomy with particular reference to the female. Pp. 349–370 in Reproductive biology and phylogeny of Cetacea (D. L. Miller, ed.). Science Publishers, Enfield, NH.

Schatten, G., H. Schatten. 1983. The energetic egg. Sciences 23:28–34.

Schiebinger, L. 1993. Why mammals are called mammals: gender politics in eighteenth-century natural history. American Historical Review 98:382–411.

Schmerler, S., G.M. Wessel. 2011. Polar bodies—more a lack of understanding than a lack of respect. Molecular Reproduction and Development 78:3–8.

Schmidt-Nielsen, K. 1984. Scaling: why is animal size so important? Cambridge University Press, Cambridge, England.

Schulz, T.M., W.D. Bowen. 2005. The evolution of lactation strategies in pinnipeds: a phylogenetic analysis. Ecological Monographs 75:159–177.

Schwanz, L.E. 2008. Chronic parasitic infection alters reproductive output in deer mice. Behavioral Ecology and Sociobiology 62:1351–1358.

Secor, W.E., et al. 2014. Neglected parasitic infections in the United States: trichomoniasis. American Journal of Tropical Medicine and Hygiene 90:800–804.

Seebacher, F., C.E. Franklin 2012. Determining environmental causes of biological effects: the need for a mechanistic physiological dimension in conservation biology. Philosophical Transactions of the Royal Society of London B 367:1607–1614.

Semb-Johansson, A., et al. 1993. Reproduction, litter size and survival in a laboratory strain of the Norwegian lemming (*Lemmus lemmus*). Pp. 329–337 in The biology of lemmings (N.C. Stenseth, R.A. Ims, eds). Academic Press, New York, NY.

Şenayli, A. 2011. Controversies on clitoroplasty. Therapeutic Advances in Urology 3:273–277.

Shah, G.M., et al. 2009. Observations on antifertility and abortifacient herbal drugs. African Journal of Biotechnology 8:1959–1984.

Sharman, G.B. 1976. Evolution of viviparity in mammals. Pp. 32–70 in Reproduction in mammals. Book 6: the evolution of reproduction (C.R. Austin, R.V. Short, eds.) Cambridge University Press, Cambridge, England.

Sharp, J.A., et al. 2011. Milk of monotremes and marsupials. Pp. 553–562 in Encyclopedia of dairy sciences, volume 1–2, 2nd edition (J.W. Fuquay, et al., eds.) Academic Press, New York, NY.

Sharp, S.P., T.H. Clutton-Brock. 2010. Reproductive senescence in a cooperatively breeding mammal. Journal of Animal Ecology 79:176–183.

Shaw, E., J. Darling. 1985. Female strategies. Walker, New York, NY.

Sherman, I.W. 2007. Twelve diseases that changed our world. American Society for Microbiology, Washington, DC.

Sherman, P.W. 1981. Kinship, demography, and Belding's ground squirrels. Behavioral Ecology and Sociobiology 8:251–259.

Sherman, P.W. 1985. Alarm calls of Belding's ground squirrels to aerial predators: nepotism or self-preservation? Behavioral Ecology and Sociobiology 17.4:313–323.

Sherman, P.W., J.U.M. Jarvis. 2002. Extraordinary life spans of naked mole-rats (*Heterocephalus glaber*). Journal of Zoology, London 258:307–311.

Sherman, P.W., et al. 1999. Litter sizes and mammary numbers of naked mole-rats: breaking the one-half rule. Journal of Mammalogy 80:720–733.

Sheriff, M.J., et al. 2010. The ghosts of predators past: population cycles and the role of maternal programming under fluctuating predation risk. Ecology 91:2983–2994.

Shindo, J., et al. 2008. Morphology of the tongue in a newborn Stejneger's beaked whale (*Mesoplodon stejnegeri*). Okajimas Folia Anatomica Japan 84:121–124.

Shome, B., A.F. Parlow. 1977. Human pituitary prolactin (hPRL): the entire linear amino acid sequence. Journal of Clinical Endocrinology and Metabolism 45:1112–1115.

Shoop, W.L., et al. 2002. Transmammary transmission of *Strongyloides stercoralis* in dogs. Journal of Parasitology 88:536–539.

Sikes, R.S. 1995. Costs of lactation and optimal litter size in northern grasshopper mice (*Onychomys leucogaster*). Journal of Mammalogy 76:348–357.

Silk, J.B. 2007a. The adaptive value of sociality in mammalian groups. Philosophical Transactions of the Royal Society B 362:539–559.

Silk, J.B. 2007b. Social components of fitness in primate groups. Science 317:1347–1351.

Silk, J.B., et al. 2004. Patterns of coalition formation by adult female baboons in Amboseli, Kenya. Animal Behaviour 67:573–582.

Sillén-Tullberg, B., A.P. Møller. 1993. The relationship between concealed ovulation and mating systems in anthropoid primates: a phylogenetic analysis. American Naturalist 141:1–25.

Silva, M., et al. 2014. Ovulation-inducing factor (OIF/NGF) from seminal plasma origin enhances corpus luteum function in llamas regardless [sic] the preovulatory follicle diameter. Animal Reproduction Science 148:221–227.

Simmons, N.B. 1993. Morphology, function, and phylogenetic significance of pubic nipples in bats (Mammalia: Chiroptera). American Museum Novitates 3077:1–37.

Skibiel, A.L., et al. 2013. The evolution of the nutrient composition of mammalian milks. Journal of Animal Ecology 82:1254–1264.

Slijper, E.J. 1966. Functional morphology of the reproductive system in Cetacea. Pp. 277–318 in Whales, dolphins, and porpoises (K.S. Norris, ed.). University of California Press, Berkeley, CA.

Smale, L., K.E. Holekamp. 1993. Growing up in the clan. Natural History 102(1):43–45.

Smale, L., et al. 1993. Ontogeny of dominance in free-living spotted hyaenas: juvenile rank relations with adult females and immigrant males. Animal Behaviour 46:467–477.

Smale, L., et al. 1995. Competition and cooperation between litter-mates in the spotted hyaena, *Crocuta crocuta*. Animal Behaviour 50:671–682.

Smale, L., et al. 1999. Siblicide revisited in the spotted hyaena: does it conform to obligate of facultative models? Animal Behaviour 58:545–551.

Smale, L., et al. 2005. Behavioral neuroendocrinology in nontraditional species of mammals: things the "knockout" mouse CAN'T tell us. Hormones and Behavior 48:474–483.

Smith, A.T. 1988. Patterns of pika (genus *Ochotona*) life history variation. Pp. 233–256 in Evolution of life histories of mammals: theory and pattern (M.S. Boyce, ed.). Yale University Press, New Haven, CT.

Smith, J.E., et al. 2008. Social and ecological determinants of fission-fusion dynamics in the spotted hyaena. Animal Behaviour 76:619–636.

Smith, J.E., et al. 2010. Evolutionary forces favoring intragroup coalitions among spotted hyenas and other animals. Behavioral Ecology 21:284–303.

Smith, K.K. 2001. Early development of the neural plate, neural crest and facial region of marsupials. Journal of Anatomy 199:121–131.

Smuts, B.B., R.W. Smuts. 1993. Male aggression and sexual coercion of females in nonhuman primates and other mammals: evidence and theoretical implications. Advances in the Study of Behavior 22:1–63.

Södersten, P., et al. 2006. Psychoneuroendocrinology of anorexia nervosa. Psychoneuroendocrinology 31:1149–1153.

Songsasen, N., et al. 2006. Patterns of fecal gonadal hormone metabolites in the maned wolf (*Chrysocyon brachyurus*). Theriogenology 66:1743–1750.

Sowls, L.K. 1966. Reproduction in the collared peccary (*Tayassu tajacu*). Symposia of the Zoological Society of London 15:155–172.

Spady, T.J., et al. 2007. Evolution of reproductive seasonality in bears. Mammal Review 37:21–53.

Speakman, J.R. 2013. Measuring energy metabolism in the mouse—theoretical, practical, and analytical considerations. Frontiers in Physiology 4:34.

Speakman, J.R., E. Krol. 2010. Maximal heat dissipation capacity and hyperthermia risk: neglected key factors in the ecology of endotherms. Journal of Animal Ecology 79:726–746.

Speakman, J.R., E. Krol. 2011. Limits to sustained energy intake. 13. Recent progress and future perspectives. Journal of Experimental Biology 214:230–241.

Staedler, M., M. Riedman. 1993. Fatal mating injuries in female sea otters (*Enhydra lutris nereis*). Mammalia 57:135–139.

Stallman, R.R., A.H. Harcourt. 2006. Size matters: the (negative) allometry of copulatory duration in mammals. Biological Journal of the Linnean Society 87:185–193.

Stearns, S.C. 1993. The evolution of life histories. Oxford University Press, Oxford, England.

Stegmaier, T., et al. 2009. Bionics in textiles: flexible and translucent thermal insulations for solar thermal applications. Philosophical Transactions of the Royal Society A 367:1749–1758.

Stenseth, N.C., R.A. Ims. 1993. The evolutionary history and distribution of lemmings—an introduction. Pp. 37–43 in The biology of lemmings (N.C. Stenseth, R.A. Ims, eds.). Academic Press, New York, NY.

Stenseth, N.C., et al. 1997. Population regulation in snowshoe hare and Canadian lynx: asymmetric food web configurations between hare and lynx. Proceedings of the National Academy of Science 94:5147–5152.

Stensland, E., et al. 2003. Mixed species groups in mammals. Mammal Review 33: 205–223.

Stephenson, P.J., P.A. Racey. 1995. Resting metabolic rate and reproduction in the Insectivora. Comparative Biochemistry and Physiology 112A:215–223.

Sterck, E.H.M., et al. 1997. The evolution of female social relationships in nonhuman primates. Behavioral Ecology and Sociobiology 41:291–309.

Stevens, N.M. 1905. Studies in spermatogenesis with especial reference to the accessory chromosome. Carnegie Institution of Washington 36:1–74.

Stevenson, M.F. 1976. Birth and perinatal behaviour in family groups of the common marmoset (*Callithrix jacchus jacchus*), compared to other primates. Journal of Human Evolution 5:365–381.

Stevenson, T.J., et al. 2015. Disrupted seasonal biology impacts health, food security and ecosystems. Proceedings of the Royal Society B 282:20151453.

Stockley, P. 2003. Female multiple mating behaviour, early reproductive failure and litter size variation in mammals. Proceedings of the Royal Society B 270:271–278.

Stockley, P., et al. 2002. Female multiple mating behaviour in the common shrew as a strategy to reduce inbreeding. Proceedings of the Royal Society B 254:173–170.

Story, H.E. 1945. The external genitalia and perfume gland in *Arctictis binturong*. Journal of Mammalogy 26:64–66.

Strassmann, B.I. 1981. Sexual selection, parental care, and concealed ovulation in humans. Ethology and Sociobiology 2:31–40.

Strassmann, B.I. 1996. The evolution of endometrial cycles and menstruation. Quarterly Review of Biology 71:181–220.

Suarez, S.S. 2015. Mammalian sperm interactions with the female reproductive tract. Cell Tissue Research 363:185–194.

Sukumar, R., et al. 1997. Demography of captive Asian elephants (*Elephas maximus*) in southern India. Zoo Biology 16:263–272.

Sun, Y., et al. 2012. Lethally hot temperatures during the early Triassic greenhouse. Science 338:366–370.

Sun, Y., et al. 2014. Occurrence of estrogenic endocrine disrupting chemicals concern in sewage plant effluent. Frontiers in Environmental Science and Engineering 8:18–26.

Swanson, E.M., et al. 2011. Lifetime selection on a hypoallometric size trait in the spotted hyena. Proceedings of the Royal Society B 278:3277–3285.

Swanson, W.J., et al. 2002. Positive Darwinian selection drives the evolution of several female reproductive proteins in mammals. Proceedings of the National Academy of Science 98:2509–2514.

Tague, R.G. 2016. Pelvic sexual dimorphism among species monomorphic in body size: relationship to relative newborn body mass. Journal of Mammalogy 97:503–517.

Tanabe, S., et al. 1982. Transplacental transfer of PCBs and chlorinated hydrocarbon pesticides from the pregnant striped dolphin (*Stenella coeruleoalba*). Agricultural and Biological Chemistry 46:1249–1254.

Tardif, S.D. 1994. Relative energetic cost of infant care in small-bodied Neotropical primates and its relation to infant care patterns. American Journal of Primatology 34:133–143.

Taylor, B.L., et al. 2013. *Orcinus orca*. The IUCN Red List of Threatened Species 2013 e. T15421A44220470.

Taylor, M.L., et al. 2014. Polyandry in nature: a global analysis. Trends in Ecology and Evolution 29:276–383.

Theis K.R., et al. 2013. Symbiotic bacteria appear to mediate hyena social odors. Proceedings of the National Academy of Science 110:19832–19837.

Thompson, R.C.A., et al. 2010. Parasites, emerging disease and wildlife conservation. International Journal of Parasitology 40:1163–1170.

Thonhauser, K.E., et al. 2013. Why do female mice mate with multiple males? Behavioral Ecology and Sociobiology 67:1961–1970.

Thorburn, G.D., J.R.C. Challis. 1979. Endocrine control of parturition. Physiological Reviews 59:863–918.

Thornhill, R. 1983. Cryptic female choice and its implications in the scorpionfly *Harpobittacus nigriceps*. American Naturalist 122:765–788.

Tilly, J.L., et al. 2009. The current status of evidence for and against postnatal oogenesis in mammals: a case of ovarian optimism versus pessimism? Biology of Reproduction 80:2–12.

Timm, R.M. 1989. Migration and molt patterns of red bats, *Lasiurus borealis* (Chiroptera: Vespertilionidae) in Illinois. Bulletin of the Chicago Academy of Sciences 14:1–7.

Tingen, C., et al. 2009. The primordial pool of follicles and nest breakdown in mammalian ovaries. Molecular Human Reproduction 15:795–803.

Toesca, A., et al. 1996. Immunohistochemical study of the corpora cavernosa of the human clitoris. Journal of Anatomy 188:513–520.

Torres, B. 1993. Sexual behavior of free-ranging Amazonian collared peccaries (*Tayassu tajacu*). Mammalia 610–613.

Toufexis, D., et al. 2014. Stress and the reproductive axis. Journal of Neuroendocrinology 26:573–586.

Treves, A. 1997. Primate natal coats: a preliminary analysis of distribution and function. American Journal of Physical Anthropology 104:47–70.

Tripp, H.R.H. 1971. Reproduction in elephant-shrews (Macroscelididae) with special reference to ovulation and implantation 26:149–159.

Trivers, R.L. 1974. Parent-offspring conflict. American Zoologist 14:249–264.

Trott, J.F., et al. 2003. Maternal regulation of milk composition, milk production, and pouch young development during lactation in the tammar wallaby (*Macropus eugenii*). Biology of Reproduction 68:929–936.

Tsutsui, T., et al. 2002. Factors affecting transuterine migration of canine embryos. Journal of Veterinary Medical Science 64:1117–1121.

Tuckey, R.C. 2006. Progesterone synthesis by the human placenta. Placenta 28:273–281.

Tulsiani, D. (ed.). 2003. Introduction to mammalian reproduction. Kluwer Academic, Boston, MA.

Tung, C.K., et al. 2015. Microgrooves and fluid flows provide preferential passageways for sperm over pathogen *Tritrichomonas foetus*. Proceedings of the National Academy of Sciences 112:5431–5436.

Tyndale-Biscoe, C.H. 1973. Life of marsupials. Elsevier, New York, NY.

Tyndale-Biscoe, C.H., M. Renfree. 1987. Reproductive physiology of marsupials. Cambridge University Press, Cambridge, England.

Tyndale-Biscoe, C.H., J.C. Rodger. 1978. Differential transport of spermatozoa into the two sides of the genital tract of a monovular marsupial, the tammar wallaby (*Macropus eugenii*). Journal of Reproduction and Fertility 52:37–43.

Uhen, M.D. 2007. Evolution of marine mammals: back to the sea after 300 million years. Anatomical Record 290:514–522.

Uriarte, N., et al. 2012. Different chemical fractions of fetal fluids account for their attractiveness at parturition and their repulsiveness during late-gestation in the ewe. Physiology and Behavior 107:45–49.

Vanden Brink, H., et al. 2013. Age-related changes in ovarian follicular wave dynamics. Menopause 20:1243–1254.

Van der Horst, C.J., J. Gillman. 1942. Pre-implantation phenomena in the uterus of *Elephantulus*. South African Journal of Medical Science 7:47–71.

Van Horn, R.C., et al. 2004. Behavioural structuring of relatedness in the spotted hyena (*Crocuta crocuta*) suggests direct fitness benefits of clan-level cooperation. Molecular Ecology 13:449–458.

Van Jaarsveld, A.S., J.D. Skinner. 1987. Spotted hyaena monomorphism: an adaptive "phallusy"? South African Journal of Science 83:612–615.

Van Jaarsveld, A.S., et al. 1988. Growth, development and parental investment in the spotted hyaena, *Crocuta crocuta*. Journal of Zoology, London 216:45–53.

Van Jaarsveld, A.S., et al. 1992. Changes in concentration of serum prolactin during social and reproductive development of the spotted hyaena (*Crocuta crocuta*). Journal of Reproduction and Fertility 95:765–773.

Van Kesteren, F. 2011. Reproductive physiology of Ethiopian wolves (*Canis simensis*). MSc. thesis, University of Oxford, Oxford, England.

Van Kesteren, F., et al. 2013. The physiology of cooperative breeding in a rare social canid: sex, suppression and pseudopregnancy in female Ethiopian wolves. Physiology and Behavior 122:39–45.

Van Noordwijk, M.A., et al. 2013. Multi-year lactation and its consequences in Bornean orangutans (*Pongo pygmaeus wurmbii*). Behavioral Ecology and Sociobiology 67:805–814.

Vanpé, C., et al. 2009. Multiple paternity occurs with low frequency in the territorial roe deer, *Capreolus capreolus*. Biological Journal of the Linnean Society 97:128–139.

Van Tienhoven, A. 1983. Reproductive physiology of vertebrates. Cornell University Press, Ithaca, NY.

Vaughan, J. 2011. Ovarian function in South American camelids (alpacas, llamas, vicunas, guanacos). Animal Reproduction Science 124:237–243.

Vernon, R.G., C.M. Pond. 1997. Adaptations of maternal adipose tissue to lactation. Journal of Mammary Gland Biology and Neoplasia 2:231–241.

Verstegen-Onclin, K., J. Verstegen. 2008. Endocrinology of pregnancy in the dog: a review. Theriogenology 70:291–299.

Veyrunes, F., et al. 2008. Bird-like sex chromosomes of platypus imply recent origin of mammal sex chromosomes. Genome Research 18:965–973.

Veyrunes, F., et al. 2010. A novel sex determination system in a close relative of the house mouse. Proceedings of the Royal Society B 277:1049–1056.

Visser, M.E., et al. 2004. Global climate change leads to mistimed avian reproduction. Advances in Ecological Research 35:89–108.

Voltolini, C., F. Petraglia. 2014. Neuroendocrinology of pregnancy and parturition. Handbook of Clinical Neurology, third series 124:17–36.

Vonhof, M.J., et al. 2006. A tale of two siblings: multiple paternity in big brown bats (*Eptesicus fuscus*) demonstrated using microsatellite markers. Molecular Ecology 15:241–247.

Vorbach, C., et al. 2006. Evolution of the mammary gland from the innate immune system? BioEssays 28:606–616.

Vyas, A., 2013. Parasite-augmented mate choice and reduction in innate fear in rats infected by *Toxoplasma gondii*. Journal of Experimental Biology 216:120–126.

Wagner, G.P., et. al. 2012. An evolutionary test of the isoform switching hypothesis of functional progesterone withdrawal for parturition: humans have a weaker repressive effect of PR-A than mice. Journal of Perinatal Medicine 40:346–351.

Wahaj, S.A., K.E. Holekamp. 2006. Functions of sibling aggression in the spotted hyaena, *Crocuta crocuta*. Animal Behaviour 71:1401–1409.

Wahaj, S.A., et al. 2004. Kin discrimination in the spotted hyena (*Crocuta crocuta*): nepotism among siblings. Behavioral Ecology and Sociobiology 56:237–247.

Wahaj, S.A., et al. 2011. Reconciliation in the spotted hyena (*Crocuta crocuta*). Ethology 107:1057–1074.

Wai-Ping, V., M.B. Fenton. 1988. Nonselective mating in little brown bats (*Myotis lucifugus*). Journal of Mammalogy 69:641–645.

Walker, K.Z., C.H. Tyndale-Biscoe. 1978. Immunological aspects of gestation in the tammar wallaby, *Macropus eugenii*. Australian Journal of Biological Sciences 31:173–182.

Wang, X., et al. 2013. Indo-Pacific humpback dolphin (*Sousa chinensis*) assist a finless porpoise (*Neophocaena phocaenoides sunameri*) calf: evidence from Ziamen waters in China. Journal of Mammalogy 94:1123–1130.

Want, Z., et al. 2008. Sperm storage, delayed ovulation, and menstruation of the female Rickett's big-footed bat (*Myotis ricketti*). Zoological Studies 47:215–221.

Wasser, S.K., M.L. Waterhouse. 1983. The establishment and maintenance of sex biases. Pp. 19–35 in Social behavior of female vertebrates (S.K. Wasser, ed.). Academic Press, New York, NY.

Watts, H.E., et al. 2011. Genetic diversity and structure in two spotted hyena populations reflects social organization and male dispersal. Journal of Zoology 285:281–291.

Weidt, A., et al. 2014. Communal nursing in wild house mice is not a by-product of group living: females choose. Naturwissenschaften 101:73–76.

Weigl, R. 2005. Longevity of mammals in captivity; from the living collections of the world. A list of mammalian longevity in captivity. Klein Senckenberg-Reihe, Frankfurt, Germany.

Weil, Z.M., et al. 2006. Photoperiod differentially affects immune function and reproduction in collared lemmings (*Dicrostonyx groenlandicus*). Journal of Biological Rhythms 21:384–393.

Weir, B.J. 1971a. The reproductive physiology of the plains viscacha, *Lagostomus maximus*. Journal of Reproduction and Fertility 25:355–363.

Weir, B.J. 1971b. The reproductive organs of the female plains viscacha, *Lagostomus maximus*. Journal of Reproduction and Fertility 25:365–373.

Weir, B.J., I.W. Rowlands. 1973. Reproductive strategies of mammals. Annual Review of Ecology and Systematics 4:139–163.

Welbergen, J.A., et al. 2008. Climate change and the effects of temperature extremes on Australian flying-foxes. Proceedings of the Royal Society B 272:419–425.

Weller, T.J., et al. 2009. Broadening the focus of bat conservation and research in the USA for the 21st century. Endangered Species Research 8:129–145.

Wells, J.C.K. 2007. Sexual dimorphism of body composition. Best Practice and Research Clinical Endocrinology and Metabolism 21:415–430.

Wells, R.S., et al. 2005. Integrating life-history and reproductive success data to examine potential relationships with organochlorine compounds for bottlenose dolphins (*Tur-*

siops truncatus) in Sarasota Bay, Florida. Science of the Total Environment 249: 106–110.

Werneburg, I., et al. 2016. Evolution of organogenesis and the origin of altriciality in mammals. Evolution and Development 18:229–244.

West, S.D. 1982. Dynamics of colonization and abundance in central Alaskan populations of the northern red-backed vole, *Clethrionomys rutilus*. Journal of Mammalogy 63:128–143.

Whateley, A. 1980. Comparative body measurements of male and female spotted hyaenas from Natal. Lammergeyer 28:40–43.

White, P.A. 2005. Maternal rank is not correlated with cub survival in the spotted hyena, *Crocuta crocuta*. Behavioral Ecology 16:606–613.

White, P.P. 2008. Maternal response to neonatal sibling conflict in the spotted hyena, *Crocuta crocuta*. Behavioral Ecology and Sociobiology 62:353–361.

White, Y.A.R., et al. 2012. Oocyte formation by mitotically active germ cells purified from ovaries of reproductive-age women. Nature Medicine 18:413–422.

Whitten, W.K., et al. 1968. Estrus-inducing pheromone of male mice: transport by movement of air. Science 161:584–585.

Wibbelt, G., et al. 2010. Emerging diseases in Chiroptera: why bats? Biology Letters 6:438–440.

Wildt, D.E., C. Wemmer. 1999. Sex and wildlife: the role of reproductive science in conservation. Biodiversity and Conservation 8:965–976.

Wilhelm, K., et al. 2003. Characterization of spotted hyena, *Crocuta crocuta*, microsatellite loci. Molecular Ecology News 3:360–362.

Williams, C.T., et al. 2013. Communal nesting in an "asocial" mammal: social thermoregulation among spatially dispersed kin. Behavioral Ecology and Sociobiology 67:757–763.

Wilsher, S., et al. 2013. Ovarian and placental morphology and endocrine functions in the pregnant giraffe (*Giraffa camelopardalis*). Reproduction 145:541–534.

Wilson, D.E., D.M. Reeder. 2005. Mammal species of the world, 3rd edition. Johns Hopkins University Press, Baltimore, MD.

Wilson, D.E., et al. 1991. Reproduction on Barro Colorado Island. Smithsonian Contributions to Zoology 511:43–52.

Wilson, D.S. 1979. Structured demes and trait-group variation. American Naturalist 113:606–610.

Wimsatt, W.A. 1975. Some comparative aspects of implantation. Biology of Reproduction 12:1–40.

Winternitz J., et al. 2012. Parasite infection and host dynamics in a naturally fluctuating rodent population. Canadian Journal of Zoology 90:1149–1160.

Wislocki, G. B. 1928. Observation on the gross and microscopic anatomy of the sloths (*Bradypus griseus* Gray and *Choloepus hoffmanni* Peters). Journal of Morphology and Physiology 46:317–397.

Wójcik, J.M., et al. 2003. The list of the chromosome races of the common shrew *Sorex araneus* (updated 2002). Mammalia 68:169–178.

Wolff, J.O. 1993. What is the role of adults in mammalian juvenile dispersal? Oikos 68:173–176.

Woodside, B., et al. 2012. Many mouths to feed: the control of food intake during lactation. Frontiers in Neuroendocrinology 33:301–314.

Wootton, J.T. 1987. The effects of body mass, phylogeny, habitat, and trophic level on mammalian age at first reproduction. Evolution 41:732–749.

Wourms, J.P., I.P. Callard. 1992. A retrospect to the symposium on evolution of viviparity in vertebrates. American Zoologist 32:251–255.

Wu, J., et al. 2014. Reproductive toxicity on female mice induced by microcystin-LR. Environmental Toxicology and Pharmacology 37:1–6.

Wurster, D.H., et al. 1970. Determination of sex in the spotted hyaena *Crocuta crocuta*. International Zoo Yearbook 10:143–144.

Yagil, R., Z. Etzion. 1980. Effect of drought condition on the quality of camel milk. Journal of Dairy Research 47:159–166.

Yamaguchi, N., et al. 2006. Female receptivity, embryonic diapause, and superfetation in the European badger (*Meles meles*): implications for the reproductive tactics of males and females. Quarterly Review of Biology 81:33–48.

Yordy, J.E., et al. 2010. Life history as a source of variation for persistent organic pollutant (POP) patterns in a community of common bottlenose dolphins (*Tursiops truncatus*) resident to Sarasota Bay, FL. Science of the Total Environment 408: 2163–2172.

Young, A.J., et al. 2006. Stress and the suppression of subordinate reproduction in cooperatively breeding meerkats. Proceedings of the National Academy of Sciences 103:12005–12010.

Young, J.M., A.S. McNeilly. 2010. Theca: the forgotten cell of the ovarian follicle. Reproduction 140:480–504.

Young, W.C., et al. 1964. Hormones and sexual behavior. Science, New Series 143: 212–218.

Youngman, P.M. 1990. *Mustela lutreola*. Mammalian Species 362:1–3.

Zainal-Zahari, Z., et al. 2002. Gross anatomy and ultrasonographic images of the reproductive system of the Sumatran rhinoceros (*Dicerorhinus sumatrensis*). Anatomia, Histologia, Embryologia 31:350–354.

Zamudio, S. 2003. The placenta at high altitudes. High Altitude Medicine and Biology 4:171–191.

Zaviačič, M. 1987. The female prostate: non vestigial organ of the female. A reappraisal. Sex and Marital Therapy 13:148–152.

Zenuto, R.R., et al. 2002. Bioenergetics of reproduction and pup development in a subterranean rodent (*Ctenomys talarum*). Physiological and Biochemical Zoology 75:469–478.

Zerbe, P., et al. 2012. Reproductive seasonality in captive wild ruminants: implications for biogeographical adaptation, photoperiodic control, and life history. Biological Reviews 87:965–990.

Zhang, G., et al. 2004. Evaluation of behavioral factors influencing reproductive success and failure in captive giant pandas. Zoo Biology 23:15–31.

Zhang, H., et al. 2009. Delayed implantation in giant pandas: the first comprehensive empirical evidence. Reproduction 138:979–986.

Zhang, X., et al. 2007. Wild fulvous fruit bats (*Rousettus leschenaulti*) exhibit human-like menstrual cycle. Biology of Reproduction 77:358–364.

Zhu, B., et al. 2003. The origin of the genetical diversity of *Microtus mandarinus* chromosomes. Hereditas 139:90–95.

Zoubida, B. 2009. Behavior at birth and anatomo-histological changes studies of uteri and ovaries in the postpartum phase in rabbits. European Journal of Scientific Research 34:474–484.

Zuckerman, S. 1951. The number of oocytes in the mature ovary. Recent Progress in Hormone Research 6:63–108.

Zuk, M. 2002. Sexual selections: what we can and can't learn about sex from animals. University of California Press, Berkeley, CA.

Common Name Index

European badger, *Meles meles*, 43, 111, 135–136, 210

European hamster, *Cricetulus, Cricetus*, 123, 130

European hare, *Lepus europaeus*, 220

European mink, *Mustela lutreola*, 110

European mole, *Talpa europaea*, 43, 51, 59, 61–62

European rabbit, *Oryctolagus cuniculus*, 20, 43, 55, 85, 99, 129, 162, 178, 244, 254, 258

European red squirrel, *Sciurus vulgaris*, 258

Fairy armadillo, *Chlamyphorus truncatus*, 104

Fallow deer, *Dama dama*, 101

False killer whale, *Pseudorca crassidens*, 185, 235

False vampire bats, Megadermatidae, 67, 178

Fennec fox, *Vulpes zerda*, 84

Ferret, *Mustela putorius*, 163

Field vole, *Microtus agrestis*, 219

Finless porpoise, *Neophocaena phocaenoides*, 229

Fishing cat, *Prionailurus viverrinus*, 257

Flying fox, Pteropodidae, Pteropod, 19, 55, 141, 161, 251, 266

Fossa, *Cryptoprocta ferox*, 63, 110, 151

Four-toed hedgehog, *Atelerix albiventris*, 54

Fox, e.g., *Urocyon, Vulpes*, 70, 133, 148, 183, 192, 201–202, 258

Fox squirrel, *Sciurus niger*, 159

Francois' leaf monkey, *Trachypithecus francoisi*, 151

Free-tailed bat, Molossidae, Molossid, 131, 147, 151, 238, 266

Fur seal, *Arctocephalus, Callorhinus*, 21, 43, 134, 136, 168, 173, 177, 192, 195–196, 210, 243, 277

Galago, *Galago*, 102

Galápagos fur seal, *Arctocephalus galapagoensis*, 177

Galápagos sea lion, *Zalophus wollebaeki*, 153, 177

Gelada baboon, *Theropithecus gelada*, 236, 267

Gerbil, jerboa, many species, including *Meriones*, 19, 63, 70, 132, 165, 172, 212, 214

Giant anteater, *Myrmecophaga tridactyla*, 136, 151, 220–222

Giant panda, *Ailuropoda melanoleuca*, 21, 136, 169, 257–258

Giant pouched rat, *Cricetomys*, 55

Gibbon, *Hylobates*, 181

Giraffe, *Giraffa*, 20, 100–102, 104, 141, 147, 150, 238

Goat, domestic, *Capra hircus*, 70, 132, 147, 207, 253

Golden hamster, *Mesocricetus auratus*, 19, 70, 99, 126, 129, 210

Golden marmot, *Marmota caudata*, 243

Golden mole, *Amblysomus*, 59, 154

Golden mouse, *Ochrotomys nuttalli*, 151, 176

Gopher, Geomyidae, 19, 55

Gorilla, *Gorilla*, 263, 277

Grasshopper mouse, *Onychomys leucogaster*, 159

Gray seal, *Halichoerus grypus*, 50, 136, 167, 213, 251

Gray tree-squirrel, *Sciurus carolinensis*, 103, 233, 254, 258

Gray whale, *Eschrichtius robustus*, 194

Great ape, ape, Hominidae, collectively chimps, bonobos, orangutans, gorillas, humans, 5, 70, 77, 109, 163, 181, 241, 247, 260, 262–265, 267–269, 277 (humans listed separately)

Greater white-toothed shrew, *Crocidura russula*, 159

Ground squirrels, e.g., *Ammospermophilus, Marmota, Tamias, Urocitellus*, 110–111, 152, 159, 176, 178, 183, 192, 209, 219–220, 225, 233, 235, 237, 242–243

Guanacos, *Lama guanicoe*, 108

Guinea pig, *Cavia porcellus*, 47, 49, 55, 61, 70, 111, 129–130, 159, 166, 244

Hairy armadillo, *Chaetophractus*, 59, 104

Hairy-tailed bat, *Lasiurus*, 151

Hamster, several genera, including *Mesocricetus, Phodopus*, 19, 70, 99, 123, 126, 129–130, 147, 159, 172, 185, 210, 214

Harbor seal, *Phoca vitulina*, 51, 135, 152, 251

Hare, *Lepus*, 59, 107, 130, 133, 144, 154–155, 162, 166, 178, 192, 199, 202, 204, 206–207, 209, 215–216, 219–220, 227–228

Hare wallaby, *Lagorchestes*, 28, 222

Harp seal, *Pagophilus groenlandicus*, 136, 152, 167, 202

Harwood's gerbil, *Dipodilus hardwoodi*, 214

Hawaiian monk seal, *Monachus schauinslandi*, 112, 259

Hedgehog, *Erinaceus*, 62–63, 110, 113, 176, 266

Hedgehog tenrec, *Setifer setosus*, 53, 103

Hippopotamus, *Hippopotamus* (hippos), 20, 40, 59, 62, 141, 146, 150, 153–154, 262

Hog badger, *Arctonyx collaris*, 136

Hog-nosed skunk, *Conepatus*, 136

Hollow- or slit-faced bat, *Nycteris*, 63

Honey possum, *Tarsipes rostratus*, 110

Hooded seal, *Cystophora cristata*, 136, 146, 154, 163–165, 167, 180, 219

Hopping mouse, *Notomys*, 160

Horse, mare, *Equus caballus*, 4, 20–21, 43, 47, 50–51, 56, 70, 79, 85, 101, 110, 113, 123–125, 128–129, 132, 147, 150, 185, 258–259

Horseshoe bat, *Rhinolophus*, 67

Human, *Homo sapiens*, ix–xi, 2, 5, 8, 11, 20, 26, 28–29, 32, 35, 42–43, 45, 47–48, 50, 57–58, 61–62, 66, 73, 79, 81, 87, 97, 101, 104, 116, 120, 129, 134, 141, 144, 153–154, 158–159, 166, 168, 177–178, 182, 184–186, 197, 222, 224, 228–231, 239, 247–260, 262–273, 275

Hyrax, dassie, order Hyracoidea, family Procaviidae, 21, 44, 104, 130, 166, 217, 229, 238, 262

Impala, *Aepyceros melampus*, 131
Indian muntjak, *Muntiacus muntjak*, 28
Indian Ocean dolphin, *Tursiops aduncus*, 165
Indian rhino, *Rhinoceros unicornis*, 20, 253, 256
Indo-Pacific humpback dolphin, *Sousa chinensis*, 229

Jamaican fruit bat, *Artibeus jamaicensis*, 135, 198
Javan rhino, *Rhinoceros sondiacus*, 20

Kangaroo, wallaby, *Macropus*, 14, 18, 20, 29, 44, 49, 57, 67, 83, 104, 116, 132–133, 135, 142–143, 149, 154, 159–160, 162–163, 168, 185, 198, 210, 220, 222, 252, 258–259, 263
Kangaroo rat, *Dipodomys*, 55, 181, 192, 214
Kiang, *Equus kiang*, 153
Killer whale, orca, *Orcinus*, 21, 133, 146, 185–186, 235, 252, 255–256, 272–273
Kirindy sifaka, Verreaux's sifaka, *Propithecus verreauxi*, 159, 180
Kit fox, *Vulpes macrotis*, 192
Koala, *Phascolarctos cinereus*, 14, 176, 220, 223, 259

Laboratory mouse, *Mus musculus*, 19, 33, 50, 70, 83, 85, 107–108, 111, 129, 135, 153, 161, 166, 171, 183, 214, 225, 231, 240, 244, 252, 258
Laboratory rat, *Rattus norvegicus*, 19, 55, 63, 70, 107, 129, 135, 143, 171, 214, 227, 252
Langur monkey, e.g., *Semnopithecus*, 5, 267
Large-headed rice rat, *Hylaeamys megacephalus*, 227
Leaf-nosed bat, *Macrotus*, 135
Least shrew, *Cryptotis parva*, 54
Least weasel, *Mustela nivalis*, 184, 201
Lemming, several genera, including *Dicrostonyx*, *Lemmus*, 19, 30, 107, 193, 199, 201–207, 209
Leopard, *Panthera pardus*, 63
Lesser anteater, *Tamandua*, 59
Lesser false vampire bat, *Megaderma spasma*, 178
Lesser hedgehog tenrec, *Echinops telfairi*, 58, 103, 159
Lion, *Panthera leo*, ix, 63, 111, 160, 185, 217–218, 232, 259, 262, 268
Little brown bat, *Myotis lucifugus*, 66, 135, 159
Llama, *Lama glama*, 70, 108, 192
Long-beaked echidna, *Zaglossus*, 17, 58, 255
Long-finned pilot-whale, *Globicephala mclaena*, 185
Long-nosed bandicoot, *Perameles nasuta*, 49
Long-nosed potoroo, *Potorous tridactylus*, 28
Lowland streaked-tenrec, *Hemicentetes semispinosus*, 105

Macaque, a monkey, e.g., *Macaca mulatta*, 70, 172–173, 176, 230, 235, 267
Maluku myotis, *Myotis moluccarum*, 133

Manatee, *Trichechus*, 21, 103, 146, 163, 178
Maned wolf, *Chrysocyon brachyurus*, 84, 151, 221
Marbled polecat, *Vormela peregusna*, 136
Marmoset, e.g., *Callithrix*, 48, 109, 147, 162, 220
Marsupial, metatherian, 9, 12, 14, 16–19, 21–22, 28–29, 31–32, 34–36, 43, 46, 49, 54, 56–57, 59, 61, 66–67, 81–83, 104–106, 110–111, 113, 115, 118, 121–123, 130, 133, 135, 138, 141–143, 148, 153–155, 158, 163–164, 168, 170, 172–173, 211, 222, 238–239, 242, 244, 277
Marten, *Martes*, 43, 112, 136
Masked flying fox, *Pteropus capistratus*, 161
Mastiff bat, Molossidae, molossid, 131, 147, 151, 238, 266
Meerkat, *Suricata suricatta*, 9, 146, 160, 177, 184, 218, 220, 240
Merriam's chipmunk, *Tamias merriami*, 110
Merriam's kangaroo rat, *Dipodomys merriami*, 214
Mexican deer mouse, *Peromyscus nudipes*, 214
Mexican free-tailed bat, *Tadarida brasiliensis*, 147, 238
Mink, *Neovison vison* (American), *Mustela lutreola* (European), 55, 70, 107, 110 (European), 136
Mole, Talpidae, e.g., *Condylura*, *Scalopus*, *Talpa*, 43, 51, 54, 59, 62–63, 113, 130, 136, 146
Mole voles, *Ellobius*, 28
Mongolian gerbil, *Meriones unguiculatus*, 70, 132
Mongoose, *Herpestes*, 130
Monjon, *Petrogale burbidgei*, 153
Monkey, many New and Old World primates, 62, 70, 88, 151, 153, 172–173, 176, 181, 186, 219, 221, 230, 235, 267
Monk seal, *Monachus*, 112, 259
Monotreme, Monotremata, Prototheria, 12–15, 17, 19, 21–22, 28–32, 34–35, 43, 45, 53, 56, 58, 61, 65, 67, 104, 118–122, 138, 141–143, 153, 155, 158, 163, 170–171, 255, 278
Montane vole, *Microtus montanus*, 214, 224
Moose (US), elk (UK), *Alces alces*, 20, 101, 155, 176, 207, 218, 220
Mouflon, *Ovis orientalis*, 257
Mountain goat, *Oreamnos americanus*, 192, 218
Mountain lion, *Puma concolor*, 258
Mountain viscacha, *Lagidium*, 131
Mouse, general term for small rodents from multiple families often Muridae. *See individual common names*
Mouse-eared bat, *Myotis*, 66, 133, 135, 151, 159, 178, 235
Mouse lemur, *Microcebus*, 252
Mule deer, *Odocoileus hemionus*, 159
Muskox, *Ovibos moschatus*, 202, 207, 218, 238

Nail-tail wallaby, *Onychogalea*, 222
Naked mole-rat, *Heterocephalus glaber*, 19, 21, 44, 66, 70, 160, 181, 239, 245
Naked-tailed armadillo, *Cabassous*, 104

Walrus, *Odobenus rosmarus*, 20–21, 136, 148, 151, 163, 178, 194, 202

Wapiti, red deer, *Cervus elaphus*, 20, 70, 80, 84, 110, 134, 159, 185, 219, 259

Water buffalo (Cape buffalo), *Bubalus bubalis*, 259

Water opossum, *Chironectes minimus*, 17–18

Water shrew, *Neomys*, 177

Weasels, *Mustela*, 20, 107, 110, 136, 163, 179, 184, 201–202, 210

Weddell seal, *Leptonychotes weddellii*, 136, 145, 168

Whale, cetacean, 19–21, 50, 67, 70, 86, 97, 110, 112, 131–133, 140–141, 146, 154, 158, 163–165, 173–174, 179, 182, 185–186, 192, 194–196, 202, 212, 225, 229, 235, 252, 255–256, 267–269, 272–273

White-faced capuchin, *Cebus capucinus*, 153

White-footed mouse, *Peromyscus leucopus*, 151, 159, 214

White rhino, *Ceratotherium simum*, 20, 149, 163

White-tail deer, *Odocoileus*, 101, 111, 152, 159, 254, 258–259, 264

White-tail deer, *Odocoileus virginianus*, 101, 152, 254, 264

Wild dogs, *Lycaon pictus*, 84, 146, 160, 236, 259

Wildebeest, gnu, *Connochaetes*, 56, 132, 217, 238

Wolf, *Canis lupus*, 84, 218, 220, 236, 254, 258

Wolverine, *Gulo gulo*, 136, 152–153

Wombat, *Lasiorhinus, Vombatus*, 181

Women, human, *Homo sapiens*, ix–xi, 2, 3, 5, 8, 11, 20, 26, 28–29, 32, 35, 42–43, 45, 47–48, 50, 57–58, 61–66, 73, 79, 81, 87, 97, 101, 104, 116, 120, 129, 134, 141, 144, 153–154, 158–159, 166, 168, 177–178, 182, 184–186, 197, 222, 224, 228–231, 239, 247–260, 262–273, 275

Woodchuck, marmot, *Marmota*, 159, 178, 219, 243

Woodland jumping mice, *Napaeozapus insignis*, 51

Wood lemming, *Myopus schisticolor*, 30

Woodrat, pack rat, *Neotoma*, 244

Yak, *Bos grunniens*, 14, 20

Yellow baboon, *Papio cynocephalis*, 236

Yellow-bellied marmots, *Marmota flaviventris*, 219

Zebra, *Equus*, 14, 149, 165, 181, 217–218

Scientific Name Index

Dasyprocta leporine, red-rumped agouti, 55, 227

Dasypus, armadillo, 44, 47, 59, 105, 136

Dasypus kappleri, greater long-nosed armadillo, 44

Dasypus novemcinctus, nine-banded armadillo, 47, 51, 56, 59, 85

Dasypus septemcinctus, seven-banded armadillo, 44

Dasyuridae, Australian marsupial family, ~20 genera, ~70 species, 19, 105, 135, 142, 145, 250, 266

Dasyuromorphia, order for Australian, often carnivorous, marsupials, e.g., quoll, 3 families, ~20 genera, ~70 species, 14, 19, 105, 135, 142, 145, 250, 253, 266

Dasyurus, quoll, native cat (a marsupial carnivore), 105, 135, 142, 145, 266

Dasyurus hallucatus, northern quoll, 142, 145

Dasyurus viverrinus, eastern quoll, 105, 135

Daubentonia madagascariensis, aye-aye, 62, 102

Delphinapterus leucas, beluga, 202

Delphinus delphis, common dolphin, 229

Dendrohyrax, tree hyrax, 104, 166

Dermoptera, colugos, 1 family, 2 genera, *Cynocephalus*, *Galeopterus*, 2 species, 19–20, 55, 151, 266

Desmodus, vampire bat, 223

Dicerorhinus sumatrensis, Sumatran rhino, 20

Diceros bicornis, black rhino, 20, 167, 218, 256

Dicrostonyx, collared lemming, 19, 30, 204–205

Didelphidae, North and South American opossums, ~15 genera, ~90 species, 10, 14, 17–18, 83, 114, 173

Didelphimorphia, order for New World opossums, 1 family, ~15 genera, ~90 species, 10, 14, 17–19, 83, 114, 173

Didelphis, opossum, 10, 14, 18, 83, 114

Dipodidae, jerboas, jumping mice, ~15 genera, ~50 species, 51

Dipodillus harwoodi, Harwood's gerbil, 214

Dipodomys, kangaroo rat, 55, 181, 192, 214

Dipodomys merriami, Merriam's kangaroo rat, 214

Dipodomys ordii, Ord's kangaroo rat, 214

Diprotodontia, order of Australian, often herbivorous marsupials, e.g., koala, kangaroo, 11 families, ~40 genera, ~140 species, 14, 18–20, 28–29, 44, 49, 57, 67, 83, 104, 110, 116, 132–133, 135, 142–143, 149, 153–154, 159–160, 162–163, 168, 170, 176, 185, 198, 210, 220, 222–223, 252, 258–259, 263

Dorcopsis, dorcopsis, 222

Dugong dugon, dugong, 14, 20–21, 62, 104, 163

Dyacopterus spadiceus, Dayak fruit bat, 161

Echinops telfairi, lesser hedgehog tenrec, 58, 103, 159

Eidolon, straw colored fruit bat, 136

Elephantidae, family for elephants, 2 genera, *Elephas*, *Loxodonta*, 14, 20–21, 44, 50, 59–60, 62, 86, 97, 103–104, 109, 123, 133, 147, 154, 163–164, 176–177, 180, 185–186, 212, 220, 235, 238, 258–259, 262–263, 268, 272–273

Elephantulus, elephant shrew, 59, 105, 128, 130, 244

Elephantulus myurus, eastern rock elephant shrew, 105

Elephas maximus, Asian elephant, 20, 59, 62, 86, 109, 147, 186

Emballonuridae, sheath-tailed bats, ~13 genera, ~50 species, 63

Enhydra lutris, sea otter, 112, 136, 140, 146, 194, 253

Eptesicus, house bats, 105, 110, 177, 235

Eptesicus fuscus, big brown bat, 105, 110, 177

Equidae, family for asses, horses, and zebra, 1 genus, *Equus*, 4, 14, 20–21, 43, 47, 50–51, 56, 70, 79, 85, 101, 110, 113, 123–125, 128–129, 132, 147, 149–150, 153, 165, 181, 185, 217–218, 258–259

Equus, zebra or horse, 14, 20–21, 43, 47, 50–51, 56, 70, 79, 85, 101, 110, 113, 123–125, 128–129, 132, 147, 149–150, 153, 165, 181, 185, 217–218, 258–259

Equus caballus, horse, 20–21, 43, 47, 50–51, 56, 70, 79, 85, 101, 110, 113, 123–125, 128–129, 132, 147, 150, 185, 258–259

Equus kiang, kiang, 153

Erethizon dorsatum, North American porcupine, 58, 165

Erethizontidae, New World porcupines, ~5 genera, ~15 species, 19, 21, 44, 58, 165

Erignathus barbatus, bearded seal, 136

Erinaceidae, hedgehogs, gymnures, ~10 genera, ~25 species, 19, 54, 62–63, 110, 113, 176, 266

Erinaceomorpha, order for hedgehogs, 1 family, ~10 genera, ~25 species, 19, 54, 62–63, 110, 113, 176, 266

Erinaceus, hedgehog, 62–63, 110, 113, 176, 266

Eschrichtius robustus, gray whale, 194

Eudorcas thomsonii, Thomson's gazelle, 153

Eumetopias jubatus, Steller sea lion, 136

Eupleridae, 7 genera, 10 species, 63, 110, 151

Eutheria, eutherian, a subclass of mammals comprising ~20 orders, 19 fig. 2.1

Felidae, cat family, ~15 genera, ~40 species, 9, 14, 20, 43, 47, 52, 56, 63, 70, 83, 107, 111, 160, 185, 220, 227–228, 232, 249, 256–259, 268, 278

Feliformia, a suborder of Carnivora with 7 families of cat-like carnivores. *See* Eupleridae; Felidae; Herpestidae; Hyaenidae; Viverridae

Felis catus, domestic cat, 43, 47, 52, 56, 70, 83, 107, 185, 232, 258, 278

Fukomys damarensis, Damaraland mole rat, 239–241

Mustela, weasels, 20, 107, 110, 136, 163, 179, 184, 201–202, 210

Mustela erminea, ermine, stoat, short-tailed weasel, 107, 179, 184, 202

Mustela lutreola, European mink, 110

Mustela nivalis, least weasel, 184, 201

Mustela putorius, ferret, 163

Mustelidae, weasel, mink, badger, otter, etc. family, ~20 genera, ~60 species, 20, 43, 55, 70, 107, 110–112, 135–136, 140, 146, 152, 163, 179, 184, 194, 201–202, 210, 219, 253

Myocastor coypus, coypu, nutria, 19, 159, 165

Myodes, red-backed vole, 19, 151, 220, 223, 236

Myodes glareolus, bank vole, 220, 223

Myodes rutilus, northern red-backed vole, 236

Myopus schisticolor, wood lemming, 30

Myotis, mouse-eared bat, 66, 133, 135, 151, 159, 178, 235

Myotis lucifugus, little brown bat, 66, 135, 159

Myotis moluccarum, Maluku myotis, 133

Myotis nigricans, black myotis, 178

Myrmecophaga tridactyla, giant anteater, 136, 151, 220–222

Myrmecophagidae, New World anteaters, 2 genera, *Myrmecophaga*, *Tamandua*, 3 species, 59, 136, 151, 220–222

Mysticeti, baleen whales, 4 families, ~5 genera, ~10 species, 20–21, 50, 67, 70, 86, 97, 110, 132–133, 146, 149, 154, 158, 163–164, 179, 182, 185–186, 194–195, 202, 212, 235

Napaeozapus insignis, woodland jumping mice, 51

Nasua, coati, 236

Neomys, water shrew, 177

Neophoca cinerea, Australian sea lion, 178, 243

Neophocaena phocaenoides, finless porpoise, 229

Neotoma, packrat, woodrat, 244

Neovison vison, American mink, 55, 70, 107, 136

Nesomyidae, e.g., pouch rat, ~20 genera, ~60 species, 55

Nesotragus (*Neotragus*), dwarf antelope, 131

Notomys, hopping mouse, 160

Notoryctemorpha, monogeneric order for 2 species of marsupial moles, *Notoryctes*, 19 fig. 2.1

Nyctalus, noctule bat, 177

Nyctereutes procyonoides, raccoon dog, 84

Nycteridae, slit- or hollow-faced bat, 1 genus, *Nycteris*, ~15 species, 63

Nycteris, slit- or hollow-faced bat, 63

Nyctinomops, a free-tailed bat, 151

Nyctomys sumichrasti, Sumichrast's vesper rat, 177

Ochotona, pika, 58, 133, 176, 178, 182, 251

Ochotonidae, pikas, 1 genus, *Ochotona*, ~30 species, 58, 133, 176, 178, 182, 251

Ochrotomys nuttalli, golden mouse, 151, 176

Odobenidae, monotypic walrus, 20–21, 136, 148, 151, 163, 178, 194, 202

Odobenus rosmarus, walrus, 20–21, 136, 148, 151, 163, 178, 194, 202

Odocoileus, white-tailed deer, 101, 111, 152, 159, 254, 258–259, 264

Odocoileus hemionus, mule deer, 159

Odocoileus virginianus, white-tailed deer, 101, 152, 254, 264

Odontoceti, toothed whales, 7 families, ~35 genera, ~70 species, 20–21, 67, 70, 112, 131, 133, 140–141, 146, 154, 165, 173–174, 185–186, 192, 194–195, 225, 229, 235, 252, 255–256, 267–269, 272–273

Ommatophoca rossi, Ross seal, 136

Onychogalea, nail-tail wallaby, 222

Onychomys leucogaster, grasshopper mouse, 159

Orcinus orca, orca, killer whale, 21, 133, 146, 185–186, 235, 252, 255–256, 272–273

Oreamnos americanus, mountain goat, 192, 218

Ornithorhynchus anatinus, platypus, 10, 12, 17, 20, 22, 28–30, 65, 118–121, 131, 142, 170, 176, 255, 263

Orycteropus afer, aardvark, 14, 20, 44, 62, 77

Oryctolagus cuniculus, common European rabbit also domesticated, 20, 43, 50, 55, 57, 61, 70, 83–85, 98–99, 107, 112–114, 129–131, 133, 143–144, 147–148, 155, 162, 172, 178, 185, 224, 244, 254, 258–259

Oryx, oryx, gemsbok, 56

Otaria byronia, southern sea lion, 136, 243

Otariidae, otariid, eared seals (fur seals, sea lions), ~7 genera, ~15 species, 20–21, 43, 62, 134, 136, 146, 152–153, 158, 168, 173, 177–178, 192, 194–196, 210, 243, 277

Otospermophilus beecheyi, California ground squirrel, 110

Ovibos moschatus, muskox, 202, 207, 218, 238

Ovis, sheep, 11, 21, 47, 50, 70, 79, 81–82, 84, 98–99, 112–113, 129–130, 132, 143, 155, 207, 257

Ovis aries, domestic sheep, ewe, 11, 21, 47, 50, 70, 79, 81–82, 84, 98–99, 112–113, 129–130, 132, 143, 155, 207

Ovis orientalis, mouflon, 257

Pagophilus groenlandicus, harp seal, 136, 152, 167, 202

Pan, chimpanzee (*P. troglodytes*), bonobo (*P. paniscus*), 109, 163, 241, 267–269

Panthera leo, lion, ix, 63, 111, 160, 185, 217–218, 232, 259, 262, 268

Panthera pardus, leopard, 63

Papio, baboon, 38, 43, 109, 220, 235–236, 241

Papio anubis, olive baboon, 241

Papio cynocephalis, yellow baboon, 236

Paraxerus cepapi, African bush squirrel, 110

Paucituberculata, order for South American shrew opossums, 1 family, 3 genera, 6 species, 19 fig. 2.1

Peramelemorpha, Australian, often herbivorous marsupials, e.g., bilby, bandicoots, 3 families, ~10 genera, ~20 species, 19, 28, 49, 128, 142

Perameles, bandicoot, 49, 128, 142

Perameles nasuta, long-nosed bandicoot, 49

Perissodactyla, order for even-toed ungulates, horses, tapirs, and rhinos, 3 families, ~5 genera, ~15 species, 14, 19–21, 43, 47, 50–51, 56, 70, 79, 85, 101, 110, 113, 123–125, 128–129, 132, 147, 149–152, 163, 165, 181, 185, 217–218, 221, 253, 256, 258–259

Peromyscus, deer mouse, 89, 111, 151, 159, 172, 176–177, 192, 214, 221, 224

Peromyscus leucopus, white-footed mouse, 151, 159, 214

Peromyscus nudipes, Mexican deer mouse, 214

Petrodromus tetradactylus, four-toed elephant shrew, four-toed sengi, 59

Petrogale, rock wallaby, 153, 222

Petrogale burbidgei, monjon, 153

Phalangeridae, phalanger, e.g., brushtail possum, cuscus, ~5 genera, ~20 species, 44, 142

Phascolarctos cinereus, koala, 14, 176, 220, 223, 259

Phoca, harbor or spotted seal, 51, 135–136, 152, 251

Phocarctos hookeri, New Zealand sea lion, 243

Phoca vitulina, harbor seal, 135, 152, 251

Phocidae, earless seals, ~13 genera, ~20 species, 20–21, 50–51, 62, 110–112, 135–136, 145–146, 152, 154, 158–159, 163–165, 167–168, 171, 176, 178, 180, 192, 194–196, 202, 208, 213, 215, 219, 251, 259

Phocoena, porpoise, 21, 195

Phodopus, hamster, including dwarf and Siberian hamster, 19, 147, 159, 210, 214

Phodopus sungorus, Djungarian hamsters, 214

Pholidota, order for pangolins, 1 genus, *Manis*, ~8 species, 19, 44, 244

Phyllostomidae, New World fruit bats, >50 genera, >150 species, 67, 125, 127, 135, 151, 178, 198, 223, 264

Physeter macrocephalus, sperm whale, 21, 173–174, 185, 192, 225, 235

Pilosa, order for sloths, anteaters, 2 families, ~5 genera, ~10 species, xi, 19–21, 44, 56, 58–59, 88, 136, 141, 146, 151, 153, 165, 217, 220–222, 264

Pinnipedia, pinniped, fin-footed carnivorans. *See* Odobenidae; Otariidae; Phocidae

Pipistrellus, pipistrelle, 105, 141, 177

Pipistrellus subflavus, eastern pipistrelle, 105

Pongo, orangutan, 163, 263, 267

Potorous tridactylus, long-nosed potoroo, 28

Praomys, soft-furred mouse, 63

Primates, order for lemurs, monkeys, apes, humans, etc., ~20 families, ~70 genera, ~375 species (excludes *Homo*), ix, 5, 19, 38, 43–44, 48, 51, 56, 62, 65–66, 70, 77, 88, 100, 102, 109, 112–113, 132, 143, 147, 151–153, 159, 162–164, 168, 172–173, 176, 180–181, 185–186, 219–221, 229–230, 233–236, 238–239, 241–242, 247, 252, 260, 262–269, 273, 277

Proboscidea, order for elephants, 1 family, 2 genera, *Elephas, Loxodonta*, 3 species, 14, 19–21, 44, 50, 59–60, 62, 86, 97, 103–104, 109, 123, 133, 147, 154, 163–164, 176–177, 180, 185–186, 212, 220, 235, 238, 258–259, 262–263, 268, 272–273

Procavia capensis, rock hyrax, 130, 166, 229

Procyon, raccoon, 9, 20, 107, 130

Procyonidae, raccoon family, ~5 genera, ~15 species, 9, 20, 107, 130, 236

Proechimys cuvieri, Cuvier's spiny rat, 227

Proechimys guairae, spiny rat, 163

Pronolagus, red rock hare, 178

Propithecus verreauxi, Kirindy sifaka, 159, 180

Proteles, aardwolf, 9

Prototheria, a subclass of mammals with one order, the egg-laying mammals. *See* Monotremata

Pseudomys, an Australian murid, 160, 177

Pseudorca crassidens, false killer whale, 185, 235

Pteropodidae, Old World fruit bats, >40 genera, >175 species, 19, 55, 135–136, 141, 161, 198, 238, 251, 266

Pteropus, flying fox, 141, 161, 238

Pteropus capistrastus, masked flying fox, 161

Pteropus rodricensis, Rodrigues fruit bat, 238

Puma concolor, mountain lion, 258

Pusa, e.g., Baikal, ribbon, ringed seal, 136, 145, 251

Pusa hispida, ringed seal, 251

Rangifer, reindeer, caribou, 20, 101, 134, 160, 193, 199, 202–204, 206–207

Rattus, rat, 19, 55, 63, 70, 107, 129, 135, 143, 171, 214, 227, 252, 258

Rattus norvegicus, Norway rat, laboratory rat, 19, 55, 63, 70, 107, 129, 135, 143, 171, 214, 227, 252

Rattus rattus, black rat, introduced worldwide, 258

Redunca, reedbuck, 264

Rhabdomys, African four-striped mouse, 113

Rhinoceros sondiacus, Javan rhino, 20

Rhinoceros unicornis, Indian rhino, 20, 253, 256

Rhinocerotidae, rhino family, 4 genera, *Ceratotherium, Dicerorhinus, Diceros, Rhinoceros*, 20, 51, 67, 149–150, 163, 167, 218, 253, 256, 262

Rhinolophidae, horseshoe bats, 1 genus
(*Rhinolophus*),~70 species, 67
Rhinolophys, horseshoe bat, 67
Rodentia, rodents,~30 families,~480 gen-
era,~2,300 species. *See* Castoridae; Cricetidae;
Dipodidae; Geomyidae; Gliridae; Heteromy-
idae; Hystricomorpha; Muridae; Nesomyidae;
Sciuridae
Rousettus aegyptiacus, Egyptian fruit bat, 198
Ruminantia, ruminant, a suborder of
Artiodactyla, 11, 20, 44, 65, 143, 207. *See also*
Antilocapridae; Bovidae; Cervidae; Giraffidae
Rupicapra rupicapra, alpine chamois, 225, 253

Saguinus, tamarin, 162
Sarcophilus harrisii, Tasmanian devil, 14, 253
Scalopus aquaticus, eastern mole, 54, 130
Scandentia, order for tree shrews, 2 families,~5
genera,~20 species, 59, 100, 162, 165,
220–222, 266–267
Sciuridae, squirrel family,~50 genera,~275
species, 19, 44, 55, 87, 103–104, 107, 110–111,
113, 152, 159, 165, 176, 178, 180, 182–183, 192,
201, 209, 219–220, 225, 233, 235, 237, 242–243,
254, 258
Sciurus, squirrel, 55, 103, 111, 159, 233, 254, 258
Sciurus carolinensis, gray tree-squirrel, 103, 233,
254, 258
Sciurus niger, fox squirrel, 159
Sciurus vulgaris, European red squirrel, 258
Scotinomys teguina, Neotropical singing mouse,
66
Semnopithecus, langur monkey, 5, 267
Setifer setosus, hedgehog tenrec, 53, 88, 103–104
Setonix brachyurus, quokka, 222
Sigmodon, cotton rat, 159
Sirenia, order for dugong and manatee, 2
families, 2 genera, *Dugong, Trichechus,* 4
species, 14, 19–21, 44, 62, 103–104, 146, 163,
178, 250
Sorex, shrew, 28, 133, 159
Sorex araneus, Eurasian shrew, 28, 133
Sorex coronatus, crowned shrew, 159
Sorex minutus, Eurasian pygmy shrew, 159
Soricidae, shrews,~25 genera,~375 species, 28,
44, 54, 56, 88–89, 103, 107, 110, 123, 130,
132–133, 159, 166, 177–180, 185, 212, 218, 238
Soricomorpha, order for shrews, moles, etc., 4
families,~45 genera,~425 species, 19 fig. 2.1.
See also Soricidae; Talpidae
Sousa chinensis, Indo-Pacific humpback dolphin,
229
Spalacopus cyanus, cururo, 245
Speothos venaticus, bush dog, 84
Spilogale, spotted skunk, 43, 136, 210
Stenella attenuata, spotted dolphin, 185
Stenella frontalis, Atlantic spotted dolphin, 229
Stenella longirostris, spinner dolphin, 185, 229

Suidae, pig family,~5 genera,~20 species, 21, 47,
50, 55, 70, 79, 85, 99, 107, 113, 117, 127–129,
132, 153, 155, 172, 180, 227, 244–245, 258
Suncus, musk shrew, 54, 103
Suncus murinus, Asian musk shrew, 54
Suricata suricatta, meerkat, 9, 146, 160, 177, 184,
218, 220, 240
Sus scrofa, domestic pig, sow, 21, 47, 50, 55, 70,
79, 85, 99, 107, 113, 117, 127–129, 132, 153, 155,
172, 180, 227, 244–245, 258
Sylvicapra grimmia, duiker, 131
Sylvilagus, cottontail, 59, 63, 107, 133, 144, 155,
178
Syncerus caffer, Cape buffalo, 217, 218

Tachyglossidae, echidna family, 2 genera,
Tachyglossus, Zaglossus, 3 species, 17, 22, 28,
30, 32, 53, 65, 67, 118–119, 121, 142, 170, 255
Tachyglossus, short-beaked echidna, 17, 30, 58,
121, 170, 255
Tadarida brasiliensis, Mexican free-tailed bat,
147, 238
Talpa altaica, Siberian mole, 136
Talpa europaea, European mole, 43, 51, 59,
61–62
Talpidae, mole family,~15 genera,~40 species,
43, 51, 54, 59, 62–63, 113, 130, 136, 146
Tamandua, lesser anteater, 59
Tamias, chipmunk, 110, 178, 192
Tamiasciurus hudsonicus, American red squirrel,
180, 233
Tamias merriami, Merriam's chipmunk, 110
Tamias obscurus, chaparral chipmunk, 110
Tamias sibiricus, Asian chipmunk, 110
Taphozous, sheath-tailed or tomb bat, 63
Tapirus, tapir, 20, 51, 151–152, 221
Tarsipes rostratus, honey possum, 110
Taxidea taxus, American badger, 136
Tayassu, peccary, 20, 112–113
Tayassuidae, family for peccaries, 3 genera,
Pecari, Tayassu, Catagonus, 3 species, 20,
112–113
Tenrec, tenrec, 53, 62, 103–104
Tenrecidae, tenrec family,~10 genera,~30
species, 53, 58, 62–63, 88–89, 103–105, 114,
130, 159
Theropithecus gelada, gelada baboon, 236, 267
Thylacinus cynocephalus, Tasmanian wolf, 250
Thylogale, pademelon, 222
Tokudaia muenninki, spiny rat, 30
Tokudaia osimensis, spiny rat, 30–31
Tokudaia tokunoshimensis, spiny rat, 30–31
Tolypeutes, three-banded armadillo, 57, 104
Trachypithecus francoisi, Francois' leaf monkey,
151
Tremarctos ornatus, spectacled bear, 136
Trichechus, manatee, 21, 103, 146, 163, 178
Trichosurus vulpecula, brushtail possum, 142

Tubulidentata, monospecific order for the aardvark, *Orycteropus afer*, 14, 19–20, 44, 62, 77

Tupaia, tree shrew, 19, 59, 100, 162, 165, 220–222, 266–267

Tursiops, dolphin, 21, 112, 146, 154, 165, 229, 236, 238, 252, 255, 268

Tursiops aduncus, Indian Ocean dolphin, 165

Tursiops truncatus, bottlenose dolphin, 112, 146, 229, 236, 238, 252, 255, 268

Tylonycteris, bamboo bat, 177

Ungulates, hoofed mammals. *See* Artiodactyla; Perissodactyla

Urocitellus, a ground squirrel, 111, 152, 159, 219, 225, 233, 237, 242

Urocitellus beldingi, Belding's ground squirrel, 111, 219, 233, 237, 242

Urocitellus columbianus, Columbian ground squirrel, 225

Ursidae, bear family, ~5 genera, 10 species, 9, 20–21, 45, 55, 69, 107, 136, 144, 146, 152, 158, 166, 168–169, 171, 180, 185, 193–194, 199–200, 202–204, 208–209, 215, 220, 232–233, 235

Ursus, bear, 55, 107, 136, 144, 146, 152, 169, 171, 180, 185, 193–194, 199–200, 202–204, 209, 215, 220, 232–233

Ursus americanus, black bear, 107, 169, 171, 232–233

Ursus arctos, brown or grizzly bear, 136, 144, 169, 171

Ursus maritimus, polar bear, 136, 146, 152, 169, 180, 185, 193–194, 199–200, 202–204, 208–209, 215, 220

Vespertilionidae, e.g., vesper, evening, common bats, ~50 genera, ~400 species, 66, 105, 110, 133, 135–136, 151, 159, 177–178, 235

Vicugna pacos, alpaca, 70, 108–109

Viverridae, 15 genera, 35 species, 59

Vombatus ursinus, common wombat, 181

Vormela, marbled polecat, 136

Vulpes, e.g., Arctic, red fox, 70, 84, 133, 183, 192, 201, 258

Vulpes lagopus, Arctic fox, 70, 84, 183, 201

Vulpes macrotis, kit fox, 192

Vulpes vulpes, red fox, 70, 84, 258

Vulpes zerda, fennec fox, 84

Wallabia bicolor, swamp wallaby, 28, 222

Xenarthra, xenarthran, a super order comprising New World anteaters, armadillos, and sloths. *See* Bradypodidae; Dasypodidae; Megalonychidae; Myrmecophagidae

Zaedyus, dwarf armadillo, 104

Zaglossus, long-beaked echidna, 17, 58, 255

Zalophus californianus, California sea lion, 136

Zalophus wollebaeki, Galápagos sea lion, 153, 177

Subject Index

abdominal cavity, 49–50, 57; pouch, 17; teats, 66

abiotic, 275; cues, 106, 109, 183–184, 270; environment, xi, 72, 145, 189, 231, 247, 278; selection pressure, 74, 154, 191–217, 246, 269; stress, 72

abortion. *See* embryo rejection

acrosome, 115

adaptation, 11, 22, 36, 63, 65, 70, 83, 87, 131, 163, 168–169, 192–194, 198, 202, 206, 208–209, 215, 276–277, 279; antimicrobial, 226–227; anti-predator, 219–222

adipose tissue, 148, 160, 173, 275

afterbirth, 140–141, 144, 147–148, 225

allantois, 11, 45–47, 120–121, 275, 277

allele, 27, 35–36, 186, 271, 275

alliance, 234, 236

allometry, 70–71, 86, 91. *See also* body size

allonursing, 240

alloparental care, 221, 234, 236–240, 245, 275

altitude, 139, 169, 191–193

altricial. *See* neonate

altruism, 236, 239

amnion, 11, 45, 47, 120, 275, 277; Amniota, 11, 13, 275; amniotic egg, 11, 45; amniotic fluid, 142, 197, 230; anamniote, 13

androcentric terminology, 3–8

androgen (testosterone), 4, 32, 49, 52, 69, 72, 75–77, 91, 225, 259, 265, 271, 275

angiogenesis, 79

anisogamy, 7

anthropogenic influences, 250–254, 258

antrum, 52–53, 103

Arctic, 152, 160, 169, 193–194, 199–204, 206–210, 215, 251, 256

assisted reproduction, ix, 100–102, 117, 249–250, 254–258, 273

atresia, 40, 52, 102–104, 275; atretic follicle, 96, 100, 102–103

autosome, 26, 28–30, 35, 276

bacteria, 42, 114, 217, 223, 225–226, 230, 251, 270, 278

basal lamina, 52, 98–99, 105

basal metabolic rate (BMR), 88

β-catenin, 31

bicornuate uterus, 55–56, 130–131

biome, 193, 199, 275

biotic, 275; community, 193; cues, 183, 194, 213–215, 270; environment, xi, 145, 189, 215–216, 231, 247, 278; selection pressure, 74, 154, 246, 279

birth, 11, 140, 146, 155–156; anatomy, 37, 63, 141, 264; after birth, 53, 93, 99, 101, 105, 116, 132, 137, 142, 147–150, 152, 158, 165, 167, 197–198, 206–207, 218, 221, 230–231, 257, 269; before birth, 51, 93, 98–99, 103, 106, 143, 147–148, 151, 186, 194, 206, 220, 230; control, 252, 259, 262; eutherian birth, 143–147; as an event or transition, xi, 3, 15, 25, 31, 45, 62, 67, 73, 82, 91, 98, 118, 122–123, 132, 134, 136, 148, 157, 159, 163, 177, 179, 182, 187, 199, 206, 209, 219, 231, 241; give birth, 47, 60, 63, 83, 85, 87, 111, 128, 136, 149–150, 152, 155, 160, 164, 169, 180, 194–195, 203, 205, 207–208, 219–220, 238, 256, 258, 263, 269; litter size, 34, 44, 47, 66, 105, 111, 129–130, 139, 141, 166, 225, 227, 244; metatherian, 18, 142–143, 244; monotreme, 119, 142; process, 37, 57, 71, 74–75, 77, 140–141, 252, 259, 266; rate, 219, 261, 270, 272–273; social facilitation, 238; stillbirth, 3, 140, 241, 251, 269, 276; synchronous, 134, 145, 209, 213, 220, 229, 238, 243; viviparity (live birth), 9, 14, 118

blastocyst, 32, 34, 41, 44–45, 53, 80, 82, 84–85, 105, 119, 124–131, 134–135, 243–244, 257, 275

blubber, 88, 154, 164, 173, 194–196, 252, 255, 275

body size, 50, 54, 69–70, 86–88, 91, 133, 139, 144, 155, 163–164, 179–180, 182, 195, 203, 206, 209, 212, 215, 264

brown adipose tissue (BAT), 148, 275

calcium, 162, 173, 180, 191, 198, 227

capacitation, 96, 115, 275

captive breeding, 100, 250, 254–255, 257–258

carbohydrates, 27, 158, 167, 198

central limitation hypothesis, 161

cervix, 3–4, 18, 37, 40, 53, 55–58, 60–62, 85, 113–116, 124, 144, 227, 253, 267

mate choice. *See* female choice; selection; sperm/spermatocyte

maternal / mother care, 9–10, 15–16, 74, 79, 134, 155, 165, 181, 228, 263; control, 48, 84, 125, 142–143, 155; genome, 25, 33, 35–36, 96, 99, 116, 122, 271; input, 16, 35, 42, 45, 95, 120–122, 127; instinct, 5; microbiome, 230; offspring interaction, 11, 14, 43, 46, 48, 53, 125, 129, 133, 137–138, 144, 246; pregnancy recognition, 64, 80–85, 90, 127, 265; resources, 3, 6, 11, 13, 47, 130, 243; size, 14, 87, 143–144, 149, 166, 178

mating. *See* coitus

matriline, 42, 192, 232, 234–237, 241

meiosis, 41, 45, 97, 99, 278

menopause, 87, 101, 176, 184–187, 262, 264, 271

menstrual cycle, 8, 87, 100–101, 185; menstruation, 57, 73, 83, 100, 262, 266–267

mesotocin, 170

metabolism, 71–72, 86, 88, 91, 127, 137, 192, 197–198, 241; hormones, 171; metabolic ceiling, 161–162, 174; metabolites, 183, 204, 214, 228; output, 89, 91, 266–267; rates, xi, 88–89, 166, 180

microbiome, 58, 67, 144, 168, 217–218, 228–231, 269–270, 278

migration, 164–165, 181, 196; blastocyst, 44, 64, 67, 130–131; primordial germ cell, 104

milk, 122, 167–172; colostrum, 170–171; composition, 18, 160, 164, 167–169, 171–172, 194, 196, 198, 207–208, 244, 266; function, 152, 158, 160, 162, 164–166, 172; marsupial, 18, 148, 168–169; monotreme, 17, 170–171; origin, 16, 83, 120, 157, 171; parasites/pollutants/microbiome, 225, 228, 230–231, 251–252; production/synthesis, 64, 79, 159, 161–162, 173, 213; release, 17, 65–66, 75; use by offspring, 173–174

minerals, 26, 139, 162, 165, 167, 171, 173, 191–192, 198, 215, 241

miscarriage. *See* embryo: rejection

mitochondria 27, 41–42, 96, 116, 127, 231, 271; disorder, 27

monogamy, 6–7, 239, 242, 268, 278

monotocous, 129, 131–132, 265, 278

monovular/polyovular, 44, 52, 104–105, 279

mucosa, 54–55

Müllerian ducts, 38–39

mycoestrogen, mycotoxins, 228, 278

natal nest, 147, 220, 224

nematode, 225

neonate, 148–155, 158, 178, 197, 205; altricial/precocial, 132–133, 143–144, 152–156, 164–166, 180; behavior, 65–66, 150–151, 172, 221–222; marsupial, 18, 142, 168; monotreme, 65, 121; pelage, 151–152, 221; size, 73, 87, 123, 136–137, 149–150, 169, 180, 207, 221

neo-oogenesis, 100, 102, 186, 278

neuroendocrine system. *See* endocrine (neuroendocrine) system

neurosteroids, neurotransmitters, 75

nipple/teat, 11, 17–18, 38, 65–67, 104–106, 130, 142, 148–149, 153, 158, 163, 166, 168, 170, 172–173, 240, 244–245

nursing. *See* suckling/nursing

nutrients, 11, 40, 45–48, 53, 70, 72, 89, 93, 113, 120–122, 124, 127, 134, 140, 147, 161–162, 166, 170, 180, 183, 191, 194–195, 198, 201, 241, 243

olfaction, 18, 109–110, 126, 142, 153–154, 220, 222, 252–253

oocyte (female gamete), ovum (ova, pl.), 4, 27, 35, 38, 41, 51, 99, 278; anatomy, 40–45, 96; composition, 27, 95, 97, 104, 122, 125; conception, 7, 115–116, 124; cryopreservation, 256–257; degeneration, 95, 100; litter size, 44, 52, 104–105; location, 52–53, 96, 103–104; numbers, 100, 102, 186; senescence, 97, 186–187, 273; size, 97–98, 103–104; transport, 105, 114, 124. *See also* oogenesis

oogenesis, 41, 43, 45, 49, 93, 95–104, 120, 186–187, 265, 278

oogonia, 41, 96–97, 99–100, 186, 278

oolemma, 42, 115

organochlorines, 251–252

orgasm, 4, 6, 8, 62, 107, 109, 113–114

oscaruncle, 121

ovary, 18, 32; anatomy, 42, 49–53, 264; development, 31, 265; dynamics, 38, 40, 50, 79, 101, 103; function, 42, 49; hormones, 49, 57–58, 75–77, 100, 102, 124, 265; litter size, 44, 128, 130–132; size, 50–51, 251. *See also* follicle; oogenesis; ovulation

oviduct, 18, 136; anatomy, 49–50, 53–55, 58, 64, 105, 111, 114–115. *See also* conception; gamete: transport; ovulation

oviparity, 14, 278

ovulation, 40, 44, 49, 53, 75, 82, 96, 99, 101–102, 104–109; concealed, 8, 267–268; cues, 61; as an event or transition, xi, 3, 73, 91, 100, 182, 266; facultative (reflex, induced), 4, 6, 100, 106–109, 113, 130–132, 180, 205, 259, 265; inhibition, 64, 77, 186, 217, 219; location, 51; timing, 84

ovulatory cycles. *See* estrus: estrous cycle

oxidative damage, 162

oxytocin, 65, 75–79, 81–82, 84–85, 147, 157, 167, 170–171

paracrine hormones, 75

parasites (endoparasites, ectoparasites), 11, 89, 106, 112, 148, 155, 187, 217, 218, 222–226, 233, 240, 246, 253, 270, 276

parathyroid hormone, 171

parent-of-origin gene expression. *See* epigenetics